Numerical Constants

FUNDAMENTAL PHYSICAL CONSTANTS

Name	Symbol	Value
|Charge of electron|	e	1.602×10^{-19} C
Avogadro's number	N_A	6.022×10^{23} mol^{-1}
Planck's constant	h	6.626×10^{-34} J•s
Atomic mass unit	u	1.661×10^{-27} kg
Speed of light in vacuum	c	2.998×10^{8} m/s
Mass of electron	m_e	9.109×10^{-31} kg
Mass of proton	m_p	1.673×10^{-27} kg
Mass of neutron	m_n	1.675×10^{-27} kg
Boltzmann's constant	k	1.381×10^{-23} J/K
Molar gas constant	R	8.315 J•mol^{-1}•K^{-1}
Stefan's constant	σ	5.671×10^{-8} W•m^{-2}•K^{-4}

OTHER USEFUL CONSTANTS

Name	Value
Density of water	1.00×10^{3} kg/m^3
Specific heat of water	4186 J•kg^{-1}•K^{-1}
Latent heat of vaporization of water	2.26×10^{6} J/kg
Latent heat of fusion of water	3.33×10^{5} J/kg
Average radius of Earth	6.37×10^{6} m
Average Earth-Sun distance	1.50×10^{11} m

MOLAR MASSES

Element	Molar Mass
C	12
O	16
S	32
Ca	40

Energy, Physics, & the Environment

Ernest L McFarland
James L Hunt
John L Campbell

Department of Physics
University of Guelph
Guelph, Ontario

Wuerz Publishing Ltd
Winnipeg, Canada

phone (204) 453 7429
fax (204) 453 6598

Wuerz Publishing Ltd
Winnipeg, Canada

Energy, Physics, & the Environment
McFarland, Ernest L; Hunt, James L; and Campbell, John L

ISBN 0-920063-62-4

Printed in Canada

Preface

In the last quarter of the twentieth century, there has been a growing public concern about the dependence of our way of life on plentiful supplies of energy. The first event to shake compacency was the 1974 'energy crisis'. That event was caused by the overnight quadrupling of prices by the suppliers of a large portion of the world's oil. It had many outcomes beyond the immediate shock to the global economy and the resulting acceleration of measures designed to conserve energy in industry, commerce, transport and the home. One of these was the belated realization that many of the energy supplies on which humankind had become dependent were finite in extent. Energy, and the environmental impact of both its generation and its use, have become central issues with ramifications that are international in scope.

The great physicist Lord Kelvin wrote: "I often say that when you can measure what you are speaking about, and express it in numbers, you know something about it." Most certainly this is true of the energy question. Most citizens and many decision-makers, with little or no scientific background, find most of their information on energy in media articles that are far from quantitative, and at times erroneous. The aim of this book is to provide a quantitative account of energy for university students in science who have a first-year preparation in Physics. There are already available a number of excellent books which take a qualitative approach; these are directed in the main towards students in disciplines other than the sciences. In contrast to many of these treatments, we want to focus on the numbers that are involved, in such a way that the reader acquires a quantitative grasp of the various dimensions of the overall energy problem. To this end we have included a large number of exercises and problems.

We have tried to provide a balanced account of the advantages and disadvantages of the various energy types now used and those that may become available. However, it is not our intent to cover every single aspect of such a vast subject — the content has been chosen to fit into a standard half-year or one-semester course. The book is therefore a foundation on which an expanding knowledge of energy problems and issues may be built. It is our hope that the book will stimulate a continuing interest in its readers, and that it will prepare them to digest the type of in-depth articles that appear in journals such as *Scientific American, Science,* etc., and that offer opportunity to stay abreast of energy issues.

Since many environmental issues are directly linked to energy, we have tried to explore environmental ramifications, again using a quantitative approach. There has certainly been emotional public debate on many issues, and a mini-industry of advocacy groups has developed. Every energy decision in the end however will be made on the basis of a complex set of considerations including geographic, economic, political and social factors; while we recognize that the quantitative understanding that we advocate here is not the only factor, we strongly assert that it is indispensable as background knowledge.

Guelph, Ontario
November 1993

E L McFarland, J L Hunt, & J L Campbell

Contents

... to the memory of Rosemary McFarland

Acknowledgements

The authors gratefully acknowledge the assistance of the following individuals, who kindly provided useful information and advice: Colin Hunt of the Canadian Nuclear Association; Norm Davis of Energy, Mines, and Resources Canada; and David Anderson of Ontario Hydro. The authors wish also to thank the Department of Physics of the University of Guelph for providing the time and resources for the development of this book, and the publisher Steve Wuerz for his constructive editorial comments. In addition, we are grateful to the University of Guelph students enrolled in Physics 308 (*Energy*) during 1993 for their many suggestions on improving the manuscript.

1 ENERGY AND POWER IN TODAY'S WORLD

1.1 SOURCES OF ENERGY

Various sources of energy are available to the human race for day-to-day functions such as cooking, transportation, and heating. It might seem at first glance that there are many energy sources; for instance, buildings are often heated with oil, gas, wood, solar energy, or electricity generated from coal, falling water, nuclear energy, or wind. Nonetheless, we will see that on the most fundamental level there are only three energy sources available on Earth. In the next few pages, we give a brief introductory discussion of energy sources, all of which will be described in more detail in later chapters.

Fossil Fuels

About 90% of the energy used in the world today is provided by fossil fuels: oil, natural gas, and coal. The energy stored in these fuels — often referred to as chemical energy — is a combination of electric potential energy of the electrons and nuclei that constitute atoms and molecules, and the kinetic energy (energy of motion) of these electrons.

Where did this fossil-fuel energy come from? As you probably know, fossil fuels were created by the action of pressure and heat on the remains of plants and animals that lived hundreds of millions of years ago. The fundamental source of energy for these plants and animals was the sun; plants use the energy of sunlight to convert carbon dioxide and water to carbohydrates (sugars and starches), some animals eat plants, other animals eat these animals, etc. So when we use gasoline to power our automobiles, natural gas to heat our buildings, and coal to generate electricity, we are actually using solar energy that has been stored for a long time.

But what provides the sun's energy? In the sun, hydrogen nuclei are fused together in a multi-step process, creating larger nuclei of helium and releasing energy. The process of combining small nuclei into larger ones

— *nuclear fusion* — is the subject of much research in the hope of developing commercial fusion-energy sources here on earth, but this prospect appears to be far in the future.

Wind and Wave Energy

Although people are now starting to use the wind to generate electricity, it has been used for hundreds of years for sailing-ships and windmills (Fig. 1-1). Like fossil fuels, the wind is a form of stored solar energy: the sun heats the earth unevenly, thus developing regions of high and low air pressure, and the resulting air movements often cause high winds at ground level. Of course, solar energy is stored in the wind for a much shorter time than in fossil fuels.

Another source of energy, derived primarily from the wind, is wave energy. Except for wave-powered navigation buoys, wave energy is not yet economically competitive with other energy sources (especially inexpensive fossil fuels), and small pilot projects are progressing slowly.

(a)   (b)

Figure 1-1:   (a) In the Netherlands, wind has been used for centuries as a source of energy for small industries such as mills.
(b) Some electric generators in the Netherlands are now driven by the wind.

Wind and wave energy are both examples of *kinetic energy*, or energy of motion. Mathematically, the kinetic energy (KE) of a moving object is defined as:

$$KE = \frac{1}{2} mv^2 \qquad (1\text{-}1)$$

where m and v represent the mass and speed, respectively, of the object.

Biomass Energy

There are many different energy sources that are categorized as biomass. Biomass refers to carbon-based material produced by plants and animals in recent years (up to decades). The largest contributor at present is wood, which is primarily burned for heating or cooking. Also included in this category are:

- crops such as sugarcane whose sugars can be fermented to produce ethanol, which can be used as a fuel;
- manure or garbage which produce methane gas through anaerobic digestion by bacteria;
- woody biomass to produce either combustible gases (methane, hydrogen, and carbon monoxide) through a gasification process, or methanol, which is a liquid fuel.

Since all biomass results directly or indirectly from photosynthesis, biomass energy is yet another form of stored solar energy. Biomass provides medium-term storage of solar energy, whereas wind gives short-term storage, and fossil fuels represent long-term storage.

Hydro Energy

Another example of short-term stored solar energy is represented by falling water, often called hydro energy. The sun evaporates water from bodies of water and soil, the water forms clouds and eventually falls as rain, snow, etc. Where local geography happens to provide a large drop in elevation, then we have the possibility of using falling water to provide energy, usually for the generation of electricity.

The energy of the water (before it falls) is an example of *gravitational potential energy* (PE). As the water falls, this gravitational PE is converted to KE of the water, and in the case of a hydroelectric station, this KE is used to turn a turbine generator, which produces electricity.

The gravitational PE of an object of mass m at an elevation h is defined as:

$$\text{gravitational PE} = mgh \qquad\qquad (1\text{-}2)$$

where g is the magnitude of the gravitational acceleration (or gravitational field). Near the surface of the earth, the value of g is:

$$g = 9.80 \text{ m/s}^2$$

(Notice that when we say that an object has an elevation h, we mean that the *centre of mass* of the object has this elevation; for a symmetrical object such as a sphere or a rectangular box, the centre of mass is at the geometric centre.)

Direct Solar Energy

So far we have been discussing various types of stored solar energy, but of course solar energy itself can be used directly to provide heat or electricity. Solar energy is in the form of electromagnetic radiation, emitted from the sun in many regions of the electromagnetic spectrum: ultraviolet, visible, infrared, microwave, and radio. The energy of this radiation is quantized in packets called photons.

Nuclear Energy

In common parlance, nuclear energy refers to the energy released in the fission, or breaking apart, of a large nucleus (typically uranium) to produce two mid-sized nuclei. This *nuclear fission*, which occurs in a controlled way in a nuclear reactor, generates large amounts of heat that can be used to make steam and turn a turbine for generation of electricity or for motive power, as in a nuclear-powered submarine.

Uranium is a naturally-occurring element in the earth, and its formation was not due to the sun; thus, nuclear energy is the first energy source discussed here that is not directly derived from solar energy.

Geothermal Energy

Most people are aware that temperature increases with depth in the earth, and that there are regions where the surface of the earth is extremely hot, producing hot springs, geysers, etc. This surface heat can be exploited as an energy source. But what generates this heat? Some elements in the earth —— uranium, thorium, and radium, for example —— are naturally radioactive and decay by emitting high-energy particles that heat the surrounding material. And so geothermal energy is actually a derivative of natural nuclear energy.

Tidal Energy

The final energy source in this discussion is the tides. In a few locations around the world, the tides are high enough on a regular basis that it is feasible to consider trapping the water with a dam at high tide, and allowing the water to run out through turbines at low tide to generate electricity.

Tidal energy is neither solar nor nuclear in origin, but comes from the gravitational PE and the KE of the earth-moon-sun system. The gravitational PE is associated with the separation of the three bodies, and the KE is associated with the rotation of the earth and the orbital motion of the bodies about each other.

Electrical Energy

You might be wondering where electricity fits into the energy picture. Electricity itself is *not* a source of energy — it must be generated by a source such as a nuclear reactor, falling water, or fossil-fuel burning. Electricity is sometimes referred to as an *energy currency*, which (like monetary currency) serves as a medium of exchange between a raw source of energy (such as coal) and a convenient end-use (such as lighting). Another term used to describe electricity is *energy carrier* — that is, something that carries energy from a producer to a user. Another energy currency or carrier is hydrogen, which is discussed in a later chapter.

Fundamental Sources of Energy

You can now understand why we stated at the beginning of this Section that there are only three fundamental sources of energy that are available to us. All the sources that we have mentioned — fossil fuels, geothermal energy, tides, etc. — are derived either from solar energy, or nuclear energy, or (in the case of the tides) the gravitational PE and the KE of the earth, moon, and sun. Table 1-1 summarizes the connection between the various derived sources and their fundamental sources. Of course, since solar energy results from nuclear fusion, we could classify solar energy as nuclear energy, and list only two fundamental sources. However, because of the dominant importance of solar energy to life on earth, it has been considered separately as a fundamental source.

Table 1-1: Fundamental Energy Sources Corresponding to Various Derived Energy Sources

Derived Energy Source	Fundamental Energy Source
Fossil fuels	Solar energy (nuclear fusion)
Wind	"
Waves	"
Biomass	"
Hydro	"
Direct solar energy	"
Nuclear energy (reactors)	Nuclear energy (fission)
Geothermal energy	Nuclear energy (radioactive decay)
Tides	KE and grav. PE of earth-moon-sun

1.2 ULTIMATE FATE OF EARTH'S ENERGY

Section 1.1 briefly listed several energy sources. It is important to consider what eventually happens to the energy consumed, whether for transportation, heating, communications, rock concerts, or whatever. The electricity generated in a coal-fired power plant is a representative example.

In the electric power plant, coal is burned in a furnace, thereby producing heat to convert water to steam; some of the *heat is lost* up the chimney. The hot high-pressure steam is used to turn turbine generators to produce electricity, and is then condensed back to liquid water (to begin the cycle again) by being cooled by water from, say, a lake or river. This cooling water is then returned to the lake or river, which *becomes hotter* as a result.

The electricity that is generated is transmitted by power lines to the customer; some of the energy is lost as *heat* in the transmission lines themselves. This lost energy is typically about 10% of the total energy transmitted.

The consumer now uses the electricity in a lightbulb. For an incandescent lightbulb, only about 5% of the electrical energy goes into light; the rest appears as *heat*. (Put your hand near a lightbulb if you don't believe this.) The light is radiated outward from the bulb, and is absorbed by surrounding objects: walls, furniture, people, etc. As a result of this absorption, the surrounding objects *become slightly hotter*.

To recap, you can see that *all* the energy in the coal is eventually converted to heat. This complete conversion to heat is true for all energy processes here on earth. (As another example, consider the kinetic energy of an automobile travelling down a highway: when the brakes are applied, all the kinetic energy is converted to heat in the brakes.)

And what happens to this heat? It is radiated away into space as electromagnetic radiation. Any object at a temperature above absolute zero emits such radiation; for objects at typical temperatures on earth (about 20°C), this radiation is primarily in the infrared region of the spectrum, although there is also some in the microwave and radio regions. This invisible radiation is sent out into space and lost to us forever. In other words, we have only one chance to use the energy available to us, and it is then irretrievably gone.

1.3 UNITS OF ENERGY

One of the unfortunate facts of life in the energy field is the plethora of units in common use. Energy values can be quoted in calories, barrels of oil, electron-volts, British thermal units, etc., and in order to make sense of all these units, you will need to be adept in converting one unit to another. Example 1-1 (in this Section) shows a sample unit conversion.

In this book as throughout the scientific and technological community, the units used are generally those of the SI (Système International d'Unités). The SI unit of energy may be determined by using the definition of, say, kinetic energy:

$$KE = \frac{1}{2} mv^2$$

On the right-hand side of this definition, the SI unit for mass m is kg, and the unit for speed v is m/s. (The "½" is unitless.) Thus, the unit for $\frac{1}{2}mv^2$ is $kg \cdot m^2/s^2$. By convention, this unit is called the *joule* (J), named after James Prescott Joule (1818-1889), an outstanding British experimental physicist who was one of the leaders in the development of the principle of conservation of energy.

$$kg \cdot m^2/s^2 = joule \ (J)$$

Although we developed this unit using kinetic energy, the same unit arises regardless of the type of energy considered. You might wish to use the definition of gravitational potential energy given in Section 1.1 to show that its SI unit also works out to be $kg \cdot m^2/s^2$.

Another energy unit is the calorie (cal), originally defined as the heat required to raise the temperature of one gram of water from 14.5°C to 15.5°C. The calorie was introduced as a unit of heat about two centuries ago, when it was thought that heat and energy were different quantities. Today the U.S. National Bureau of Standards defines the calorie in terms of joules, with no reference to the heating of water:

$$1 \ cal = 4.184 \ J \ (exact)$$

You have undoubtedly encountered another type of calorie in connection with food consumption: the food calorie, or Calorie (Cal). (Notice the uppercase "C".) This Calorie is 1000 cal, or 1 kcal (kilocalorie), which is equivalent to 4184 J, or 4.184 kJ.

$$
\begin{aligned}
1 \ Cal \ &= 1 \ \text{food calorie} \\
&= 1 \ \text{kcal} \\
&= 4184 \ J \\
&= 4.184 \ kJ
\end{aligned}
$$

Note here that we have used the standard SI prefix "k" for kilo or 10^3.

Figure 1-2: SI units are taken seriously in Australia.

In North America, food energy is still commonly measured in the archaic Calories, but metrification is proceeding at a more rapid pace in some other countries, where joules are being used (Fig. 1-2). Another unit of energy frequently used in North America is the British thermal unit (Btu); furnaces and air conditioners are often rated in terms of how many Btu of heat energy they can supply or remove per hour. The definition of the Btu is similar to the original definition of the calorie: a Btu is the amount of heat energy required to raise the temperature of one pound of water from 63°F to 64°F. In joules, this is[1]:

$$1 \text{ Btu} = 1055 \text{ J}$$

A huge quantity of energy is consumed in the world each year, and to express this amount of energy handily, we need correspondingly large units. One such unit is the quad:

$$1 \text{ quad} = 1 \text{ quadrillion Btu's} = 10^{15} \text{ Btu}$$

World energy consumption in 1991 was 312 quads, of which 78 quads was consumed in the U.S.A. The U.S.A., which has only about 5% of the world's population, is responsible for approximately 25% of the world's energy consumption. Canada consumed 8 quads of energy in 1991, which although obviously less than the total U.S. consumption, is actually somewhat greater on a per capita basis. Indeed, Canadians are the biggest energy 'consumers' on Earth. This is perhaps easily explained by Canada's distances and by its climate, which are more extreme than those of other nations. (In addition, Canadian resource industries such as aluminum-smelting are very energy-intensive.) Still, no other nation uses a greater amount of energy (per capita) than Canada.

Although the terms calories, Btu's, and quads are still in use, they are gradually being replaced by SI units. For example, a quad and an exajoule ($1 \text{ EJ} = 10^{18}$ J) are roughly equal (as shown below), and exajoules are now being used more and more frequently instead of quads.

$$\begin{aligned} 1 \text{ quad} \quad &= 10^{15} \text{ Btu} \\ &= 1.055 \times 10^{18} \text{ J} \quad \text{(using 1 Btu = 1055 J)} \\ &= 1.055 \text{ EJ} \end{aligned}$$

Oil is a dominant energy source today, whether measured in calories of energy, in dollars of value, or in importance to international

---

[1] There are also other definitions of the Btu, which result in values from 1054 J to 1060 J.

development. The price of crude oil in $U.S. per barrel is an important general indicator of world energy prices, and is widely published in daily newspapers. As a result, consumption of energy resources other than oil is often quoted in terms of the amount of oil that would be required to provide the same amount of energy. The quantity of oil is usually given in barrels or in tonnes. (1 tonne = 1 metric tonne = 1000 kg ≈ 7.3 barrels.) For example, Canada's coal consumption in 1989 is tabulated in a variety of publications as "33 million tonnes of oil equivalent." Conversion to more conventional energy units is straightforward using the following (approximate) conversion factors:

$$1 \text{ tonne of (crude) oil} \approx 4.0 \times 10^7 \text{ Btu}$$

$$1 \text{ barrel of (crude) oil} \approx 5.5 \times 10^6 \text{ Btu}$$

A final energy unit, which is important in the study of subatomic particles, and which will be useful in our later discussion of nuclear reactors, is the electron-volt (eV). This unit is defined as the energy gained by an electron in passing through a potential difference of one volt. In joules:

$$1 \text{ eV} = 1.60 \times 10^{-19} \text{ J}$$

Energies in the keV to MeV range are common for single subatomic particles.

Example 1-1 *(This example reviews a general method for performing unit conversions.)*
Nuclear energy in Western Europe in 1989 provided 158 million tonnes of oil equivalent. Convert this quantity to joules.

A conversion factor from tonnes of oil directly to joules was not provided in the preceding section. However, we have a conversion factor from tonnes of oil to Btu (1 tonne of oil ≈ 4.0 × 10⁷ Btu), and another one from Btu to joules (1 Btu = 1055 J). We apply each of these factors in turn:

$$158 \times 10^6 \text{ tonnes of oil} \times \frac{4.0 \times 10^7 \text{ Btu}}{1 \text{ tonne of oil}} \times \frac{1055 \text{ J}}{1 \text{ Btu}} = 6.7 \times 10^{18} \text{ J}$$

Notice the approach: for each conversion, we multiply by a fraction in which the numerator and denominator are equivalent. In the first fraction, tonne of oil was placed in the denominator to "cancel" the tonnes of oil that we started with, leaving us with Btu, and then in the second fraction, Btu was placed in the denominator to "cancel" Btu and give us joules.

Note also that the final answer is given to two significant digits, since one of the conversion factors had only two significant digits. It is important to state numerical answers to the appropriate number of digits. Appendix I provides rules and sample exercises for determining the correct number of significant digits.

1.4 ENERGY CONSUMPTION AND SOURCES

In Section 1.3 we mentioned that world energy consumption in 1991 was 312 quads. Figure 1-3 shows a breakdown of this energy according to source. You can see that fossil fuels (oil, gas, and coal) account for 91% of world energy. This heavy dependence on fossil fuels is rather disturbing —— the reserves of these fuels are finite, and there are serious environmental problems (greenhouse effect, air pollution, etc.) associated with using them.

How is all this energy used? About 38% is used by industry, another 38% by the residential and commercial sectors (including public buildings such as schools and government offices), and 24% in transportation. Oil is the dominant source for transportion, of course, and road vehicles alone account for half of the annual consumption of oil.

The hydroelectric and nuclear energy components in Fig. 1-3 together provide 10% of world energy as electricity. However, do not be misled into thinking that electricity constitutes only 10% of world energy —— fossil fuels are also used to produce electricity. Approximately 45% of annual coal consumption generates electricity, along with 10% of gas, and 5% of oil. Combining this fossil fuel use with hydro and nuclear energy, a total of about 30% of world energy is used for the production of electricity.

World Energy Consumption
1991

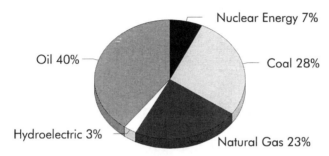

Figure 1-3: Fossil fuels account for 91% of world energy consumption.
(The total of all the items is greater than 100% because of
rounding.)
(Source: BP Statistical Review of World Energy, June 1992)

Canadian Energy Consumption
1991

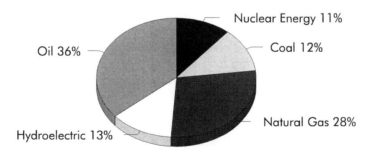

Figure 1-4: Canadian energy comes less from coal, and more from
hydroelectricity, when compared to world data of Fig. 1-3.
(Source: BP Statistical Review of World Energy, June 1992)

Canadian energy consumption by source is illustrated in Fig. 1-4. Fossil fuels account for 76% of Canadian energy, compared with 91% for the world, largely because of a smaller dependence on coal (12% for Canada versus 28% for the world). Hydroelectricity plays a larger role in Canada, supplying 13% of the energy, in contrast with only 3% in the world.

Figures 1-3 and 1-4 provide information only on commercially traded fuels. While other energy sources (wood, animal waste, wind, etc.) can be locally important in various regions of the world, their consumption is poorly documented and contributes only a small amount to the global energy total. In Canada, other energy sources are estimated to account for about 5% of the country's energy consumption. These sources are primarily solid wood wastes in the forestry and pulp and paper industries, wood used in residential heating, and "black liquor" (also known as pulping liquor), which is a burnable substance made up of wood constituents and chemicals that are by-products in the manufacture of pulp.

1.5 POWER

In everyday speech, the words "energy" and "power" are often used interchangeably, but the scientific meanings of these words are different. Energy has already been discussed above, and is related to the conept of power. The SI unit of power, the watt, is one that you have undoubtedly heard of: for example, hairdryers and lightbulbs are labelled according to their wattage.

Power is the rate at which energy is used or provided, that is, power (P) is simply energy (E) divided by the time (t) during which the energy was used. For example, if a car's KE increases, the power input to the car is the increase in the KE divided by the time taken. Mathematically, power (P) can be written as:

$$\text{Power P} \quad = \quad \frac{E}{t} \tag{1-3}$$

With energy in joules and time in seconds, the unit of power is joule per second (J/s), which is conventionally referred to as a *watt* (W):

$$W \text{ (watt)} = \frac{J}{s}$$

Watts and kilowatts (kW) are often used to indicate the power consumption of home appliances such as lightbulbs and hair-dryers, and megawatts (MW) are used in describing power output of electrical plants.

> **Did You Know?** The watt is named after James Watt (1736-1819) who did not, contrary to common opinion, invent the steam engine. Using scientific principles he improved the efficiency of the engine, invented by Thomas Newcomen (1663-1729), making it a commercial success.

A 60-watt lightbulb uses 60 J of electrical energy per second (but as seen in Section 1.2, only 5% of this energy is converted to light). As you sit reading this book, you use about 100 W of power, that is, each second you convert 100 J of food energy into other forms of energy (mainly thermal energy, which is then radiated away). A typical Canadian nuclear power reactor produces about 600 to 800 MW of electrical power, enough for 300 000 to 400 000 homes. At Niagara Falls, Ontario Hydro produces a total of 1800 MW of power at five hydro-electric stations. The electrical power demand in the Province of Ontario has a peak of about 23 000 MW or 23 GW, which occurs in January or February. The solar power striking the earth's surface is 178 000 TW, or 178 PW.

Another unit of power, often used in the automotive industry, is the horsepower:

$$1 \text{ horsepower} = 1 \text{ hp} = 746 \text{ W}$$

■

Example 1-2 *The speed of an automobile (mass 1.1 × 10³ kg) increases from 15 m/s (54 km/h) to 25 m/s (90 km/h) during a time interval of 5.5 s. Neglecting friction and thermal losses, what power output is required during this time?*

To determine the power, P, we consider the increase in the kinetic energy of the car, and divide by the time taken:

$$P = \frac{\frac{1}{2}mv_2^2 - \frac{1}{2}mv_1^2}{t} = \frac{\frac{1}{2}m(v_2^2 - v_1^2)}{t}$$

where v_2 and v_1 represent the final and initial speeds of the car, respectively, and t is the time. Substituting numerical values:

$$P = \frac{\frac{1}{2}(1.1 \times 10^3 \text{ kg})((25 \text{ m/s})^2 - (15 \text{ m/s})^2)}{5.5 \text{ s}} = 4.0 \times 10^4 \text{ W}$$

Thus, the power required is 4.0×10^4 W, or 40 kW. (As will be seen in Chapter 4, the actual power used by this automobile would be roughly four times this amount, with most of the additional energy going into waste heat.)

Example 1-3   A small hydroelectric plant produces 250 kW of electrical power from water falling through a vertical height of 18 m. What volume of water passes through the turbine each second?   (Assume complete conversion of the water's energy to electrical energy.)

In a hydro plant, the gravitational potential energy of the water is converted to kinetic energy as it falls, and then to electrical energy. Hence, the electrical power can be written as the gravitational potential energy lost by the water divided by the time:

$$P = \frac{mgh}{t}$$

where m is the mass of water falling through a vertical height h in time t, and g is the magnitude of the gravitational acceleration. Since the answer requires the volume of water, replace the mass m with the product of density, ρ, and volume, V:

$$P = \frac{\rho V g h}{t}$$

Re-arranging to solve for volume, and substituting numerical values:

$$V = \frac{Pt}{\rho g h} = \frac{(250 \times 10^3 \text{ W})(1.0 \text{ s})}{(1.00 \times 10^3 \text{ kg/m}^3)(9.80 \text{ m/s}^2)(18 \text{ m})} = 1.4 \text{ m}^3$$

Hence, a volume of 1.4 m^3 of water passes through the generator each second.

Kilowatt·hour —— An Energy Unit

One of the most confusing units for the general public (and even for physics students) is the kilowatt·hour (kW·h). Because part of the unit is kilowatt, which is a unit of power, many people believe that the kilowatt·hour is also a power unit. However, since the kilowatt is multiplied by a time unit (hour), the kilowatt·hour is actually a unit of energy, as shown below.

Since power is energy divided by time ($P = E/t$), then energy can be written as:

$$E = Pt$$

This relationship states that the product of power and time is energy (regardless of the particular units used). If SI units are used, power in W multiplied by time in s gives energy in J, that is, $J = W \cdot s$. Since a watt is just a joule per second, this should make sense. But power and time can also be expressed in other units; if power has units of kW and time has units of h (hours), then the product —— still an energy —— has units of kW·h. An energy of 1 kW·h can be expressed in J through a unit conversion, using $1 \text{ kW} = 10^3 \text{ W}$, $1 \text{ W} = 1 \text{ J/s}$, and $1 \text{ h} = 3600 \text{ s}$:

$$1 \text{ kW·h} \times \frac{10^3 \text{ W}}{1 \text{ kW}} \times \frac{1 \text{ J/s}}{1 \text{ W}} \times \frac{3600 \text{ s}}{1 \text{ h}} = 3.6 \times 10^6 \text{ J} = 3.6 \text{ MJ}$$

Hence, 1 kW·h is equivalent to 3.6 MJ.

The kW·h is a very handy unit of electrical energy consumption, since the total power requirement of a house is often in the kW range, and time can easily be measured in hours. Using joules for electrical energy would result in extremely large numbers. Example 1-4 provides a sample calculation of kW·h. For electrical energy consumption at the national and world levels, gigawatt·hours (GW·h) and terawatt·hours (TW·h) are often used.

Table 1-2: Electricity Prices in the Residential Sector, January 1992

City	Price (Cdn. ¢ per kW. h)
Seattle, U.S.A.	4.7
Sydney, Australia	6.8
Montreal, Canada	7.2
Vancouver, Canada	7.9
Houston, U.S.A.	11.7
Stockholm, Sweden	12.8
Los Angeles, U.S.A.	13.0
Chicago, U.S.A.	14.2
Paris, France	14.8
Mexico City, Mexico	16.1
New York, U.S.A.	16.5
London, England	17.0

Example 1-4   *A 100-W lightbulb is turned on for 12 h.*

(a) What is the energy consumption in kW·h?

(b) How much does this energy consumption cost?

(a) To determine the energy in kW·h, we multiply the power in kW by the time in h. Since the power is given in W, we first convert this to kW:

$$100 \ W \ \times \ \frac{1 \ kW}{10^3 \ W} = 0.10 \ kW \quad (assuming \ 2 \ significant \ digits)$$

Then, $\quad E = Pt = (0.10 \ kW) \cdot (12 \ h) = 1.2 \ kW \cdot h$

Thus, the electrical energy consumed is 1.2 kW·h.

(b) To calculate the cost of 1.2 kW·h, we need to know the price per kW·h, which varies from place to place. In Guelph, Ontario, as of January 1, 1994, the price is 12.44¢ per kW·h for the first 500 kW·h used in a two-month billing period, and 7.33¢ per kW·h for all remaining kW·h in the billing period. A typical household uses between 2000 and 3000 kW·h of energy in two months, giving an average price of about 8¢ per kW·h. (See Problem 1-4

at the end of this chapter for a calculation of average price.) Therefore, the cost of 1.2 kW·h is approximately:

$$1.2 \ kW \cdot h \ \times \ \frac{8¢}{kW \cdot h} \ \approx \ 10¢ \quad \textit{(to the nearest cent)}$$

Thus, leaving a 100-W lightbulb on for 12 h costs about 10¢. A resident of London, England, would pay roughly twice this amount. (See Table 1-2 for the price of electricity at selected locations around the world.)

■

1.6 ELECTRICAL ENERGY

Now that the kilowatt· hour has been introduced, we turn our attention to electrical energy in Canada and around the world. Canada has a rich history in the production of electricity, summarized in Table 1-3. The world's largest hydroelectric producer is Canada, which is also a leader in long-distance transmission of electricity.

Table 1-4 provides an international comparison of electricity generation (in TW· h) by fuel type for various countries in the year 1989, showing Canada's unique position in hydroelectricity. As you can see from the table, about 60% of Canada's electricity is generated from hydro, compared to only 18.3% in the world overall. Notice that over 60% of the world's electricity is generated by burning fossil fuels, primarily coal.

The U.S.A. leads all countries in total electricity generation (as it does in total energy consumption). Canada occupies fifth position in electricity generation in the world — this is astounding, considering that Canada has only 0.5% of the world's population. On a per capita basis, Canada is second only to Norway in electrical consumption — on average, each Canadian uses about 18000 kW· h of electrical energy per year, compared to a world average of approximately 2000 kW· h, and 12000 kW· h in the U.S.A.

Table 1-3: Significant Developments in Electricity Production in Canada
(Source: *Electric Power in Canada 1990*, Energy, Mines and Resources Canada)

YEAR	EVENT
1846	Toronto, Hamilton, Niagara, and St. Catharines Electro-Magnetic Telegraph Company formed
1876	Alexander Graham Bell made first long-distance telephone call from Paris, Ontario, to Brantford, Ontario, via battery power in Toronto (a total distance of 218 km)
1878	Experiments with electric street lighting in Victoria
1878	Canada's first electric light company (American Electric and Illuminating Company) formed in Montreal
1881	Toronto's first electric generator built by J.J. White for customer T. Eaton
1882	One of North America's first hydroelectric generating facilities constructed at Chaudière Falls on the Ottawa River
1883	Canada's first street lights installed in Hamilton, Ontario
1884	Canada's first electric utility set up in Pembroke, Ontario
1885	First Canadian electric streetcar enters service in Toronto
1897	First long-distance, high-voltage transmission line (11 kV) carried power 29 km to Trois-Rivières, Quebec
1900	Hydroelectric power distributed in all provinces except P.E.I. and Saskatchewan
1921	Sir Adam Beck No. 1 generating plant, then the largest in the world, opened in Niagara Falls
1930	Canada's installed generating capacity reached 4700 MW
1967	First commercial-scale (220 MW) CANDU nuclear generating station in service at Douglas Point, Ontario
1971	First two of four 542-MW nuclear generating units commissioned at Pickering, Ontario

Table 1-4: Electricity Generation in TW·h by Fuel Type, 1989
(Source: *Electric Power In Canada 1991*, Energy, Mines and Resources, Canada)

Country	Conventional Thermal (primarily coal)	Hydro	Nuclear	Geothermal	Total
U.S.A.	2163	272	529	17	2981
Japan	518	98	183	1	800
China	473	110	0	0	582
Canada	128	291	80	0	500
France	52	51	304	0	407
United Kingdom	240	7	65	0	312
Australia	133	14	0	1	148
Mexico	92	23	0	5	120
World total	7409 (64.8%)	2094 (18.3%)	1884 (16.5%)	39 (0.3%)	11427 (100.0%)

Table 1-5: Canadian Electrical Energy Production in TW·h by Fuel Type, 1990
(Source: *Electric Power in Canada 1990*, Energy, Mines and Resources Canada)

	Coal	Oil	Gas	Nuclear	Hydro	Other	Total
Nfld.	0	1.9	0	0	34.9	0	36.8
P.E.I.	0.1	<0.1	0	0	0	0	0.2
N.S.	5.7	2.5	0	0	1.2	0.1	9.4
N.B.	1.1	6.3	0	5.3	3.5	0.3	16.6
Quebec	0	1.9	0	4.1	129.4	0	135.5
Ontario	26.4	1.1	2.0	59.4	40.2	0.3	129.3
Manitoba	0.3	<0.1	<0.1	0	19.7	<0.1	20.1
Sask.	8.6	<0.1	0.5	0	4.2	0.2	13.5
Alberta	35.2	0	5.1	0	2.1	0.5	42.9
B.C.	0	0.7	1.6	0	57.2	1.2	60.7
Yukon	0	<0.1	0	0	0.4	0	0.5
N.W.T.	0	0.2	0	0	0.3	0	0.5
Canada	77.4	14.7	9.2	68.8	293.1	2.6	466.0

The mix of energy sources used for electricity generation varies greatly from province to province within Canada. Table 1-5 shows electrical energy production in Canadian provinces by fuel type in the year 1990. There are a number of interesting differences between provinces:

- In Newfoundland, Quebec, Manitoba, and British Columbia, hydro generation accounted for more than 94% of the total, but in Alberta about 82% came from coal. (Coal played an important role in Saskatchewan and Nova Scotia as well, producing 64% and 62% of the electricity, respectively.)

- Only three provinces —— New Brunswick, Quebec, and Ontario —— have nuclear power, with Ontario being by far the largest nuclear user, producing 46% of its electricity with nuclear reactors.

- Natural gas is not used at all for electricity generation in Quebec and the Maritime Provinces.

Total electricity generation in Canada actually decreased in 1989 and 1990, dropping by a total of 7.5% from the 1988 value. This decline was due to economic recession and tighter regulations on emissions of acid gases. Between 1975 and 1990, Canadian electricity generation increased by an average of 3.8% annually.

Recommended Further Reading

References to additional reading for most of the chapters are provided in Appendix V.

EXERCISES

Please refer to the Appendices and inside bookcovers for useful information.

1-1 A truck of mass 4500 kg brakes to a halt from a speed of 82 km/h in a distance of 77 m.
(a) How much energy (in megajoules) must be dissipated? [Ans. 1.2 MJ]
(b) What happens to this energy? [Ans. is given in Section 1.2]

1-2 By using conversion factors provided in Section 1.3 or inside bookcovers, perform the following unit conversions:
(a) 5.82 Btu to Calories [Ans. 1.47 Cal]
(b) 22 tonnes of crude oil to quads [Ans. 8.8×10^{-7} quads]
(c) 52 billion barrels of crude oil to quads [Ans. 2.9×10^2 quads]

1-3 A sprinter, starting from rest, has a power output of 5.1×10^2 W for a time of 7.2 s. Neglecting losses due to production of heat, etc., determine his final speed. The sprinter's mass is 67 kg.
[Ans. 1.0×10^1 m/s]

1-4 (a) By using the basic definition of a watt (and the meaning of kilo), determine how many joules there are in one kilowatt·hour.
[Ans. 3.6×10^6 J]
(b) Convert 581 kW·h to joules. [Ans. 2.09×10^9 J]

1-5 The annual solar energy striking the earth's surface is 178 000 TW·y. Convert this to joules. [Ans. 5.61×10^{24} J]

1-6 (a) Given below are the electrical power requirements for five household appliances. Determine the number of kilowatt·hours of electrical energy consumed if all these appliances are running simultaneously for two hours in a house. [Ans. 2.67 kW·h)]

  solid state color TV: 145 W
  automatic washing machine: 512 W
  furnace fan: 500 W
  clock: 2 W
  humidifier: 177 W

(b) If the average residential cost for electricity is 8¢ per kW·h, what would be the cost (to the nearest cent) for the energy calculated in (a)?
[Ans. 21¢]

PROBLEMS

Please refer to the Appendices and inside bookcovers for useful information.

1-1 In a 770-kW hydrolectric plant, 300 m³ of water pass through the turbine each minute. Assuming complete conversion of the water's initial gravitational potential energy to electrical energy, through what distance does the water fall? Assume two significant digits. [Ans. 16 m]

1-2 Calculate the power in megawatts available from a tidal power scheme where the incoming tide fills up a catchment area enclosed by concrete walls, and then at low tide this water is allowed to fall through openings at the bottom of one of the walls to spin turbines. The square catchment area has 1.2-km sides, and the tide rises by 3.7 m. Assume that the process of emptying takes 1.0 h, and that all the energy of the water is converted to energy of the turbines. Hint: when using the expression "mgh" for gravitational potential energy, remember that the "h" represents the height of the centre of mass. [Ans. 27 MW]

1-3 (a) Suppose that an automobile is travelling at a constant speed on a horizontal road; the engine is running and obviously producing energy. Where does this energy go? (It does not go into kinetic energy of the auto, because the kinetic energy does not change if the speed is constant.)
(b) An automobile moving at 90 km/h ascends a hill of gradient 1 in 25 (i.e., a vertical rise of 1 unit for a horizontal distance of 25). Its mass is 1300 kg. What power (in kilowatts) is needed for the climb up the hill over and above the normal power used when moving horizontally? Assume two significant digits. [Ans. 13 kW]

1-4 (a) The residential rates for electricity in Guelph (as of January 1994) are 12.44¢ per kW·h for the first 500 kW·h (in a two-month billing period), and 7.33¢ for each additional kW·h. If a family has a consumption of 2875 kW·h in a billing period, what is its average cost per kW·h? [Ans. 8.22¢ per kW·h]
(b) Does this rate structure promote conservation?

1-5 The intensity of radiation from a distant source like the sun varies inversely with the square of the distance from the source. At the top of the atmosphere the solar intensity (power per unit area) is 1.35 kW/m^2. If a space ship with a solar panel of area 125 m^2 were halfway between the sun and earth, determine the total solar energy that the panel would receive in a day. Express your answer in kilowatt·hours and in joules.

[Ans. 1.62×10^4 kW·h, 5.83×10^{10} J]

ENERGY CONSUMPTION: PAST, PRESENT, AND FUTURE

The data shown in Chapter 1 demonstrated that almost 90% of world energy consumption comes from fossil fuels. A question of obvious importance is: How long will our fossil-fuel resources last? An accurate answer to this question is important; it requires an understanding of trends in energy consumption, and some mathematical "crystal-ball-gazing" as we make projections for the future.

2.1 HISTORICAL ENERGY CONSUMPTION

First, consider the question, 'How much energy does one need for daily survival?' Ignore, for a moment, the energy consumed by people in driving automobiles, in manufacturing and using all sorts of commercial products, in heating, cooling, and lighting, etc., and all we *really* need is food energy. A typical person uses about 2000 to 3000 food calories (or roughly 10 MJ) per day. This was evidently the per capita energy consumption of primitive humans.

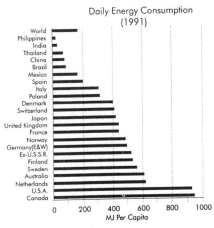

Figure 2-1: North Americans use about 50% more energy per capita than people in other industrialized nations.
(Source: BP Statistical Review of World Energy, June 1992)

Throughout history and even in pre-history, humans have been cleverly inventive in tapping energy from many sources and for many ends. Primitive people learned to use fire in about 100,000 B.C., they domesticated animals for agriculture and transportation around 5000 B.C., and began to exploit the wind in about 1000 B.C. At about the time of Christ, the Romans developed the waterwheel, and a thousand years later, coal started as an energy source. The Industrial Revolution of the late 18th century, with steam engines driving machinery, greatly increased the consumption of energy. Nowadays, there is scarcely a single facet of our daily activities that does not use some external source of energy. Oil is used for transportation; gas for heating and cooking; and coal, hydro, and nuclear energy to generate electricity for lighting, heating, cooking, cooling, entertainment, computing, etc. In addition, fossil fuels and other energy sources are used in a variety of ways in manufacturing and in resource-based industries such as forestry and mining.

As a result, the 10 MJ daily per capita energy consumption of our primitive ancestors has now increased to a world average of almost 200 MJ. Figure 2-1 shows the per capita energy consumption for a number of countries and the world. North Americans use about 50% more energy per capita than people in other industrialized countries, who in turn use considerably more than people in developing countries.

Notice that per-capita energy consumption is highest in North America, with Canadian consumption higher than U.S.A. consumption by a small margin. It should also be noted that the data shown represent only 'measurable' energy, that is, the sorts of energy which are commonly bought and sold, such as gas, oil, and electricity. In less-developed countries, the amount of traditional fuels used — peat, animal waste, and so forth — is not easily determined. If such fuels were measurable, Fig. 2-1 would show a somewhat larger energy consumption in countries such as the Philippines and India.

Total world energy consumption depends not only on per capita consumption, but also on world population. Per capita consumption and world population have both been increasing, and the result is an escalating world energy consumption (Fig. 2-2). Although this graph shows a couple of downward segments, the general trend is steadily upward. The slight dips are themselves interesting: notice the drop due to the economic depression in the early 1930s, and the decrease in the early 1980s because of rising oil prices and the consequent economic recession.

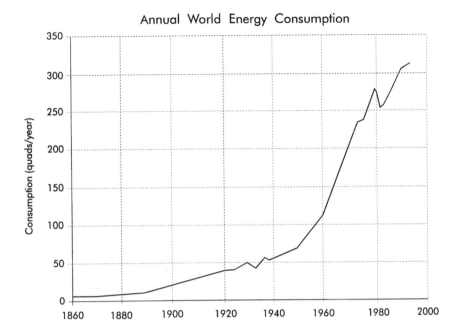

Figure 2-2: World energy consumption has generally shown an upward trend.
(Sources: BP Statistical Review of World Energy, June 1992; J. Fowler, Energy & the Environment, 1984)

Table 2-1: Average Annual Growth Rates, 1974 - 1989
(Source: *A Summary of Energy, Electricity, and Nuclear Data 1991*, AECL)

Country Group	Annual Growth Rate (%) in Population	Annual Growth Rate (%) in Total Energy Consumption
North America	0.9	0.4
Latin America	2.3	3.0
Western Europe	0.5	1.0
Eastern Europe	0.9	3.2
Africa	2.9	4.2
Middle East and South Asia	2.1	6.0
South-East Asia and the Pacific	2.1	4.3
Far East	1.5	3.9
World Average	1.7	2.3

Growth in energy consumption and population is not uniform around the world. Table 2-1 shows the average annual growth rates for population and energy consumption for various geographical regions during the period 1974 - 1989. North America and Western Europe show the smallest growth rates in both population and energy use, whereas developing countries tend to have large growth rates in both. Of course, as shown in Fig. 2-1, countries in North America and Western Europe already consume very large amounts of energy, compared to developing nations. One of the serious problems facing the world today is the friction between industrialized nations — with an already-high standard of living and large energy consumption — and developing nations, which often have a desire to increase their energy consumption quickly in order to improve their standard of living. Unfortunately, there are major environmental consequences associated both with the large energy consumption of the industrialized nations and with the rapidly increasing consumption in the developing countries. These environmental effects are discussed in later chapters.

2.2 THE MATHEMATICS OF EXPONENTIAL GROWTH

Population and energy consumption have been increasing. In order to set the stage for making predictions of future population and energy consumption, and for calculating how long fossil fuels might last, we need to consider the mathematics of *constant-percentage growth*, or *exponential growth*.

As an example of exponential growth, suppose that you invest $5.00 in a bank account which pays 10% interest per year, compounded annually. After one year, you have the original $5.00, plus $0.50 interest, for a total of $5.50. In essence, the original $5.00 has been multiplied by a factor of 1.10, where the 0.10 represents the 10% interest. The $5.50 is then invested for a second year, and is multiplied by another factor of 1.10, giving a total of $6.05 after two years. The original $5.00 has now been multiplied twice by 1.10, i.e., by $(1.10)^2$. Table 2-2 shows how the original deposit of $5.00 grows with time, with each year contributing a multiplication by 1.10. After n years, the total is $5.00 times $(1.10)^n$. Notice that the *exponent* of the 1.10 is growing — after 2 years we have $(1.10)^2$, after 3 years $(1.10)^3$, etc. — hence the name *exponential growth*.

Exponential growth results whenever a quantity grows by a certain percentage (such as 10%) per unit time (such as a year).

Table 2-2: Investing $5.00 at 10% interest

Time (yr.)	Amount ($)
0	5.00
1	$5.00 \, (1.10) = 5.50$
2	$5.00 \, (1.10)^2 = 6.05$
3	$5.00 \, (1.10)^3 = 6.66$
n	$5.00 \, (1.10)^n$

In order to write down a mathematical equation for exponential growth, begin by defining:

N = quantity that is growing (such as money in a bank account, energy consumption, etc.)

t = time

N_o = quantity at time $t = 0$ (beginning of time interval of interest)

C = constant multiplying factor (such as the 1.10 used above)

Following our example of the money invested in the bank account, it is straightforward to write an expression for the quantity N at any time t:

$$N = N_o C^t \qquad (2\text{-}1)$$

Although Eqn. 2-1 is a perfectly good equation for exponential growth, it is not in the form normally used — instead of using the base C, it is much more common to write the equation with the base "e," where e has the usual value of 2.718 . . . We can change to base e by first writing C as:

$$C = e^{\ln C}$$

where ℓn C is the natural logarithm of C. Now, since C is a constant, then ℓn C is just another constant, which we can write as k:

$$\ell n\ C = k$$

Thus, C can be expressed as: $\quad C = e^{\ell n\ C} = e^k$

Substitution of $C = e^k$ into Eqn. 2-1 gives:

$$N = N_o(e^k)^t$$

Or, more simply: $\qquad N = N_o e^{kt}$ \hfill (2-2)

Equation 2-2 is the usual way that exponential growth is expressed mathematically.

The constant k is referred to as the *growth constant*, and it is useful to relate k to the percent growth rate, symbolized by R. (If the growth rate is 10% per year, then R is 10.) C is related to R by $C = 1 + R/100$ (for $R = 10$, then $C = 1.10$), and since $k = \ell n\ C$, then:

$$k = \ell n\ (1 + R/100)$$ \hfill (2-3)

Figure 2-3 shows a graph of $N = N_o e^{kt}$. Notice that as t increases, the curve rises ever more steeply — that is, the rate of change of N increases with time. In other words, with exponential growth, the growth becomes bigger and bigger as time progresses. For a larger value of k, the curve would rise even more steeply. Notice the similarity between Fig. 2-3 and Fig. 2-2, which shows total world energy consumption as a function of time.

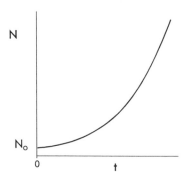

Figure 2-3: Graph of $N = N_o e^{kt}$.

The equation $N = N_0 e^{kt}$ can be written in an alternative way, which is also very useful. Start by re-arranging $N = N_0 e^{kt}$ as:

$$\frac{N}{N_0} = e^{kt}$$

Taking natural logarithms of both sides:

$$\ell n \frac{N}{N_0} = \ell n \, e^{kt}$$

But $\ell n \, e^{kt}$ is just kt. Therefore, $\ell n \dfrac{N}{N_0} = kt$ $\qquad\qquad$ (2-4)

This logarithmic form of the equation for exponential growth is very handy when dealing with situations in which N and N_0 are known (or the ratio N/N_0 is known), and either one of k or t is known while the other is to be determined.

■

Example 2-1   *Between the years 1950 and 1975, world energy consumption grew from 70 quads/yr to 230 quads/yr, with a growth that was approximately exponential.*

(a) Determine the growth constant.
(b) What was the average annual percent growth rate?
(c) What would the consumption be in the year 2000, assuming continued exponential growth at the same rate?

(a) *The given quantities are N = 230 quads/yr, N_0 = 70 quads/yr, and t = 25 yr. In order to determine k, start with Eqn. 2-4:*

$$\ell n \frac{N}{N_0} = kt$$

Re-arranging and substituting: $k = \dfrac{1}{t} \, \ell n \dfrac{N}{N_0}$

$$= \frac{1}{25 \; yr} \, \ell n \, \frac{230 \; quads/yr}{70 \; quads/yr}$$

$$= 4.8 \times 10^{-2} \; yr^{-1}$$

Note that the units of k are yr^{-1}, if time t is in yr. This means that the exponent *(kt) in $N = N_o e^{kt}$ is dimensionless, as all exponents must be.*

(b) To calculate the average annual percent growth rate, R, we use Eqn. 2-3, knowing k:

$$k = \ell n \,(1 + R/100)$$

To remove R from inside the logarithm, perform the exponentiation using base e:

$$e^k = e^{\ell n \,(1 + R/100)}$$
$$= 1 + R/100$$

Re-arranging to solve for R, and substituting the value of k:

$$R = 100 \,(e^k - 1) = 100 \,(e^{4.8 \times 10^{-2}} - 1) = 4.9$$

Hence, the average annual growth rate is 4.9% per year.

(Perhaps you are wondering about units for e^k, since it appears that the exponent is not dimensionless. However, the k in this exponent is implicitly multiplied by a time of one year, since R is the growth rate for one year.)

(c) To determine the projected energy consumption in the year 2000, use Eqn. 2-2:

$$N = N_o e^{kt}$$

To find N in this particular question, either

1. *set t = 0 in the year 1950 and use N_o = 70 quads/yr; then t = 50 yr for the year 2000*

or 2. *set t = 0 in the year 1975 and use N_o = 230 quads/yr; t = 25 yr.*

Using the latter set of values and substituting:

$$N = (230 \text{ quads/yr}) \, e^{(4.8 \times 10^{-2} \text{ yr}^{-1})(25 \text{ yr})} = 7.6 \times 10^2 \text{ quads/yr}$$

Thus, if exponential growth at the 1950-1975 rate had continued until the year 2000, the annual consumption in that year would have been 7.6×10^2 quads/yr. ∎

2.3 DOUBLING TIME

One significant feature of exponential growth is that for a given growth constant, there is a constant time required for the growing quantity to double. This time, called the *doubling time*, is inversely related to the growth constant.

To show this relationship, start with the equation $\ell n \, \dfrac{N}{N_o} = kt$. For a doubling, $\dfrac{N}{N_o} = 2$, and the time t is the doubling time, represented as T_2.

Hence, $\ell n\ 2 = kT_2$

Solving for T_2: $T_2 = \dfrac{\ell n\ 2}{k} = \dfrac{0.693}{k}$ (2-5)

Notice that if k has units of, say, yr^{-1}, then the units of T_2 are yr.

Example 2-2   *In Example 2-1, it was shown that the growth constant for world energy consumption between 1950 and 1975 was $4.8 \times 10^{-2}\ yr^{-1}$. What is the doubling time associated with this growth?*

Use Eqn. 2-5: $T_2 = \dfrac{0.693}{k} = \dfrac{0.693}{4.8 \times 10^{-2}\ yr^{-1}} = 14\ yr$

Thus, the doubling time is 14 yr.

It is important to recognize that the doubling time applies to all doublings of the growing quantity. In Example 2-2, energy consumption doubles *every* 14 yr. This means that in 14 years energy consumption doubles, after another 14 years it doubles again (so that it has quadrupled relative to the original value), then after another 14 years it doubles so that it is 8 times the original value, etc. To put this another way: when growth is exponential, the amount of the increase is itself a rapidly growing quantity.

2.4 HANDY APPROXIMATIONS

Thus far, equations used for exponential growth have all been mathematically exact. However, it is often convenient to use approximate relationships for exponential growth that permit us to do some calculations in our head. We begin by looking at Eqn. 2-3, which relates the growth constant k and the percent growth rate R:

$$k = \ell n\ (1 + R/100)$$

The right-hand side of this equation has the form of $\ell n\,(1 + x)$, where $x = R/100$. Now, if $|x| < 1$, $\ell n\,(1 + x)$ can be written in a Taylor series expansion:

$$\ell n\,(1 + x) = x - \frac{x^2}{2} + \frac{x^3}{3} - \frac{x^4}{4} + \ldots \quad \text{for } |x| < 1$$

This expansion can be simplified if $|x| << 1$; in this case, the first term, x, is much greater than any of the subsequent terms involving x^2, x^3, etc. Thus, the expansion can be approximated by x alone:

$$\ell n\,(1 + x) \approx x \quad \text{for } |x| << 1$$

Returning to $k = \ell n\,(1 + R/100)$, we use the above approximation to write the right-hand side as

$$\ell n\,(1 + R/100) \approx R/100 \quad \text{for } R/100 << 1$$

Hence, $\qquad\qquad k \approx R/100 \quad \text{for } R/100 << 1$ \hfill (2-6)

Approximation (2-6) is a very handy way to relate k and R if the growth rate is relatively small, that is, *less than about 10% per unit time*. This approximation can also be used to develop a relation between doubling time and percent growth rate. Start with Eqn. 2-5:

$$T_2 = \frac{0.693}{k}$$

Use approximation 2-6 to replace k with $R/100$:

$$T_2 \approx \frac{0.693}{R/100} \approx \frac{69.3}{R}$$

Since this is just an approximation, the value 69.3 can be replaced with 70:

$$T_2 \approx \frac{70}{R} \quad \text{for } R/100 << 1 \qquad (2\text{-}7)$$

■

Example 2-3   *World energy consumption between 1950 and 1975 grew at a rate of 4.9% per year. Determine approximate values for the growth constant and doubling time.*

Since the growth rate is less than 10% per year, use approximations 2-6 and 2-7. The growth rate of 4.9% is very close to 5%, and since approximations are being employed, it is legitimate to use 5% — that is, R ≈ 5.

$$k \approx R/100 \approx 5/100 \approx 0.05 \ yr^{-1}$$

$$T_2 \approx \frac{70}{R} \approx \frac{70}{5} \approx 14 \ yr$$

Thus, the growth constant is approximately 0.05 yr^{-1} and the doubling time is approximately 14 yr. Compare these values with the more accurate numbers determined in Examples 2-1 and 2-2: 0.048 yr^{-1} and 14 yr. The approximations are obviously very good.

■

2.5 SEMILOGARITHMIC GRAPHS

It is often useful to graph data in such a way that the graph appears as a straight line. With exponential growth, a linear graph will be obtained if the data are plotted on a semilogarithmic graph, as explained below.

Start with Eqn. 2-4, the logarithmic form of the exponential growth equation:

$$\ell n \ \frac{N}{N_o} = kt$$

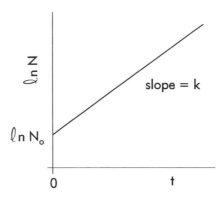

Figure 2-4:   A graph of ℓn N vs. t for exponential growth.

Rewriting the left-hand side as ℓn N - ℓn N_o , and then re-arranging to solve for ℓn N gives:

$$\ell n\ N = kt + \ell n\ N_o$$

It is now possible to plot a graph in which ℓn N is on the y-axis, and t on the x-axis, i.e., ℓn N = y, and t = x. The equation becomes, in effect,

$$y = kx + \ell n\ N_o$$

This has the form of the equation of a straight line: y = mx + b, where the slope m is the growth constant k, and the y-intercept b is ℓn N_o. In other words, under conditions of exponential growth a plot of ℓn N vs. t will yield a straight line with slope k and y-intercept ℓn N_o (Fig. 2-4). Such a graph is referred to as a *semilogarithmic (or semilog) graph*, since one of the two axes involves a logarithm.

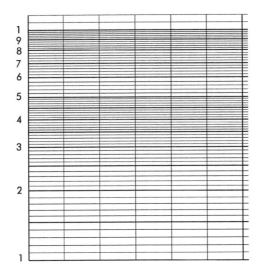

Figure 2-5: Semilogarithmic graph paper.

When a semilogarithmic graph is being plotted, it is tedious to take natural logarithms of all the numbers to be plotted on the y-axis. It would be easier just to plot the y-values in the regular way and somehow have the logarithms determined automatically. A special type of graph paper, semilogarithmic paper (Fig. 2-5), does just that for us. It has spacings along the y-axis that automatically space the data points as if we had taken logarithms. The x-axis has normal linear spacings. Notice on Fig. 2-5 that equal distances on the vertical axis correspond to *multiplication* by a given number. For example, the distances between 1 and 2, 2 and 4, and 4 and 8 are all equal, corresponding to a multiplication by two. On the linear x-axis, equal distances indicate *addition* by a given number.

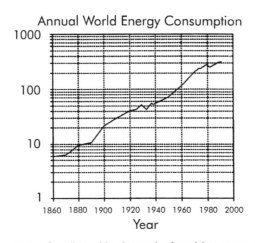

Figure 2-6: Semilogarithmic graph of world energy consumption.

Figure 2-6 shows the data of Fig. 2-2 for world energy consumption replotted on a semilog graph. If you were actually plotting this graph yourself, you would place the data points just as on a normal graph; the consumption in the year 1920 was 40 quads/yr, for example, and you can confirm the plotting of this point easily in Fig. 2-6. (You might wish to find the corresponding point in Fig. 2-2.) Although the semilog graph of Fig. 2-6 has some bumps and wiggles, it is nonetheless roughly linear, indicating that world energy consumption has indeed been approximately exponential from 1860 to the present day. Therefore, it is quite reasonable to model world energy consumption as an exponential function.

Example 2-4   *The data shown in the semilog graph of Fig. 2-6 can be approximated quite well as a straight line joining two points: 200 quads/yr in the year 1970, and (approximately) 7 quads/yr in the year 1870. (Use a straight edge to confirm this.) Determine the growth constant associated with this line.*

The growth constant (k) is just the slope of the line; normally the slope is calculated simply as rise/run, but since this is a semilog graph, it is necessary to take natural logarithms of the vertical variable (consumption N):

$$k = slope = \frac{\ln N_2 - \ln N_1}{t_2 - t_1} = \frac{\ln (N_2/N_1)}{t_2 - t_1}$$

$$= \frac{\ln (200/7)}{100 \ yr}$$

$$= 0.03 \ yr^{-1}$$

Notice that the calculation of this slope is equivalent to solving for k in the equation $\ln (N/N_o) = kt$, using $N = 200$ quads/yr, $N_o = 7$ quads/yr, and $t = 100$ yr.

2.6 LIFETIME OF FOSSIL FUELS (CONSTANT CONSUMPTION MODEL)

It is now possible to provide one answer to the question: How long will fossil fuels last? Since energy consumption has been increasing exponentially, it seems appropriate to use exponential growth as a model for fossil-fuel consumption, and the next section deals with this topic. In the present section, as a first step, consider fossil fuel lifetimes if consumption is constant.

The lifetime of fossil fuels depends on two factors: the amount of fossil fuels remaining, and the rate of consumption. The amount remaining is often quoted in two categories:

Reserves (or proven reserves) refer to fossil fuels that have already been discovered, their quantity measured, and are known to be extractable at a competitive price.

Resources (or recoverable resources) include the reserves and, in addition, fossil fuel deposits that are inferred or expected, for which recovery is anticipated to be technically and economically feasible. For the purposes of estimating fossil fuel lifetime, resources are the important quantity.

Table 2-3   Fossil Fuel Reserves and Resources (in quads)[1]

Geographic Region	Oil Reserves	Oil Resources	Gas Reserves	Gas Resources	Coal Reserves	Coal Resources
World	5500	7000	4000	8000	20,000	150,000
Canada	40	150	100	300	230	2000
U.S.A.	200	450	250	700	6000	36,000

Table 2-3 shows fossil fuel reserves and resources (in quads) for the world, Canada, and U.S.A. The data shown for resources have rather large uncertainties, and should be taken only as "ball-park" estimates. The reserves are known more accurately, but of course even they change in

---

[1] Sources: *BP Statistical Review of World Energy, June 1992; Canadian Energy Statistics 1990,* Energy, Mines, and Resources, Canada; W. Fulkerson et al, *Energy from Fossil Fuels,* in *Scientific American,* September 1990; *A Summary of Energy, Electricity and Nuclear Data 1991,* Atomic Energy of Canada, Ltd.; H.A. Bethe and D. Bodansky, *Energy Supply,* in *A Physicist's Desk Reference, The Second Edition of Physics Vade Mecum,* American Institute of Physics, 1989.

time as fuels are consumed, new discoveries are made, new extraction technologies are developed, and the prices of fuels change. It should be mentioned that 66% of world oil reserves are located in the Middle East, and only 4% in North America. Notice the large resource of coal, compared to oil and gas.

Table 2-4: Fossil Fuel Consumption in 1991 (in quads/yr)
(*BP Statistical Review of World Energy, June 1992*)

Region	Oil	Gas	Coal
World	126	71	87
Canada	3.0	2.3	1.0
U.S.A.	31	20	19

As the data in Table 2-4 demonstrate, the consumption of oil is greater than that of gas or coal. Since the resources and present consumption rate are known for these fossil fuels, it is possible to determine how long the resources would last *if* consumption remains constant at its 1991 value by simply dividing each of the resources in Table 2-3 by the appropriate consumption rate in Table 2-4. The results (rounded-off to one or two significant digits) are shown in Table 2-5. For the lifetimes of resources in Canada and U.S.A., the effect of imports and exports has been neglected, that is it has been assumed that all consumption is from domestic sources.

Table 2-5: Lifetime (in yr) of Fossil Fuel Resources
Assuming Constant Consumption (1991 Rates)

Region	Oil	Gas	Coal
World	60	110	1700
Canada	50	130	2000
U.S.A.	15	35	1900

As can be seen from Table 2-5, there is not much oil left in the world — about 60 years' worth. The situation in the U.S.A. is even worse, with only about 15 years of oil left in domestic supplies. It is easy to see why the U.S.A. has a strong interest in maintaining a steady and secure supply of oil from the Middle East. The situation for gas is somewhat better, with roughly 100 years of resource remaining. For coal, there still remains almost 2000 years' supply, yet of all the fossil fuels, coal is the one that has the most damaging environmental effects.

It is important to remember that the lifetimes presented in Table 2-5 are calculated with the assumption that consumption remains constant. It is obvious that if consumption increases, as it has for many decades in the past, then the lifetimes will be smaller. Lifetimes for exponentially increasing consumption are calculated in the next section.

Unconventional Sources of Fossil Fuels

The data for fossil fuel resources in Table 2-3 include only conventional deposits of oil and gas. However, there are other sources that might possibly double the recoverable resources. There is much uncertainty about the quantity of these unconventional sources, and how much it would cost to extract them with minimum effect on the environment. For oil, these other sources consist of:

- Tar sands, consisting of a thick hydrocarbon (bitumen) mixed with sand.
  The world's largest deposits are the Athabasca tar sands in Alberta. The energy content of the Alberta tar sands is perhaps 5000 quads, and 10% of the deposit is close enough to the earth's surface to be accessible through open-pit mining. Although there is some activity in mining, processing (heating), and refining the tar sands, the oil extracted is rather expensive, with its raw cost at present being four to five times that of Middle Eastern oil.

- Oil shale, in which a heavy hydrocarbon (kerogen) is deposited throughout rock.
  In a few places in the world, the kerogen content is large enough that the rocks can actually be burned (usually mixed with coal). In other cases, the oil shale can be processed by pyrolysis (heating without oxygen) to produce a liquid oil, along with some gas and residual carbon. At the present time, this oil from shale cannot compete economically with conventional oil.

Natural gas (primarily methane, CH_4) is normally found in association with oil deposits, and in the early days of the petroleum industry, it was considered a nuisance that was simply burned off at the well. It was not until storage and pipeline facilities became available that gas became commonly accessible as a fuel. Unconventional sources of gas comprise:

● Gas in non-porous rocks, i.e., gas which does not flow freely to a well.

● Coal-seam methane, i.e., gas found in coal deposits.

● Geopressured methane, i.e., methane trapped in hot subterranean deposits of water.

● Gas hydrates, i.e., frozen mixtures of water and natural gas found in polar regions.

There are also deposits of coal not included in the category of recoverable resources because the coal seams are too thin, too deep, or too inaccessible (under a town or lake, for example).

2.7 LIFETIME OF FOSSIL FUELS (EXPONENTIAL GROWTH MODEL)

When determining fossil fuel lifetimes in Section 2.6, we assumed constant consumption at 1991 levels. However, since energy consumption has historically risen exponentially, it is natural to wonder how long fossil fuels would last if exponential growth were to continue in the future. The mathematics of exponential growth provide a quick analysis.

Suppose that the annual consumption N of a resource is expressed as an exponential function of time t: $N = N_o e^{kt}$. To emphasize that N is a function of time, write this as $N(t) = N_o e^{kt}$. At some arbitrary time t, consider a small time interval of duration Δt (Fig. 2-7). The consumption at this time is $N(t)$, which is the "height" of the graph above the origin at time t. In Fig. 2-7, a narrow rectangle has been drawn with a base of width Δt and a height of $N(t)$. The *area* of this rectangle is $N(t) \cdot \Delta t$, and represents the *quantity of resource consumed* in the time interval Δt. Note the units: $N(t)$ has units such as quads/yr and Δt would be in years, and thus their product is in quads.

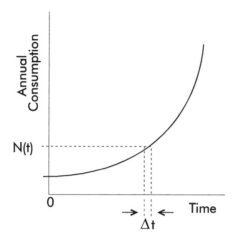

Figure 2-7: The area of the narrow rectangle is $N(t) \cdot \Delta t$, representing consumption during the time interval Δt.

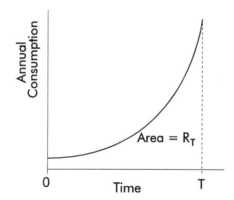

Figure 2-8: The area under the curve from time $t = 0$ to T (the resource lifetime) equals the resource consumed (Q_T) during this time.

In Fig. 2-8, consider the total area under the exponential curve from the present ($t = 0$) until some time T in the future, when all the resource will be just consumed. T is the *resource lifetime*, that is, the time required to consume the resource that remains now (at $t = 0$). The area under the curve is essentially just the sum of the areas of a series of rectangles such as the one in Fig. 2-7, and since each of these areas represents a quantity of resource consumed, the total area is the total quantity of resource consumed from now until time T. This is just the quantity of resource now remaining, which we represent as Q_T. You probably recognize that this area is also the integral of the function from $t = 0$ to $t = T$. Hence:

$$Q_T = \text{resource remaining}$$

$$= \text{quantity consumed from the present } (t = 0) \text{ to time T}$$

$$= \text{area under curve}$$

$$= \int_0^T N(t)\, dt$$

Recalling that $N(t) = N_o e^{kt}$: $\qquad Q_T = \int_0^T N_o e^{kt}\, dt$

Performing the integration: $\qquad Q_T = \dfrac{N_o}{k}\, e^{kt}\, \Big|_0^T$

$$\therefore\ Q_T = \frac{N_o}{k}\, (e^{kT} - 1)$$

This equation must be solved for the resource lifetime T. First re-arrange to isolate e^{kT} :

$$e^{kT} = \frac{kQ_T}{N_o} + 1$$

Taking natural logarithms: $\qquad kT = \ell n\, (\dfrac{kQ_T}{N_o} + 1)$

Dividing by k to solve for T: $\quad T = \dfrac{1}{k}\, \ell n\, (\dfrac{kQ_T}{N_o} + 1)$ $\qquad\qquad$ (2-8)

Equation 2-8 is extremely useful — given the resource remaining Q_T at a particular time, the consumption rate N_o at that time, and the growth constant k, it is possible to determine how long the resource will last.

■

Example 2-5 *In the two decades prior to the "oil crisis" of 1973, world oil consumption had been growing at about 7% per year.*

(a) *If consumption were to return to this 7% growth, how long would world oil last?*

(b *If the actual resource remaining were discovered to be twice the current estimate, how long would oil last, assuming 7% growth?*

(a) Use Eqn. 2-8: $\quad T = \dfrac{1}{k} \ln \left(\dfrac{kQ_T}{N_o} + 1 \right)$

Since the growth rate is 7% per year, $k \approx 0.07$ yr^{-1} from approximation 2-6. Use $Q_T \approx 7000$ quads from Table 2-3, and $N_o = 126$ quads/yr from Table 2-4. Substituting values:

$$T \approx \frac{1}{0.07 \ yr^{-1}} \ ln \left(\frac{(0.07 \ yr^{-1})(7000 \ quads)}{126 \ quads/yr} + 1 \right)$$

$$\approx 23 \ yr$$

Hence, if world oil consumption were to grow at 7% per year, the world as a whole would run out of oil in about 23 yr (or roughly 20 yr). Contrast this value with the 60 yr we calculated in Section 2.6 assuming no growth.

(b) If the resource is doubled, $Q_T \approx 14000$ quads. Re-doing the calculation with this value gives:

$$T \approx 31 \ yr$$

Doubling the resource extends the lifetime by less than 10 yr!! You can see that exponential growth has an insidious effect on resource lifetime. ∎

Table 2-6: Approximate Lifetime (in yr) of Fossil Fuel Resources Assuming Exponential Growth With Various Annual % Growth Rates

Region & Fuel	0%	2%	5%	10%	Actual Average Annual Growth Rate 1986-91
World Oil	60	37	27	19	1.5%
Canada Oil	50	35	25	18	1.0%
U.S.A. Oil	15	13	11	9	0.6%
World Gas	110	59	38	25	3.4%
Canada Gas	130	64	40	26	6.7%
U.S.A. Gas	35	27	20	15	3.7%
World Coal	1700	180	89	52	0.9%
Canada Coal	2000	190	92	53	-5.1%
U.S.A. Coal	1900	180	91	52	1.9%

Table 2-6 shows the results of applying Eqn. 2-8 to determine resource lifetimes of oil, gas, and coal for annual growth rates of 2%, 5% and 10%. Also tabulated for reference are the no-growth (constant consumption) lifetimes from Table 2-5, and the actual average annual growth rates from 1986 to 1991. It is apparent that even a modest growth rate of 2% has an enormous effect on resource lifetime, and a growth rate as large as 10% is devastating. You might even be wondering whether the coal lifetimes are correct — it does seem surprising that a no-growth lifetime of 1700 yr can be reduced to only 180 yr with a mere 2% annual growth. However, consider the following: at 2% growth per year, the doubling time is 35 yr; this means that after 35 yr of 2% growth, coal is being consumed twice as fast, and if the growth were then to cease, the resource would last only about 850 yr instead of 1700 yr. Only 35 yr of 2% growth has effectively wiped out 850 yr of coal resources. And the growth does not stop — after another 35 yr, consumption doubles yet again, and this process continues until all the coal is depleted, in only 180 yr.

It is clear that if fossil fuels are to last a long time, we must control the growth in their consumption. However, the last column in Table 2-6 indicates that consumption of all fossil fuels has been growing recently, with gas showing the largest increase (3.4% annually). The only entry showing no growth is Canadian coal consumption, showing an average annual decline of 5.1% for the period 1986-91.

The Future

Is exponential growth a good model for the future? Probably not. It would be more reasonable to expect that as a resource begins to dwindle, its price will rise and consumption will fall, and we explore this type of scenario in the Hubbert model of resource consumption in Chapter 3. Nevertheless, a knowledge of exponential growth is important in understanding how consumption usually behaves when the resource is still very abundant, and how quickly a finite resource would disappear if consumption were to increase exponentially.

EXERCISES

2-1 The current growth rate of the world's population is about 1.7% annually (down from 2.0% in 1974). The population is about 5.5 billion (as of summer 1993).

(a)   Determine the growth constant and doubling time which correspond to the present growth rate. [Ans. 0.017/yr, 41 yr]
(b)   If the population continues to grow at 1.7% per year, what will the world's population be in 30 years? [Ans. 9.1 billion]
(c)   If the average mass of a person is 70 kg, how long would it take until the mass of people equals the mass of the earth, assuming population growth of 1.7% per year? The mass of the earth is 5.98×10^{24} kg. [Ans. 1.8×10^3 yr, a short time in the history of humanity]

2-2 Between 1986 and 1991, consumption of Canadian natural gas grew with an average growth rate of 6.7% per year. Estimate (in your head) approximate values of the growth constant and doubling time. [Ans. 0.07 yr^{-1}, 10 yr]

2-3 Oil consumption in Taiwan increased steadily from 14.6 million tonnes in the year 1983 to 28.6 million tonnes in 1992, that is, consumption approximately doubled in 9 yr. Determine (in your head) approximate values of the growth constant and annual percent growth rate for Taiwanese oil consumption.   [Ans. 0.08 yr^{-1}, 8% per year]

2-4 Confirm that the slope of the straight line on the semilog graph of Fig. 3-14 is 0.07 yr^{-1}.

2-5 The following data represent exponential growth of a quantity, N, as a function of time, t.

t (yr)	0	1	2	3	4	5	6	7	8
N	2.85	4.50	6.90	9.00	14.9	22.3	35.5	51.3	90.0

(a)   Plot the data on semilog paper. (A sheet is provided in Appendix in the Appendices.)

(b)   Draw a straight line to fit the points as well as possible, and then use two well-separated points on your line to determine the growth constant. [Ans. (b) 0.40 yr^{-1} to 0.42 yr^{-1}]

2-6   Imagine that a savings account was established at the time of Christ, 2000 years ago, with a deposit equivalent to 1¢.   Assume interest, compounded annually, at 5%.   Calculate the present value of the savings account.   [Ans. 2.4×10^{40}]

PROBLEMS

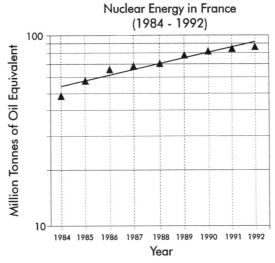

Figure 2-9: Problem 2-1.

2-1   Figure 2-9 shows nuclear-energy consumption[2] in France from 1984 to 1992.   The triangles indicate the actual data points, and the line is a computerized linear best-fit to the points.

(a) In 1990 what was the nuclear-energy consumption in France?

(b) Convert your answer in (a) to exajoules.

(c) Using the line on the graph, determine the growth constant for French

---

[2]   The unit, million tonnes of oil equivalent, is often used to quantify consumption of energy sources other than oil.

nuclear-energy consumption.

(d) Assuming exponential growth at the rate determined in (c), when would nuclear-energy consumption in France reach 200 million tonnes of oil equivalent?

[Ans. (a) 80 million tonnes of oil equivalent

(b) 3.4 EJ   (c) 0.067 yr^{-1} (d) 2004]

2-2 How long would world natural gas resources last if consumption were to continue to grow at 3.4% per year (the average growth rate for 1986-91)? Use appropriate data from Tables 2-3 and 2-4.   [Ans. 47 yr]

2-3 From 1981 to 1988, the growth in coal consumption in China was approximately exponential. The 1981 consumption was equivalent to 394 million tonnes of oil, and the 1988 consumption was equivalent to 581 million tonnes of oil.

(a) Determine the growth constant and the average annual % growth rate. [Ans. 0.055 yr^{-1}, 5.7%]

(b) Determine the doubling time of Chinese coal consumption, assuming constant exponential growth at the 1981-1988 rate. [Ans. 12 yr]

(c) Chinese coal resources were approximately 20 000 quads at the end of 1988. Determine the lifetime of coal resources in China if:

(i) consumption remains constant at the 1988 rate [Ans. 8.5×10^2 yr]

(ii) consumption continues to increase exponentially at the 1981-1988 rate. [Ans. 70 yr]

THE HUBBERT MODEL
OF RESOURCE
CONSUMPTION

The previous chapter outlined two possible models of fossil fuel consumption: constant consumption, and exponential growth. We argued that exponential growth seemed to be a good model, based on historical and present data, for the consumption of a resource when the resource is still abundant. However, it is not very reasonable to expect that consumption of oil (or any other commodity) will continue to increase until the very moment when it is depleted. As supplies start to dwindle, it is more probable that prices will increase and consumption will level off and then decline. This model was suggested in the 1950s by Dr. M. King Hubbert, a geophysicist now retired from the U.S. Geological Survey. At that time, his work was given little credence by the petroleum industry, but U.S. domestic oil production actually peaked in the mid-1970s and is now well into its decline. (Because of imports, U.S. *consumption* has been able to increase.)

3.1 INTRODUCTION TO THE HUBBERT MODEL

To gain an overall appreciation of the Hubbert model, consider the usual pattern of resource consumption. The specific factors to consider are the rate of discovery of the resource, the rate of consumption of the resource, and the size of the reserves at any time. Figure 3-1 illustrates a plausible graph of the cumulative quantity of resource consumed, Q_C, vs. time, t, for a finite resource such as oil, gas, or coal. The resource is first utilized slowly, and then more rapidly as technologies are developed to make better use of the resource's potential. However, the rate of consumption (that is, the slope of the curve) eventually declines as the resource begins to "run out," and ultimately the cumulative consumption asymptotically approaches a value that equals the total quantity of the resource that was ever available. In Fig. 3-1 this latter quantity is represented as Q_∞. A curve as smooth as the one shown would not occur in actual practice, of course; there would be bumps and wiggles along it, depending of the state of the economy, international conflicts, etc. Nonetheless, it is reasonable to assume that a graph of long-term cumulative consumption vs. time would show the general "S-shape" of Fig. 3-1.

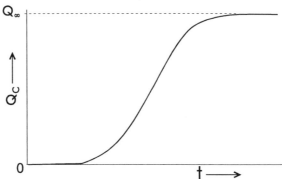

Figure 3-1: A graph of cumulative quantity of resource consumed, Q_C, vs. time, t. Q_∞ represents the total quantity of the resource.

A graph of the cumulative quantity of resource *discovered* will have a similar shape, but precedes the consumption of the resource. Figure 3-2 shows both the cumulative quantity of resource consumed, Q_C, and the cumulative quantity of resource discovered, Q_D. The Q_D-curve precedes the Q_C-curve by a fixed amount of time.

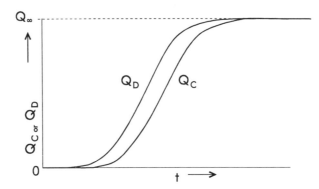

Figure 3-2: Quantity of resource discovered, Q_D, and quantity consumed, Q_C, vs. time, t. The Q_D-curve precedes the Q_C-curve.

At any given time the quantity of resource that is known to be available is simply the difference between the cumulative amount discovered and the cumulative amount consumed. This available quantity represents the reserves, and is represented as Q_R:

$$Q_R = Q_D - Q_C \qquad\qquad (3\text{-}1)$$

The three quantities, Q_C, Q_D, and Q_R, are illustrated as functions of time in Fig. 3-3. You might wish to confirm by direct measurement on the graph that, for any given time, $Q_R = Q_D - Q_C$.

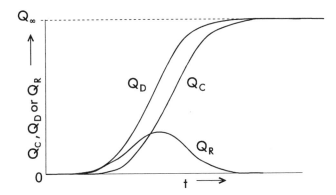

Figure 3-3: The quantity of reserves, Q_R, at any given time is the difference between the quantity discovered, Q_D, and the quantity consumed, Q_C.

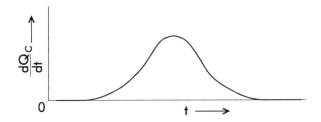

Figure 3-4: The rate of consumption of a resource, dQ_C/dt or N, vs. time, t.

Thus far the *cumulative* quantities of resource consumed and discovered, and their difference (the reserves), have been discussed. It is also useful to investigate the time rate of change of each of these quantities. First consider the rate of change of the cumulative quantity consumed, that is, we consider dQ_C/dt, which is the slope of the curve of Fig. 3-1. This slope is zero at the beginning (bottom) of the curve, reaches a positive maximum in the middle, and approaches zero again at the end (top). This slope, that is, the rate of consumption, is shown in Fig. 3-4. In Chapter 2, rate of consumption (often expressed in quads/yr) was given the symbol N, and we continue with that notation here:

$$N = \frac{dQ_C}{dt} \tag{3-2}$$

The rate at which resource is discovered (dQ_D/dt) is the slope of the curve of Q_D vs. t in Fig. 3-2, and has the same shape as the graph of rate of consumption vs. time (Fig. 3-4), but precedes it in time. Figure 3-5 illustrates both the consumption rate ($N = dQ_C/dt$) and the discovery rate (dQ_D/dt).

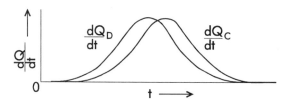

Figure 3-5: The consumption rate, dQ_C/dt, and the discovery rate, dQ_D/dt, of a resource.

Now consider the rate of change of reserves, dQ_R/dt. The graph of Q_R vs. t shown in Fig. 3-3 begins with a slope of zero, has a positive slope until the quantity Q_R reaches its peak, at which time the slope is again zero, and then the slope is negative (that is, the reserves are decreasing); the slope eventually becomes zero again at large time-values. Figure 3-6 shows a graph of dQ_R/dt vs. t, along with dQ_C/dt and dQ_D/dt.

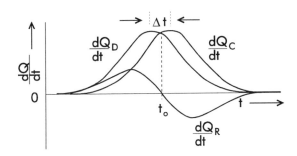

Figure 3-6:   The rate of change of reserves, dQ_R/dt, has a value of zero at time t_0. Δt is the lag time between discovery and consumption (see text).

The three curves in Figure 3-6 together illustrate a very interesting property: as the curve of dQ_R/dt passes through its central value of zero (that is, as the reserves are at their maximum), the curve of consumption rate going up crosses the curve of discovery rate going down.   This statement can be justified mathematically by taking the time derivative of Eqn. 3-1:

$$\frac{dQ_R}{dt} = \frac{dQ_D}{dt} - \frac{dQ_C}{dt}$$

Therefore, when $dQ_R/dt = 0$, we have $dQ_D/dt = dQ_C/dt$, that is, the discovery rate equals the consumption rate.

Furthermore, from the symmetry of the curves, this particular time is half-way (in time) between the peak of the discovery rate curve and the peak of the consumption rate curve. This observation provides a method of predicting in advance when the peak of the consumption rate will occur[1]. All that is required is: (1) the lag time, Δt, between discovery rate and consumption rate, which is straightforward to obtain by plotting a graph of these quantities vs. time, and (2) the time, t_0, at which $dQ_R/dt = 0$, which can be obtained from a graph of dQ_R/dt vs. t. (See Fig. 3-6.) Then the peak in consumption can easily be predicted to occur at a time of $t = t_0 + \Delta t/2$, that is, one-half lag time after the time when $dQ_R/dt = 0$. In the 1960s Hubbert plotted dQ_R/dt vs. t for U.S. domestic oil production and found that the graph crossed the t-axis around 1961. A plot of this graph, including data up to 1979, is shown in Fig. 3-7; notice that this plot corresponds only to the central, approximately linear, region of the graph of dQ_R/dt shown in Fig. 3-6. When Hubbert plotted his graph, the lag time Δt was known to be about 12 yr, and so he estimated that U.S. oil production would peak in about the year 1967 (i.e., 1961 + 12/2). It is now clear that U.S. production peaked in 1973, and so Hubbert's prediction was not far off the mark.

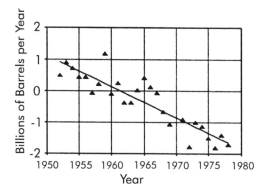

Figure 3-7: Graph of dQ_R/dt for U.S. oil production. The triangles are data points, and the line is a computerized fit to the data.

The next section focuses on consumption rate ($N = dQ_C/dt$) as a function of time, and uses Hubbert's model to predict the peak consumption rate. Hubbert's model is also used to predict the amount of remaining resource at any future time.

---

[1] A different method of predicting this peak time will be discussed in Section 3.2.

3.2 THE MATHEMATICS OF THE HUBBERT MODEL

Figure 3-4 in the previous section shows a graph of consumption rate of a resource vs. time; this rate increases, then levels off and declines in a symmetrical way. A convenient mathematical function that follows this pattern is the *Gaussian*, or *normal*, function (Fig. 3-8).

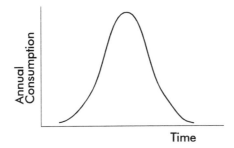

Figure 3-8: A Gaussian function.

To describe the consumption of a resource by such a function, three parameters are of particular importance:

- the time T_M when the function has its maximum value (peak)

- the height of the peak, that is, the maximum consumption rate N_M (the consumption rate at time T_M)

- the "width" of the function. There is a convention for specifying the width: the *standard deviation*, represented by σ. There are a number of ways to describe σ, but probably the easiest is to say that σ is the width (measured from the peak in either direction) at which the value of the function — the consumption rate — is 61% of the peak value $(0.61\ N_M)$.

The parameters T_M , N_M , and σ are illustrated in Fig. 3-9.

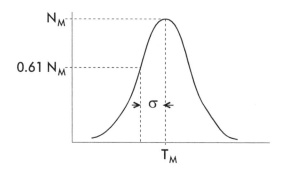

Figure 3-9: Parameters T_M, N_M, and σ of a Gaussian function.

The full mathematical expression of the consumption rate N (or dQ_C/dt) as a Gaussian function of time t is:

$$N = N(t) = N_M \, e^{-\frac{(t - T_M)^2}{2\sigma^2}} \tag{3-3}$$

Notice that when $t = T_M$, this equation becomes $N = N_M \, e^0 = N_M$, as it should. As well, when the time t differs from the mid-time T_M by σ, i.e., when $|t - T_M| = \sigma$, then $N = N_M \, e^{-1/2} = 0.61 \, N_M$ as mentioned above.

The total resource, including that already consumed and that remaining, is the area under the entire curve, which is the integral of the function; represent this total quantity of resource as Q_∞, as in Section 3.1. (Notice that in Section 2.7, Q_T represented only the quantity of resource remaining.)

$$Q_\infty = \int_{-\infty}^{+\infty} N(t) \, dt = \int_{-\infty}^{+\infty} N_M \, e^{-\frac{(t - T_M)^2}{2\sigma^2}} \, dt \tag{3-4}$$

The quantities N_M, Q_∞, and σ are related:

$$N_M = \frac{Q_\infty}{\sigma \sqrt{2\pi}} \tag{3-5}$$

Equation (3-5) will not be derived here, but its usefulness will be demonstrated in Example 3-2.

When working with Gaussian functions, it is convenient to define another variable, z:

$$z = \frac{|\, t - T_M\, |}{\sigma} \qquad\qquad (3\text{-}6)$$

With this definition of z, Eqn. 3-3 simplifies to:

$$N = N_M\, e^{-z^2/2} \qquad\qquad (3\text{-}7)$$

As illustrated in Example 3-1 below, the variable z tells us how many standard deviations a particular time t is away from the peak-time T_M.

Example 3-1 *If z = 2, what are the corresponding values of the time, t?*

Using Eqn. 3-6: $\quad 2 = \dfrac{|\, t - T_M\, |}{\sigma} \qquad \therefore\ 2\sigma = |\, t - T_M\, |$

There are two possible values of t, depending on whether the quantity $t - T_M$ is positive or negative. If $t - T_M > 0$, that is, if $t > T_M$ (the time is past the peak-time), then

$$|\, t - T_M\, | = t - T_M$$

Hence, $\qquad\qquad 2\sigma = t - T_M$

and solving for t gives $\qquad t = T_M + 2\sigma$

If $t - T_M < 0$, that is, $t < T_M$ (the time has not yet reached the peak-time), then

$$|\, t - T_M\, | = T_M - t$$

Therefore, $\qquad\qquad 2\sigma = T_M - t$

giving $\qquad\qquad t = T_M - 2\sigma$

Thus, if z = 2, then $t = T_M \pm 2\sigma$. (If z = 4, what would be the values of t?)

Example 3-1 shows that if z = 2, then the time t is two standard deviations away from T_M (Fig. 3-10). If z were 3.67, then t would be 3.67 standard deviations away from T_M. In effect, z is just a different way of specifying time t; for any arbitrary value of z, time t = $T_M \pm z\sigma$.

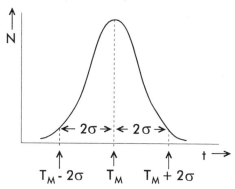

Figure 3-10:   If the variable z is equal to 2, this means that t is either 2σ to the right of T_M, or 2σ to the left of T_M.

Before proceeding to another example using a Gaussian function, consider one of its useful features. Regardless of the specific values of T_M, N_M , and σ, the area under the curve from time T_M - σ to time T_M + σ is always 0.6827 Q_∞, that is, this area is always 68.27% of the total area under the curve. Similarly, the area from T_M - 2σ to T_M + 2σ is always 95.45% of the total, i.e., 0.9545 Q_∞. In general, the area from T_M - $z\sigma$ to T_M + $z\sigma$ as a fraction of the total area depends only on z. As will soon be obvious, this property is very convenient when dealing with resource consumption. Table 3-1 shows the area under the curve from T_M - $z\sigma$ to T_M + $z\sigma$ (as a decimal fraction of the total area) as a function of z. The value of z is given in the left-hand column (to one decimal place) and the top row (to the second decimal place). Use the table to confirm that for z = 1, the fractional area is 0.6827, and for z = 2, the value is 0.9545. What fraction of the total area is contained between times T_M - 1.46σ and T_M + 1.46σ?
[Ans. 0.8557]

Table 3-1: Integral of the Gaussian Function vs. z.

This table gives the integral of the Gaussian function from $T_M - z\sigma$ to $T_M + z\sigma$, shown as the shaded area to the right, as a decimal fraction of the total area under the function.

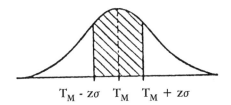

$$T_M - z\sigma \quad T_M \quad T_M + z\sigma$$

z	.00	.01	.02	.03	.04	.05	.06	.07	.08	.09
0.0	.0	.0080	.0160	.0239	.0319	.0399	.0478	.0558	.0638	.0717
0.1	.0797	.0876	.0955	.1034	.1113	.1192	.1271	.1350	.1428	.1507
0.2	.1585	.1663	.1741	.1819	.1897	.1974	.2051	.2128	.2205	.2282
0.3	.2358	.2434	.2510	.2586	.2661	.2737	.2812	.2886	.2961	.3035
0.4	.3108	.3182	.3255	.3328	.3401	.3473	.3545	.3616	.3688	.3759
0.5	.3829	.3899	.3969	.4039	.4108	.4177	.4245	.4313	.4381	.4448
0.6	.4515	.4581	.4647	.4713	.4778	.4843	.4907	.4971	.5035	.5098
0.7	.5161	.5223	.5285	.5346	.5407	.5467	.5527	.5587	.5646	.5705
0.8	.5763	.5821	.5878	.5935	.5991	.6047	.6102	.6157	.6211	.6265
0.9	.6319	.6372	.6424	.6476	.6528	.6579	.6629	.6680	.6729	.6778
1.0	.6827	.6875	.6923	.6970	.7017	.7063	.7109	.7154	.7199	.7243
1.1	.7287	.7330	.7373	.7415	.7457	.7499	.7540	.7580	.7620	.7660
1.2	.7699	.7737	.7775	.7813	.7850	.7887	.7923	.7959	.7995	.8029
1.3	.8064	.8098	.8132	.8165	.8198	.8230	.8262	.8293	.8324	.8355
1.4	.8385	.8415	.8444	.8473	.8501	.8529	.8557	.8584	.8611	.8638
1.5	.8664	.8690	.8715	.8740	.8764	.8789	.8812	.8836	.8859	.8882
1.6	.8904	.8926	.8948	.8969	.8990	.9011	.9031	.9051	.9070	.9090
1.7	.9109	.9127	.9146	.9164	.9181	.9199	.9216	.9233	.9249	.9265
1.8	.9281	.9297	.9312	.9327	.9342	.9357	.9371	.9385	.9399	.9412
1.9	.9426	.9439	.9451	.9464	.9476	.9488	.9500	.9512	.9523	.9534
2.0	.9545									
2.5	.9876									
3.0	.9973									
3.5	.9995									
4.0	.9999									

Any Gaussian function can be characterized by the three parameters T_M, N_M, and σ. In order to determine these parameters, we need three pieces of information as input. In Example 3-2 that follows, the three given quantities are:

- current consumption rate (N)
- cumulative consumption to date
- total resource (Q_∞)

Example 3-2   *In 1992, the world oil consumption (i.e., rate) was 125 quads/yr, and the cumulative consumption to the end of 1992 was 4069 quads. The resource remaining (from Table 2-3) is approximately 7000 quads. Use the Hubbert model to determine*
(a) the year of peak consumption (T_M)
(b) the peak consumption (N_M)
(c) the year when 20% of the resource will remain
(d) the consumption in the year 2025.

(a) *We first determine the total resource, Q_∞, which is the sum of the resource consumed and the resource remaining:*

$$Q_\infty \quad = (4069 + 7000)\ quads$$
$$= 11069\ quads$$
$$\approx 11000\ quads$$

(Use 11000 quads as a round number for Q_∞ since there is much uncertainty in the value of 7000 quads remaining; in the calculations that follow, all numbers — including intermediate answers — have been rounded off to two significant digits.)

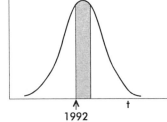

Figure 3-11:   By the end of 1992, the world had consumed 4069 quads of oil (the left-hand unshaded area), out of a total of about 11000 quads.

Now find the value of z for the year 1992, and then determine σ and T_M.

At the end of 1992, the world had consumed 4069 quads of oil (the left-hand unshaded area in Fig. 3-11), out of a total of about 11000 quads. As a fraction of the total, this is 4069/11000 = 0.37, or 37%. Because of the symmetry of the Gaussian function, in the future as oil consumption is dwindling there will be another 37% consumed, represented by the right-hand unshaded area. The two unshaded areas make up 0.74 of the total, leaving 0.26 in the centre (shaded). Knowing this area, z can be found. Look in the body of Table 3-1 for the entry closest to 0.26, and read off the z-value, 0.33.

Now use Eqn. 3-7:

$$N = N_M e^{-z^2/2}$$

Substitute the expression for N_M from Eqn. 3-5, $N_M = \dfrac{Q_\infty}{\sigma \sqrt{2\pi}}$, into Eqn. 3-7 to give:

$$N = \frac{Q_\infty}{\sigma \sqrt{2\pi}} e^{-z^2/2}$$

Re-arrange to solve for σ: $\sigma = \dfrac{Q_\infty}{N \sqrt{2\pi}} e^{-z^2/2}$

Use this equation to calculate σ; for the year 1992, N = 125 quads/yr and z = 0.33; and since $Q_\infty \approx 11000$ quads:

$$\sigma = \frac{11000 \ quads}{(125 \ quads/yr)\sqrt{2\pi}} e^{-(0.33)^2/2}$$

$$= 33 \ yr$$

Now, knowing σ, z, and t, use Eqn. 3-6 to find T_M:

$$z = \frac{|t - T_M|}{\sigma}$$

For 1992, t < T_M, since less than half of the total oil has been consumed at this time. Therefore, $|t - T_M| = T_M - t$.

Thus, $z = \dfrac{T_M - t}{\sigma}$

$$\therefore \ z\sigma = T_M - t \quad giving \quad T_M = t + z\sigma = 1992 + (0.33)(33) = 2003$$

Hence, the Hubbert model predicts that world oil consumption will peak in about the year 2003.

(b) To determine the peak consumption N_M, simply use Eqn. 3-5:

$$N_M = \frac{Q_\infty}{\sigma \sqrt{2\pi}} = \frac{11000 \ quads}{(33 \ yr)\sqrt{2\pi}} = 1.3 \times 10^2 \ quads/yr$$

Thus, the peak consumption predicted by the Hubbert model is about 1.3×10^2 quads/yr.

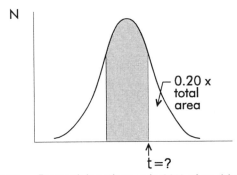

Figure 3-12:   Determining when only 20% of world oil will remain.

(c) The situation when 20% of the oil remains is illustrated in Fig. 3-12. The answer requires the time t at the beginning of the right-hand unshaded region, which contains 0.20 of the total area. By symmetry, the left-hand unshaded region also contains 0.20 of the total area, leaving 0.60 as the shaded area in the centre. From Table 3-1, this corresponds to z = 0.84. Now use Eqn. 3-6 to find t:

$$z = \frac{|t - T_M|}{\sigma}$$

For the time t when only 20% remains, $t > T_M$, since more than half of the oil has been consumed. Therefore, $|t - T_M| = t - T_M$.

Hence, $\quad z = \dfrac{t - T_M}{\sigma} \quad$ or $\quad z\sigma = t - T_M$

Thus, $\quad t = T_M + z\sigma = 2003 + (0.84)(33) = 2031$

The Hubbert model predicts that 20% of world oil will remain in about the year 2031.

(d) To find the consumption in the year 2025, first determine the z-value for that year:

$$z = \frac{|t - T_M|}{\sigma} = \frac{|2025 - 2003|}{33} = 0.67$$

Now use Eqn. 3-7:

$$N = N_M\, e^{-z^2/2} = (1.3 \times 10^2 \text{ quads/yr})\, e^{-(0.67)^2/2} = 1.0 \times 10^2 \text{ quads/yr}$$

Therefore, according to the Hubbert model, the consumption will be approximately 1.0×10^2 quads/yr in the year 2025. ∎

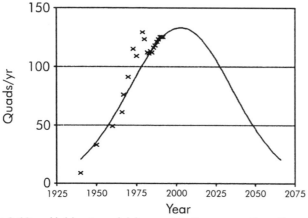

Figure 3-13:   Hubbert model for world oil consumption; the crosses correspond to actual data.

It is interesting to use the results of the above example to plot a graph of world oil consumption vs. time (Fig. 3-13). The solid line, a Gaussian function, has been determined from calculations of consumption N for various times t, just as was done in part (d) of Example 3-2 for the year 2025. The crosses show actual data points for world oil consumption. Notice that from about 1940 to 1975, consumption increased sharply and fit the Gaussian curve only moderately well, but since about 1982 the data and the curve agree extremely well.

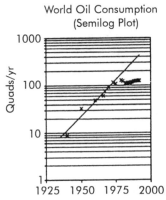

Figure 3-14: World oil consumption from 1940 to 1975 exhibited exponential growth (approximately) with a growth constant of k = 0.07 yr^{-1}.

Figure 3-14 shows the same world oil consumption data replotted on a semilog graph. The straight line on this graph has a slope 0.07 yr^{-1}, and fits the data reasonably well between 1940 and 1975. This indicates that world oil consumption was increasing exponentially with a growth of about 7% per year during this period. Since 1975, the line and the data points have very poor agreement.

3.3 FUTURE ENERGY CONSUMPTION

Chapters 2 and 3 have briefly explored various scenarios for future consumption of fossil fuel resources: constant consumption, exponential growth, and the Hubbert model. For any one fossil fuel (or any finite resource), it is likely that its early consumption will grow exponentially, and then will level off and decline, following a curve such as a Gaussian.

But what about total energy consumption, which historically has shown exponential growth with very few downturns? What will it do in the future?

This question is extremely difficult to answer, because consumption of individual energy sources and total consumption of energy depend on a large number of factors:

- population
- economic activity (GDPs of major countries)
- the price of energy
- availability of energy (e.g., a strike by coal-miners can stop the supply of coal)
- technological advances[2]
- development of alternative sources of energy[2]
- government policies
- world political stability
- public attitude

Many of these factors are inter-related — for example, public attitude can influence government policies, which in turn can determine the price of energy and, thereby, economic activity. At present, some people have strong concerns about the environmental effects of energy production, and might be willing to pay more for energy if environmental problems could be reduced. These concerns might well affect the future choice of sources, the price of energy, and the willingness of government and private industries to reduce environmental pollution and encourage greater energy efficiency.

In the long term, neither population nor energy consumption can continue to increase indefinitely — the Earth simply would be unable to withstand the environmental pressures.

---

[2]   Recent technological advances and alternative energy sources, as well as energy conservation, are discussed in Chapters 16 to 18.

EXERCISES

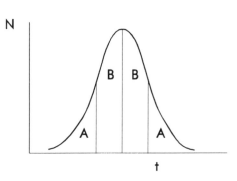

Figure 3-15: Exercise 3-1.

3-1 Use Fig. 3-15 for this exercise.
(a) If area 2B represents 75% of the total area under the curve, what is the corresponding z-value?
(b) If area 2A represents 30% of the total area under the curve, what is the corresponding z-value?
(c) If area B represents 30% of the total area under the curve, what is the corresponding z-value?
(d) If area A represents 19% of the total area under the curve, what is the corresponding z-value?
(e) If area B is twice area A, what is the corresponding z-value?
(f) If z = 1.19, then what percentage of the total area under the curve is represented by 2B?
(g) If z = 1.96, then what percentage of the total under the curve is represented by A?
[Ans. (a) 1.15 (b) 1.04 (c) 0.84 (d) 0.88 (e) 0.97 (f) 76.60% (g) 2.50%]

3-2 A Gaussian function defined by $N = N_M\, e^{-\frac{(t - T_M)^2}{2\sigma^2}}$ has the following parameters:

$$T_M = 1992 \text{ (yr)}; \quad \sigma = 45 \text{ yr}; \quad \text{and } N_M = 78 \text{ quads/yr.}$$

(a) If t = 2015 (yr), determine N.
(b) If N = 37 quads/yr, determine t.
[Ans. (a) 68 quads/yr (b) 2047 or 1937]

PROBLEMS

3-1 The following are data for Canadian oil production:

cumulative production to end of 1992 = 18.5 billion barrels
production in 1992 = 98.2 million tonnes
approximate resource remaining at end of 1992 = 150 quads

Apply the Hubbert model to Canadian oil production to predict:

(a) the year in which the production will peak [Ans. 1998]
(b) the production in that peak year [Ans. 4.1 quads/yr]
(c) the year in which only 13% of the resource will remain [Ans. 2026]
(d) the production in the year 2010. [Ans. 3.6 quads/yr]

3-2 Figure 3-16 shows the production and consumption of oil (in exajoules per year) in the U.S.A. to the end of 1992. Imports account for the difference between production and consumption. You can see that production has passed its peak and is now declining. Use the Hubbert model to predict when U.S. oil production will decline to 10 EJ/yr. Use the following data:

production in 1992 = 17.6 EJ

cumulative production to end of 1992 = 998.3 EJ

approximate resource remaining at end of 1992 = 475 EJ
[Ans. 2013]

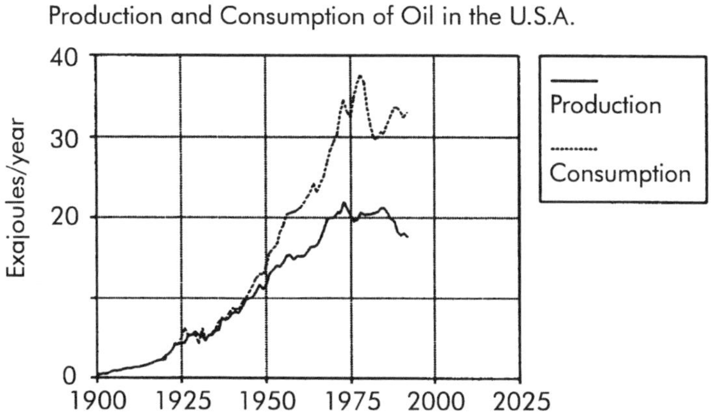

Figure 3-16:   Problem 3-2.

4 EFFICIENCY OF ENERGY CONVERSIONS

Imagine how wasteful it would be to discard more than half of the available energy when converting one form of energy to another. Unfortunately, that's exactly what we do in many energy-converters, such as coal-burning electric power plants and automobiles. Why don't engineers design these machines to operate more efficiently? This chapter explores the fundamental law of nature that dictates the high level of waste energy in such devices.

4.1 TEMPERATURE, THERMAL ENERGY, AND HEAT

The study of the efficiency of energy conversions is an intriguing one, but it is one with several specific terms and distinctions. In particular, there are differences between the specific meanings of temperature, heat, and thermal energy — synonyms in common speech — which are important in the study of efficiency.

Although the term *temperature* appears frequently in day-to-day life, its precise meaning is surprisingly difficult to define. Mathematically, the temperature (T) of an ideal gas is defined through the *ideal-gas law*:

$$PV = nRT \tag{4-1}$$

where P and V are the pressure and volume, respectively, of n moles of an ideal gas, that is, a gas at sufficiently low density that the interactions between the gas molecules can be ignored. The temperature in Eqn. 4-1 is measured on the absolute temperature scale, for which the unit is the kelvin (K). The quantity R is the universal gas constant:

$$R = 8.315 \text{ J} \cdot \text{K}^{-1} \cdot \text{mol}^{-1} \tag{4-2}$$

Although Eqn. 4-1 defines temperature, it does not provide much of an intuitive feeling for the meaning of temperature. It is useful to think of temperature as being related to the average translational[1] kinetic energy ($\frac{1}{2}mv^2$) of molecules: as temperature increases, the average molecular kinetic energy increases, that is, the molecules move more rapidly; conversely, as temperature decreases, average kinetic energy decreases and the molecules move more slowly. In fact, there is a quantitative relation between average molecular kinetic energy and temperature:

$$(\frac{1}{2} mv^2)_{AVG} = \frac{3}{2} kT \qquad (4\text{-}3)$$

where m is the mass of one molecule, v is its speed, and k is Boltzmann's constant:

$$k = 1.381 \times 10^{-23} \text{ J/K} \qquad (4\text{-}4)$$

Equation 4-3 shows that average molecular kinetic energy is linearly related to temperature T; this relationship is valid for typical gases, liquids, and solids at room temperature. (For solids, the atoms vibrate in place, but still have a kinetic energy associated with the vibration.) At low enough temperatures, quantum energy effects make Eqn. 4-3 invalid.

Whereas temperature is related to the *average* kinetic energy of a collection of molecules, *thermal energy* (sometimes called *internal energy*) is the *total* energy of all the molecules. Thus, thermal energy depends on the number of molecules as well as on the average molecular energy. A large number of molecules at low temperature can have more thermal energy than a small number of molecules at high temperature, simply because there are more of them. Thermal energy includes not only translational kinetic energy associated with the centre-of-mass motion of molecules, but also energy within the molecules themselves, such as rotational and vibrational kinetic energy, and electric potential energy.

The term *heat* is often used loosely to mean thermal energy, but there is in fact a difference between these two concepts. Heat refers to a *transfer of energy* from one object to another, as a result of a difference in

---

[1] The term "translation" refers to a change in position of the centre of mass of an object. Thus, translational KE is the KE associated with the centre-of-mass motion, i.e., the motion of the object as a whole. This contrasts with, say, rotational KE, which is associated with rotational motion of the object around the centre of mass. An object can have zero translational KE, but still have rotational KE; an example is a wheel spinning on a fixed axle.

temperature between the objects. As an example, consider a hot cup of coffee on a table in a room — as the coffee cools, its thermal energy decreases, heat is transferred from the coffee to the table and air in the room, increasing the thermal energy of the table and air. It is not correct, strictly speaking, to refer to the heat of the coffee, nor the heat of the table.

When heat is received by an object, all of it does not necessarily go into increasing the object's thermal energy. For example, a gas might be heated and hence expand to move a piston, thus doing work[2] on the piston. In this case, some of the heat received by the gas increases the thermal energy of the gas, and the rest provides the work done.

4.2 SPECIFIC HEAT AND LATENT HEAT

Suppose that the same amount of heat is applied to different objects. How much does the temperature of each object increase? This depends in part on the mass of the object: if heat is applied to a very massive object, it will experience a smaller temperature increase than a less massive object (of the same material). The temperature change also depends on what the object is made of. If the molecules have many modes of motion (such as translation, rotation, vibration), then the heat can go into increasing the energies of each mode by a small amount, and the temperature is not raised very much. On the other hand, if we consider applying heat to something such as a monatomic gas, which has only translational kinetic energy of its atoms to make up its thermal energy, then all the heat goes into increasing the energy of only one mode of motion, and the temperature increase is larger. This difference between the internal modes of molecular motion manifests itself through different *specific heats* of objects, discussed below.

When an amount of heat Q is applied to an object having mass m, then the object undergoes an increase in temperature ΔT. The quantities Q, m, and ΔT are related by:

---

2 Recall the definition of work: if a force of magnitude F is applied to an object, which undergoes a displacement of magnitude Δr, then the work W done by the force on the object is by $W = F\Delta r\cos\theta$, where θ is the angle between the force and displacement. If the force and displacement are in the same direction, $\theta = 0$, and hence $W = F\Delta r$. The SI unit of work is the joule (Exercise 4-1).

$$Q = mc\Delta T \qquad\qquad (4\text{-}5)$$

where c is the specific heat of the object, defined as the amount of heat, per unit mass and per unit change in temperature, required to change the temperature of a substance. The SI unit of specific heat is joule· kilogram^{-1}· kelvin^{-1} (J· kg^{-1}· K^{-1}). Eqn. 4-5 can be re-arranged as:

$$\Delta T = \frac{Q}{mc} \qquad\qquad (4\text{-}5b)$$

From Eqn. 4-5b it is easy to see that if the same amount of heat Q is applied to various objects having the same mass, then the object having the largest specific heat will experience the smallest increase in temperature. Notice also from Eqn. 4-5b, that if objects made of the same material receive the same heat Q, then the object with the largest mass will undergo the smallest temperature change. Although this analysis has focused on application of heat to produce temperature increases, Eqn. 4-5 works just as well for removal of heat and corresponding decreases in temperature.

Table 4-1: Selected Specific Heats

Substance	Specific Heat (J· kg^{-1}· K^{-1})
Lead	128
Mercury	139
Copper	387
Aluminum	895
Ice (-10 °C)	2.22×10^3
Ethyl alcohol	2.43×10^3
Water	4186

Table 4-1 lists the specific heat for a few substances. The values tabulated were determined at room temperature (20°C) and one atmosphere of pressure, unless otherwise noted. (The specific heat of a material depends somewhat on both temperature and pressure.) Notice the large value of specific heat for water; this makes water an excellent coolant, since a relatively small mass of water can absorb a large quantity of heat without undergoing a large increase in temperature.

Thus far, it has been assumed that heating or cooling an object will result in only an increase or decrease of temperature. However, the object might undergo a change of state, that is, it might melt, solidify, vaporize, or condense. For a solid of mass m to liquefy (or a liquid of mass m to solidify), heat Q must be added (or removed):

$$Q = mL_F \qquad\qquad (4\text{-}6)$$

where L_F is the *latent heat of fusion* of the material. (Sometimes the word "latent" is omitted.) Notice that there is no ΔT in Eqn. 4-6; changes of state occur with no temperature change.

Table 4-2: Selected latent heats of fusion and vaporization

Substance	Melting Temperature (°C)	Latent Heat of Fusion (J/kg)	Boiling Temperature (°C)	Latent Heat of Vaporization (J/kg)
Mercury	-39	1.1×10^4	357	2.96×10^5
Lead	327	2.3×10^4	1744	8.58×10^5
Nitrogen	-210	2.6×10^4	-196	2.00×10^5
Hydrogen	-259	5.8×10^4	-253	4.55×10^5
Water	0	3.33×10^5	100	2.26×10^6
Aluminum	660	4.0×10^5	2467	1.05×10^7

For a liquid to vaporize through the addition of heat, or a vapour to condense by the removal of heat, there is a similar relationship:

$$Q = mL_V \qquad\qquad (4\text{-}7)$$

where L_V is the *latent heat of vaporization*. Table 4-2 gives some typical heats of fusion and vaporization.

■ *Example 4-1* *How much heat is required to vaporize 1.3 kg of water, initially at 20°C?*

Heat is required first to increase the temperature of the water from 20°C to the boiling point (100°C, from Table 4-2), and then to change the water at 100°C to water vapour (steam) at 100°C. We use Eqn. 4-5 to calculate the heat needed to raise the temperature from 20°C to 100°C:

$$Q = mc\Delta T$$
$$= (1.3 \ kg)(4186 \ J \cdot kg^{-1} \cdot K^{-1})(80 \ K)$$
$$= 4.35 \times 10^5 \ J \quad (keeping \ one \ extra \ significant \ digit)$$

Note that we used the specific heat of water from Table 4-1, and that the increase *in temperature is:* $\Delta T = 80°C = 80 \ K$.

To determine the heat required to vaporize the water at 100°C, use Eqn. 4-7 and the latent heat of vaporization of water from Table 4-2:

$$Q = mL_V = (1.3 \ kg)(2.26 \times 10^6 \ J/kg) = 2.94 \times 10^6 \ J$$

When the two heats are added together, the total is 3.4×10^6 J required to heat and vaporize the water. ■

4.3 FIRST LAW OF THERMODYNAMICS

In leading up to a discussion of energy conversions and efficiency, it is useful to discuss the *first law of thermodynamics*. This law states that energy is conserved (although it can be converted from one form to another). This law has already been used implicitly; for example, in Section 4.1 it was seen that when a gas is heated and expands against a piston, the heat energy provided equals the increase in thermal energy of the gas plus the work done by the gas on the piston.

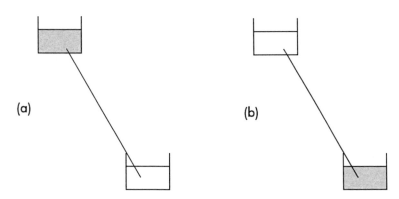

Figure 4-1: (a) Room-temperature water at rest in a tank.
(b) The water has flowed down the incline to the lower tank, with a resulting increase in temperature.

As another example, consider the situation shown in Fig. 4-1. Room-temperature water of mass m rests in a tank at the top of an incline (Fig. 4-1 (a)); a valve is then opened at the bottom of the tank, allowing the water to run down the incline into a tank at the bottom (Fig. 4-1 (b)). The gravitational potential energy of the water has decreased by mgh, where h is the vertical distance that the centre of mass of the water has moved, and has been converted into increased thermal energy of the water. The water is hotter. Conservation of energy gives:

$$mgh = mc\Delta T$$

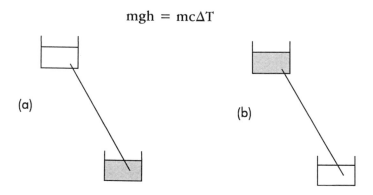

Figure 4-2: (a) Hot water is at rest in the lower tank.
(b) Conservation of energy would allow the water to flow to the higher tank, converting thermal energy to gravitational potential energy.

Consider now Figure 4-2, which shows hot water flowing up an incline, converting thermal energy into gravitational potential energy, and becoming cooler as it goes uphill. This uphill flow is, of course, impossible, but it is not conservation of energy that makes it so. In order to explain the non-reversibility of the water flow, we need to invoke another law of nature: the second law of thermodynamics.

4.4 THE SECOND LAW OF THERMODYNAMICS

The second law of thermodynamics can be stated in many ways; probably the simplest is that disorder is more probable than order. It is easy to generate disorder from order (just think of your room!), but much more difficult to create order from disorder. The thermal energy of the hot water in the lower tank in Fig. 4-2 (a) is disordered; the water molecules are moving randomly in all directions. It is highly unlikely — essentially impossible — that the molecules would all move spontaneously in the same direction up the incline in an ordered way. On the other hand, gravitational potential energy is an example of ordered energy (the vertical direction is a specific direction and gives order to this energy), and it is simple to convert this ordered energy to disordered thermal energy by allowing water to run down the incline.

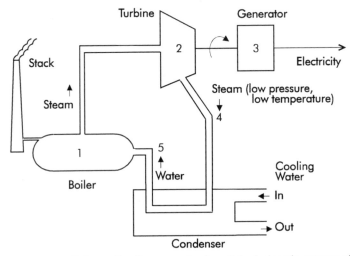

Figure 4-3: Schematic diagram of a fossil-fuel electric power plant.

These ideas can now be applied to an important example of energy production: a fossil-fuel electric power plant (Fig. 4-3). Fuel is burned to vaporize water in the boiler (item 1, see Figure 4-3), producing steam at high pressure and temperature. The steam passes over blades of a turbine (2), imparting energy to the turbine and causing it to spin. The turbine turns an electric generator (3), which produces an electric current in the external wires. Because the steam loses energy in spinning the turbine, the pressure and temperature of the steam decrease (4). The low-temperature steam is cooled and condenses back to water (5), to begin the process over again. The condensation is usually done by passing the steam through pipes that are cooled by water pumped from a nearby lake or river.

The energy of the hot steam is disordered thermal energy. However, the spinning turbine and the electric current have energy that is ordered. The electric plant is performing a conversion of disordered energy into ordered energy, which goes against the natural tendency from order to disorder, and it is impossible to convert all the disordered thermal energy in the steam into useful ordered kinetic energy of the turbine. The function of the condenser can now be understood. The heat that is discarded into the cooling water increases its thermal energy and its disorder. The disordered energy of the hot steam is partly converted into ordered energy of the turbine, and partly into disordered energy of the cooling water. If heat were not discarded to the cooling water (or elsewhere), the electric plant could not function — it cannot convert all the disordered energy of the steam into ordered energy. Not surprisingly, almost all the ordered energy of the spinning turbine can be converted into ordered electric energy.

It will be demonstrated in Section 4.6 that the cooling in the condenser improves the power-plant efficiency, which depends on the temperature of the steam produced by the boiler, and on the temperature of the cooling water in the condenser. In addition, from a very practical point of view, if the water that is being returned to the boiler were not condensed, it would require all the energy output from the turbine to pump the waste steam from the turbine back to the boiler. This waste steam is at low pressure and the boiler is at high pressure, and the waste steam would not flow to the boiler without pumping. In order to move liquid water from the condenser to the boiler, a pump is also required (not shown in Fig. 4-3). However, since the water has been condensed, its volume is roughly 1000 times smaller than the corresponding volume of steam, and hence the energy required to pump it to the boiler is about 1000 times smaller than to pump the same mass of steam. (The energy required here can be expressed as: energy = (difference in pressure) × volume.)

4.5 EFFICIENCY

Our discussion of the operation of a thermal electric plant in the previous section indicated that not all the energy available from the burning fuel is converted to useful electric energy. How much *is* converted? To answer this question, we turn to the concept of *efficiency*:

$$\text{efficiency} = \frac{\text{useful energy or work out}}{\text{total energy or work in}} \qquad (4\text{-}8)$$

Writing this with symbols:

$$\eta = \frac{\text{Useful } E_{OUT}}{\text{Total } E_{IN}} \qquad (4\text{-}8b)$$

where efficiency is represented by η (the lower-case Greek letter "eta"). Efficiencies are very often quoted as percentages:

$$\eta = \frac{\text{Useful } E_{OUT}}{\text{Total } E_{IN}} \times 100\% \qquad (4\text{-}8c)$$

A modern fossil-fuel electric plant has an efficiency of about 40%, that is, only about 40% of the energy in the coal is converted to useful electric energy. Much of the rest is discarded as waste energy to the cooling water, some is lost up the smokestack, and some to inevitable heat losses in steam pipes and friction in the turbine and generator. It is important to keep in mind that the main reason this efficiency is so low is that the electric plant is converting disordered energy into ordered energy. Older thermal electric plants have efficiencies in the range of 30% to 35%.

Table 4-3: Typical Efficiencies

Device	Efficiency
Fossil-fuel electric power plant	40%
Hydroelectric plant	95%
Automobile	25%
Home gas furnace	85%
Electric generator	98%
Large electric motor	95%
Wind generator	55%
Electric heater	100%
Fluorescent light	20%
Incandescent light	5%

Table 4-3 lists typical efficiencies of various energy-converting devices. Notice that automobiles, which — like thermal electric plants — convert disordered thermal energy to ordered mechanical energy have an efficiency of only about 25%. Thus, roughly three-quarters of the energy that you put into your gasoline tank is wasted. High efficiencies are achieved by devices that convert ordered energy to ordered energy (hydroelectric plants, electric generators, etc.), or ordered energy to disordered energy (electric heaters).

Note that the common incandescent light has an efficiency of only 5%; although both electric energy and light energy have high order, the conversion from electricity to light occurs through the intermediary of disordered thermal energy. About 95% of the energy used by an incandescent lightbulb is dissipated as waste heat. The recent interest in compact fluorescent bulbs is a result of their higher efficiency, about 20%; while this is not high, it is nonetheless four times better than the incandescent efficiency. A 60-W incandescent bulb (i.e., one that uses 60 W of electric power) can be replaced by a 15-W fluorescent bulb to give the same amount of light.

A Variation on a Unit —— MWe

The total thermal power provided by a fossil-fuel electric plant is typically of the order of 2500 MW. At an efficiency of 40%, this produces about 1000 MW of electrical power (0.40 × 2500 MW). In order to distinguish this useful electrical power from the total power, the electrical power is often written with a unit of MWe, where the "e" stands for electric.

Example 4-2 *A coal-burning electric power plant produces 950 MWe at an efficiency of 38%. What mass of coal (having energy content 28 MJ per kg) is required annually to operate this plant?*

To determine the answer, first use the definition of efficiency (Eqn. 4-8c) to determine the total power input to the plant. Since power (P) is just the rate of energy use, power can be inserted into the definition of efficiency in place of energy:

$$\eta = \frac{Useful\ P_{OUT}}{Total\ P_{IN}} \times 100\%$$

Hence,

$$Total\ P_{IN} = \frac{Useful\ P_{OUT}}{\eta} \times 100\%$$

$$= \frac{950\ MW}{38\%} \times 100\%$$

$$= 2.50 \times 10^3\ MW$$

Converting to watts: $2.50 \times 10^3\ MW \times \dfrac{10^6\ W}{1\ MW} = 2.50 \times 10^9\ W$

Now determine the total energy E produced in one year, by multiplying the power P (in watts, or joules per second) and the time t (the number of seconds in a year):

$$E = Pt$$
$$= (2.50 \times 10^9\ W)(365 \times 24 \times 60 \times 60\ s)$$
$$= 7.88 \times 10^{16}\ J$$

To calculate the mass of coal required, divide this total energy by the energy content per kg of coal:

$$\frac{7.88 \times 10^{16} \; J}{28 \times 10^6 \; J/kg} \; = \; 2.8 \times 10^9 \; kg$$

Thus, the amount of coal required per year is 2.8×10^9 kg, or 2.8 million tonnes. (That's a lot of coal, equivalent to almost 100 railroad-car loads per day!)

Capacity Factor

The solution to Example 4-2 assumed that the power plant would be running all year without stopping. However, electrical plants are often shut down for maintenance — either regular or unexpected — or because the demand for electricity is low. In addition, if demand is low, a plant might be operating at less than full capacity, that is, a 1000-MWe plant might produce only 600 MW for some period of time. These factors are incorporated into an annual *capacity factor*, which is defined as:

$$\text{capacity factor} = \frac{\text{actual electrical energy produced in one year}}{\substack{\text{maximum electrical energy producible in one year} \\ \text{running at full capacity}}}$$

Suppose that a plant's capacity factor is 65% for a given year. This means that its output is only 65% of the output that it would have produced if it had been running at 100% capacity all year. A capacity factor of 65% could mean that the plant was operating at its rated (maximum) power output for 65% of the time, and was completely shut down for 35% of the time. Or it could mean that it operated every day, but at only 65% of its maximum power. A capacity factor of 65% could also arise from any one of a number of combinations of shutdowns and operations at less than 100% power. Whatever the combination, a capacity factor of 65% means that the plant is using only 65% of the fuel that would be used by a plant with a capacity factor of 100%.

Nuclear plants are intended to run with a capacity factor of nearly

100%, because of the high construction cost and low fuel cost. As well, startup procedures at these plants are time-consuming. Fossil-fuel plants have lower capacity factors; these plants are cheaper to build than nuclear plants, but the fuel is more expensive, and startup is straightforward.

4.6 HEAT ENGINES

The automobile and the thermal electric plant are examples of a general type of machine called a *heat engine*, which is any device that converts heat energy into mechanical work. The second law of thermodynamics can be written specifically for heat engines: *no device can be constructed that, operating in a cycle (like an engine), accomplishes only the extraction of heat energy from some source and its **complete** conversion to work.*

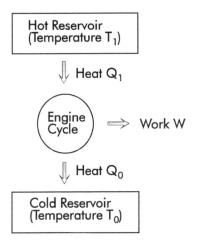

Figure 4-4: Schematic diagram of a heat engine.

Figure 4-4 shows the general structure of a heat engine. Heat energy Q_1 is extracted from a thermal reservoir at high absolute temperature T_1, some useful work W is done, and heat Q_0 is discarded to a reservoir at low absolute temperature T_0.

The efficiency of a heat engine is given by:

$$\eta = \frac{\text{Useful } E_{OUT}}{\text{Total } E_{IN}}$$

$$= \frac{W}{Q_1}$$

By the first law of thermodynamics, the input heat Q_1 must equal the sum of the useful work W and the discarded heat Q_0:

$$Q_1 = W + Q_0$$

Thus, $$W = Q_1 - Q_0$$

Hence, the expression for η can be rewritten as:

$$\eta = \frac{Q_1 - Q_0}{Q_1}$$

or $$\eta = 1 - \frac{Q_0}{Q_1}$$ (4-9)

Although Eqn. 4-9 is a correct expression for the efficiency of a heat engine, it is not particularly useful. In order to develop a more useful equation, it will be necessary to investigate the concept of entropy.

Entropy

We have already discussed disordered and ordered energy. *Entropy* is a measure of disorder. One statement of the second law of thermodynamics is that in any energy transfer or conversion within a closed system, the entropy of the system must increase (or in rare cases, remain constant). In a thermal electric plant, if all the disordered thermal energy of the hot steam were to be converted solely to electric energy, then there would be a decrease in the total entropy, in violation of the second law.

Is it possible to be more quantitative about entropy? It is certainly beyond the scope of this book to go into the fundamental, philosophical roots of the concept and its mathematical formulation in terms of the randomness, or microscopic level of disorganization, of a system. However, it is nevertheless possible to adopt the macroscopic approach to entropy of the original thermodynamicists. They stated that if heat is added to a system, then the change in entropy (ΔS) of the system is the heat energy received divided by the absolute temperature of the system[3]:

$$\Delta S = \frac{Q}{T} \qquad (4\text{-}10)$$

At first sight, there appears to be no relation between this definition and the basic one of randomness, but there are some semblances. The heat input Q is obviously an input of a disorganized form of energy, since it consists of random molecular motions. Thus, heat input increases entropy and conversely loss of heat decreases the entropy of an object. But why is temperature in the denominator? In very cold materials, there is not much molecular motion, so a given quantity of heat energy will stir things up quite noticeably, increasing the disorder a great deal. But if the same quantity of heat energy is added to a very hot object (already having a huge amount of disorder), the entropy increase is much smaller.

When an amount of heat Q is extracted from an object at temperature T, the change in entropy is negative:

$$\Delta S = \frac{-Q}{T}$$

It is possible to apply this expression for ΔS to a specific case. Consider a heat engine in which an amount of heat Q_1 is converted to work W and the remainder Q_0 is discarded. We consider the system to be the engine plus the two thermal reservoirs. The engine runs in

---

[3] This statement is valid as long as the system is so large that the addition of heat Q increases the temperature T by only an infinitesimal amount.

cycles, and at the end of a cycle, it is in the same state in which it began; that is, there is no change, and $\Delta S = 0$ for the engine alone. The entropy change due to the heat flowing out of the hot reservoir at temperature T_1 is

$$\Delta S_1 = \frac{-Q_1}{T_1}$$

The change in entropy due to heat Q_0 flowing into the cold reservoir at temperature T_0 is

$$\Delta S_0 = \frac{Q_0}{T_0}$$

Since there is no change in entropy associated with the organized mechanical work W, the total entropy change is

$$\Delta S = \frac{-Q_1}{T_1} + \frac{Q_0}{T_0}$$

According to the second law of thermodynamics, the overall entropy of a system can never decrease, that is, $\Delta S \geq 0$.

Hence,
$$\frac{-Q_1}{T_1} + \frac{Q_0}{T_0} \geq 0$$

Re-arranging,
$$\frac{Q_0}{T_0} \geq \frac{Q_1}{T_1}$$

Therefore,
$$\frac{Q_0}{Q_1} \geq \frac{T_0}{T_1} \qquad (4\text{-}11)$$

Returning to Eqn. 4-9, which gives the efficiency η of a heat engine:

$$\eta = 1 - \frac{Q_0}{Q_1}$$

Substituting relation 4-11 into this:

$$\eta \leq 1 - \frac{T_0}{T_1} \quad \text{(heat engine)} \quad \text{(4-12)}$$

Relation 4-12 is an extremely useful expression for the efficiency of a heat engine, in terms of only the temperatures of the thermal reservoirs. The equality sign in 4-12 holds only in the case of what is known as a reversible heat engine (for which $\Delta S = 0$), and gives the maximum possible efficiency of a heat engine:

$$\text{Max. } \eta = 1 - \frac{T_0}{T_1} \quad \text{(heat engine)} \quad \text{(4-13)}$$

Unfortunately, a reversible heat engine is impractical — it can operate only if the temperature difference ($T_1 - T_0$) is infinitesimal, or, in the case of a finite temperature difference, the engine requires an infinite amount of time in which to work. Nonetheless, Eqn. 4-13 serves as a convenient expression for the maximum conceivable efficiency of a heat engine. Real heat engines will always have efficiencies less than this, because of the irreversible transfer of heat from the hot reservoir to the cold reservoir, which have a finite difference in temperature, and because of energy dissipated in the engine by friction, imperfect insulation, etc.

Equation 4-13 can be used to determine the maximum possible efficiency of a fossil-fuel electric plant, as shown in Example 4-3 below.

■
Example 4-3 *In a fossil-fuel electric plant, such as that shown in Fig. 4-5, the temperature of the hot steam is about 500°C (773 K), and the low-temperature reservoir is the cooling water from a lake or river with a temperature of roughly 20°C (293 K). What is the maximum efficiency that is theoretically possible in such a plant?*

The maximum possible efficiency is given by Eqn. 4-13:

$$Max.\ \eta = 1 - \frac{293\ K}{773\ K} = 0.62,\ or\ 62\%$$

Hence, the maximum efficiency theoretically possible for a typical fossil-fuel electric plant is 62%.   ■

The answer in Example 4-3 above shows that even if one could build a reversible heat engine operating between the temperatures encountered in a typical thermal electric plant, its efficiency at converting disordered energy to ordered energy would only be 62%. Such a heat engine would still discard 38% of the available heat energy to the low-temperature reservoir. As can be seen from Equation 4-13, the efficiency could be increased by increasing T_1 or decreasing T_0. However, temperature T_1 is limited by the strength of the boiler materials, and T_0 is just the natural temperature of the cooling waters.

Figure 4-5: The maximum efficiency theoretically possible from a typical fossil-fuel electric plant is only about 62%.

In nuclear electric plants, a nuclear reactor produces heat to generate steam at high temperature and pressure. This steam is then used the same way as in a fossil-fuel plant to spin turbines. In order to have high efficiency, one wants to have the core of the reactor at the highest possible temperature, but this results in unique demands on the structural materials and the cooling fluid, which must maintain their integrity under the twin stresses of temperature and radiation damage. Consequently, T_1 has to be kept lower than in fossil-fuel plants, giving an efficiency of typically only 30%, compared to the 40% attainable by burning fossil fuels.

4.7 REFRIGERATORS

Refrigerators are one example of a heat-moving device, that is, something that transfers heat from one place to another. A refrigerator moves heat from the inside of the appliance to the outside; the heat is expelled to the air in the room via coils at the rear of the refrigerator. Heat naturally is transferred from hot objects to cold ones; refrigerators perform the reverse (unnatural) function, and in order to do so, need an external input of energy, normally provided by electricity. Another version of the second law of thermodynamics — this one specifically for refrigerators — is: no device can be constructed that (operating in a cycle) accomplishes *only* the transfer of heat from a cooler source to a hotter one.

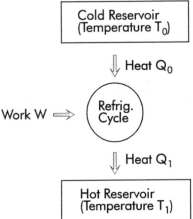

Figure 4-6: A refrigerator. Heat Q_0 is extracted from a low-temperature reservoir by using external work W, and heat Q_1 is expelled to a high-temperature reservoir.

Figure 4-6 illustrates the basic operation of a refrigerator. Heat Q_0 is extracted from a reservoir at a low absolute temperature T_0, by input of external work W. Heat Q_1 is expelled to a reservoir at high absolute temperature T_1. By the first law of thermodynamics, the sum of the heat extracted and the external work must equal the heat expelled:

$$Q_0 + W = Q_1 \qquad (4\text{-}14)$$

The "efficiency" of a refrigerator is the ratio of the useful heat transferred to the external work provided. (The reason for the quotation marks will be given shortly.) Since the function of a refrigerator is to cool the interior of the appliance, the *useful* heat transferred is the heat Q_0 removed from the interior. Thus, the "efficiency" is

$$"\eta" = \frac{Q_0}{W} \qquad (4\text{-}15)$$

From Eqn. 4-14, the work is:

$$W = Q_1 - Q_0$$

Substitution of this expression for W into Eqn. 4-15 gives

$$"\eta" = \frac{Q_0}{Q_1 - Q_0}$$

$$= \frac{1}{\dfrac{Q_1}{Q_0} - 1}$$

By using an analysis involving entropy (similar to that used in Section 4.6 for heat engines), it can be shown (Problem 4-5) that

$$\frac{Q_1}{Q_0} \geq \frac{T_1}{T_0}$$

Substituting this relation into our expression for "efficiency,"

$$"\eta" \leq \frac{1}{\dfrac{T_1}{T_0} - 1} \qquad \text{(4-16)}$$

Hence, the maximum possible "efficiency" of a refrigerator is

$$\text{Max. } "\eta" = \frac{1}{\dfrac{T_1}{T_0} - 1} \qquad \text{(4-17)}$$

A typical maximum "efficiency" of a refrigerator is now calculated. Assume temperatures of $T_0 = 263$ K (i.e., $-10\,°$C) in the freezer compartment, and $T_1 = 293$ K ($20\,°$C) in the surrounding room. Simple substitution into Eqn. 4-16 gives an "efficiency" of 8.77, or 877%.

How can an "efficiency" be greater than 100%? What has been calculated is not actually the efficiency of an energy *conversion*, but rather a parameter that indicates how well a refrigerator *moves* heat. The value of 8.77 indicates that a refrigerator could move 8.77 units of heat energy for an input of one unit of external work. It should now be clear that although the word "efficiency" has thus far appeared in quotation marks, it is not efficiency at all; it is instead something which is commonly called the *coefficient of performance (C.O.P.)*:

$$"\eta" = \text{C.O.P.}$$

and $$\text{C.O.P.} \leq \frac{1}{\dfrac{T_1}{T_0} - 1} \qquad \text{(refrigerator)} \qquad \text{(4-18)}$$

For completeness, and as a reminder of the fundamental meaning of the C.O.P. of a refrigerator, Eqn. 4-15 is now re-written with "η" replaced by C.O.P.:

$$\text{C.O.P.} = \frac{Q_0}{W} \qquad \text{(refrigerator)} \qquad \text{(4-19)}$$

Actual mass-produced refrigerators have a C.O.P. slightly less than 1, and the most efficient refrigerators have a C.O.P. of about 2. Notice that C.O.P.s are normally not quoted as percentages. The C.O.P. could be improved easily by using thicker insulation, improving door seals, improving the efficiency of the motor and compressor, and placing the motor and compressor (both of which get hot during operation) above the cool cabinet so that the heat from these devices will rise into the room instead of into the cabinet. A number of sensible design features of refrigerators in the 1950s, such as thicker insulation, were discarded in the era of cheap energy in the 1960s.

How a Refrigerator Works

The operation of a real refrigerator is rather simple. A fluid is alternately compressed and allowed to expand, and in between these processes either absorbs or gives off heat. A fluid commonly used in household refrigerators is freon (CCl_2F_2), one of the chlorofluorocarbons (CFCs) responsible for the decline of ozone in the upper atmosphere. In industry, ammonia is frequently employed as the working fluid.

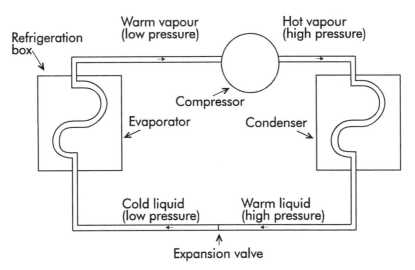

Figure 4-7: Operation of a refrigerator.

The workings of a refrigerator cycle begin at the expansion valve as shown in Fig. 4-7. To the right of the valve, the fluid is a warm liquid at high pressure. When a sensor in the refrigeration box indicates that cooling is required, the expansion valve (also known as the throttling valve) is opened. The valve has only a very small opening, and liquid sprays slowly through it into the lower-pressure region to the left. Because of the pressure decrease, some of the liquid vaporizes, thus removing energy from the liquid, and the temperature falls. The liquid is now cold, with a small quantity of it having been vaporized. This cold liquid flows through coils in the refrigeration box, removing heat from the contents of the box. The liquid vaporizes during this process, and the coils that it is travelling through are called the evaporator. At this point it is a warm, low-pressure vapour which passes into the compressor. This device is run by an electric motor and compresses the vapour, heating it somewhat. The fluid is now a hot, high-pressure vapour. It passes through coils (usually at the rear of the refrigerator), giving up heat to the surroundings. In the process, it condenses (hence the name "condenser" in Fig. 4-7) to form a warm liquid, which is ready to re-enter the cycle.

When a refrigerator is operating, the heat given off at the condenser is appreciable — put your hand on the coils and feel it. A little attention to these coils can make a refrigerator work more efficiently. If the coils are dirty, the efficiency of heat transfer from the condenser decreases, and so it is useful to clean the coils periodically. As well, if the heat from the coils cannot easily escape from the region surrounding the refrigerator, some of it will be conducted back into the refrigeration box, requiring further refrigeration cycles. Home refrigerators are often placed in a recess in a wall, hampering heat escape; it makes good sense to allow for ample air circulation around the refrigerator, especially at the top, since hot air rises. Ideally, it would perhaps be best to allow the heat to vent through the rear wall, heating another room.

Example 4-4 *A refrigerator that has a C.O.P. of 1.2 consumes 0.50 kW·h (1.8 MJ) of electrical energy while operating over a particular period of time. During this time, (a) how much heat energy (in megajoules) is removed from the refrigerator compartment, and*

(b) how much heat energy (in megajoules) is dissipated by the condenser?

(a) The C.O.P. of a refrigerator is given by Eqn. 4-19:

$$C.O.P. = \frac{Q_0}{W}$$

Given quantities are: C.O.P. = 1.2, and external work W (the electrical energy provided) = 1.8 MJ. The energy removed from the cold refrigerator compartment is Q_0, and is given by a simple re-arrangement of the above equation:

$$Q_0 = C.O.P. \times W = 1.2 \ (1.8 \ MJ) = 2.2 \ MJ$$

Thus, 2.2 MJ of heat energy is removed from the refrigerator compartment.

(b) By the first law of thermodynamics, the heat (Q_1) dissipated by the condenser must equal the sum of the external work provided and the heat extracted, as stated in Eqn. 4-14:

$$Q_1 = W + Q_0$$

$$\therefore \ Q_1 = (1.8 + 2.2) \ MJ = 4.0 \ MJ$$

Hence, the heat dissipated by the condenser is 4.0 MJ.

■

Thermoacoustic Refrigerators

Virtually all of the refrigeration and air-conditioning machines in the world operate on a compression and expansion cycle of a working gas, usually a chlorofluorocarbon (CFC). The environmental damage caused by released CFCs is so severe that the "Montréal Protocol," an international agreement, has mandated a halt to CFC production by 1995. The search for alternatives has focused on modifications to CFCs to permit rapid degradation when they are released into the environment.

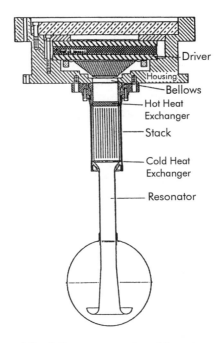

Figure 4-8: A thermoacoustic refrigerator.

However, there is one possible candidate for a completely different refrigerator technology that has no moving parts and uses only inert gases such as helium, argon, and xenon as the working fluid. This is the thermoacoustic refrigerator illustrated in Fig. 4-8. A driver, which is just a high-power loudspeaker, launches an acoustic wave into a resonator that is about one-quarter of a wavelength long. The alternate compressions and rarefactions of the gas due to the acoustic wave establishes a temperature gradient between hot and cold heat-exchangers, that is, heat is pumped from the cold stage to the hot stage where it is radiated to the surroundings. At the microscopic level the operation is complicated and its description is beyond the scope of this discussion.

A version of this refrigerator was flown on the space shuttle "Discovery" in January 1992 and was able to pump between 3 and 4 watts of heat at a C.O.P. of about 4. By 1993 improved versions were under development that would have the same power as a domestic refrigerator.

4.8 AIR CONDITIONERS AND HEAT PUMPS

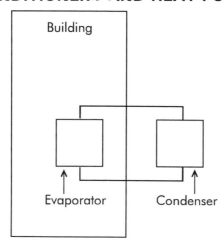

Figure 4-9: Schematic diagram of an air conditioner.

This section examines two devices — air conditioners and heat pumps — that are essentially identical in form and operation to refrigerators. An air conditioner is just a refrigerator that has the evaporator (where the cooling occurs) in the interior of a building (Fig. 4-9), and the condenser (where the heat is expelled) outside the building, or at least in a window where the heat can easily be sent outside. The C.O.P. of an air conditioner is defined in exactly the same way as that of a refrigerator:

$$\text{C.O.P.} = \frac{Q_0}{W} \leq \frac{1}{\dfrac{T_1}{T_0} - 1} \quad \text{(air conditioner)} \qquad (4\text{-}20)$$

Sometimes air conditioners or refrigerators carry a label specifying their energy-efficiency ratio (EER). This is the rate at which heat energy is removed in Btu per hour divided by the electrical power consumption in watts (an interesting, but unfortunate, combination of units).

A heat pump is virtually the same as an air conditioner, except that the placement of the evaporator and condenser are reversed (Fig. 4-10).

The evaporator is placed outside a building, where it extracts heat from the air, ground, or groundwater; the condenser is positioned inside, where it is used to heat the building. The function of the heat pump is to provide heat to the interior, and hence the *useful* heat is the heat Q_1 transferred into the warm building. Therefore, the C.O.P. is the ratio of the heat Q_1 provided to the work W required. Using an analysis similar to that performed in Section 4.6 for a heat engine, it is possible (Problem 4-7) to write the C.O.P. of a heat pump in terms of the absolute temperatures of the two heat reservoirs:

$$\text{C.O.P.} = \frac{Q_1}{W} \leq \frac{1}{1 - \dfrac{T_0}{T_1}} \quad \text{(heat pump)} \quad \text{(4-21)}$$

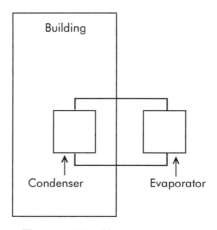

Figure 4-10: Heat pump.

Many heat pumps serve double-duty as air conditioners. In the winter, the working fluid flows in the direction shown in Fig. 4-10; in the summer, a simple flick of a switch causes the compressor to drive the fluid in the opposite direction. Thus, the evaporator becomes the condenser — and vice-versa — and the interior of the building is now cooled.

Suppose that a heat pump is operating with an exterior temperature of $0°C$ ($\therefore T_0 = 273$ K), and an interior temperature of $20°C$ ($\therefore T_1 =$

293 K). Simple substitution into expression 4-21 gives a maximum possible C.O.P. of 15. The actual C.O.P. of a typical heat pump working between these temperatures is about 3. This means that for 1 unit of electric energy as input, the pump can provide 3 units of heat energy to the building's interior. However, as the exterior temperature drops below about -10°C, the typical C.O.P. drops below 1, and it becomes more efficient simply to use an electric heater.

Notice in expression 4-21 that in order for a heat pump to have a high C.O.P., the temperatures T_1 and T_0 should be as close as possible. In contrast, consider the efficiency of a heat engine given earlier in expression 4-12:

$$\eta \leq 1 - \frac{T_0}{T_1} \qquad \text{(heat engine)}$$

For this efficiency to be large, T_0 should be small and T_1 large; that is, T_1 and T_0 should be as far apart as possible.

The use of heat pumps was more attractive a decade or two ago, when electricity prices were considerably lower. With the price of electricity now rising faster than that of natural gas or fuel oil, fewer people are installing heat pumps.

4.9 SYSTEM EFFICIENCY

Is it more efficient to heat a home using electric heaters or an oil furnace? At first glance, the answer might appear to be obvious: electric heaters convert electrical energy to thermal energy with an efficiency of 100%, whereas an oil furnace is only about 75% efficient. However, to give a more complete answer, one should also consider the overall efficiency of producing and delivering the electricity and the oil. If the electricity is produced in a coal-fired electrical plant, then the energy consumed in mining and transporting the coal should be taken into account. The *system efficiency* includes the efficiencies of *all* such steps in an energy-conversion process.

Table 4-4: Approximate System Efficiency for Electric Heat and Oil Heat

Electric Heat			Oil Heat		
Step	Step Efficiency	Cumulative Efficiency	Step	Step Efficiency	Cumulative Efficiency
Coal-mining	96%	96%	Oil extraction	97%	97%
Transporting coal	97%	93%	Transporting oil	97%	94%
Electricity generation	40%	37%	Heating	75%	70%
Electricity transmission	90%	34%	System efficiency		70%
Heating	100%	34%			
System efficiency		34%			

Table 4-4 shows the efficiency of each step for electric heat (assuming a coal-burning power plant) and oil heat. The cumulative efficiency is the product of the various step efficiencies, and the system efficiency is the final cumulative efficiency. The system efficiency of 70% for oil heat turns out to be larger than that (34%) of electric heat, indicating that it more efficient to heat with oil than with electricity from a coal-fired plant. Such system efficiencies are useful when comparing alternative ways to perform the same task.

Energy Intensity

It is well known that countries such as China, the former U.S.S.R., and the emerging democracies of eastern Europe have many factories with outmoded technologies that require more energy to produce goods than factories in western Europe and North America. This can be quantitatively expressed by the tern *energy intensity*, which is the amount of energy required for a country to produce a dollar's worth of gross domestic product (GDP). In a way, energy intensity is equivalent to system efficiency applied to an entire country. Table 4-5 shows energy intensities for a few countries; a range of energy intensity is given for each country because of different ways of measuring economic output and fluctuating monetary exchange rates.

Table 4-5: Energy Intensity[4] (1990)

Country	Energy Intensity (MJ per $ of GDP)
U.S.S.R.	31 - 48
Poland	33 - 42
Czechoslovakia	28 - 42
China	26 - 35
E. Germany	28 - 35
Hungary	22 - 29
U.S.A.	19 - 22
W. Germany	11 - 13
France	10 - 12

EXERCISES

4-1 Use the definition of work ($W = F \Delta r \cos\theta$) to show that the unit of work is kg· m^2/s^2, which is a joule. (You might find it handy to recall Newton's second law of motion: $F = ma$.)

4-2 How much heat is required to vaporize 7.8 kg of water, initially at 25 °C? [Ans. 2.0×10^7 J]

4-3 Suppose that a 60-W incandescent bulb is replaced by a 15-W fluorescent bulb. How much less electrical energy (in kilowatt· hours) is consumed in a week by the fluorescent bulb, operating for 12 h each day? [Ans. 3.8 kW· h]

---

[4] W. Chandler et al, *Sci. Am.* **263**, No. 3, p. 122 (Sept. 1990)

4-4 Suppose that an ocean-thermal-energy-conversion device were developed to produce work from the temperature difference between the warm surface waters of the ocean and the colder deeper waters. If the surface and deep-water temperatures are 22°C and 5°C respectively, what is the maximum theoretical efficiency of such an engine? [Ans. 5.8%]

4-5 What is the maximum C.O.P. that is theoretically possible for an air conditioner operating between a room at 21.0°C and exterior air at 31.0°C? [Ans. 29]

4-6 A heat pump having a C.O.P. of 2.4 consumes 0.75 kW·h (2.7 MJ) of electrical energy while operating over a certain period of time. During this time, (a) how much heat energy (in megajoules) is transferred into the high-temperature reservoir, and (b) how much heat energy (in megajoules) is transferred out of the low-temperature reservoir? [Ans. (a) 6.5 MJ (b) 3.8 MJ]

4-7 The C.O.P. of a refrigerator or an air conditioner is given by C.O.P. = Q_0/W, whereas the C.O.P. of a heat pump is given by C.O.P. = Q_1/W. Why are these expressions different?

PROBLEMS

4-1 A satellite in orbit could derive its energy from a very strong radioactive source of alpha particles. Suppose that 30 W (two significant digits) of electrical power are to be produced with 5.0% efficiency by conversion of the (kinetic) energy of the alpha particles. If the energy of each alpha particle is 5.5 MeV, how many particles must be emitted by the source per second? [Ans. 6.8×10^{14}]

4-2 It takes 2.0 million tons of coal a year to feed a 1000-MWe steam-electric plant. Assuming that the plant has a 60% capacity factor, what is its efficiency? Coal has an energy content of 22 MBtu per ton. Assume two significant digits. [Ans. 41%]

4-3 The latent heat of fusion of ice is 333 kJ/kg. In the freezer

compartment of a refrigerator, a tray holds 12 cubes of water each of mass 8.0 g. If the C.O.P. of the refrigerator is 3.3, how many kilowatt·hours of electrical energy would be required to run the refrigerator to change the water cubes (at 0°C) into ice cubes (at 0°C)?
[Ans. 2.7 × 10⁻³ kW·h]

4-4 A refrigerator is operating at a C.O.P. of 0.98, which is 13% of the maximum C.O.P. theoretically possible. The temperature of the freezer compartment of the refrigerator is -12°C. What is the temperature (to the nearest degree Celsius) of the surrounding room? [Ans. 23°C]

4-5 Use the second law of thermodynamics and the concept of entropy to show that for a refrigerator, $Q_1/Q_0 \geq T_1/T_0$. (The symbols are defined in Section 4.7.)

4-6 A particular heat pump uses 1.0 kW of electrical power. Every second, it is able to remove 2.5×10^3 J of energy from a low-temperature reservoir. What is its C.O.P.? [Ans. 3.5]

4-7 Derive expression 4-20 for the C.O.P. of a heat pump.

4-8 Calculate the system efficiency of a fluorescent light powered by electricity from a coal-fired power plant. Use Tables 4-3 and 4-4 as necessary. [Ans. 7%]

4-9 A 15-W compact fluorescent bulb produces the same amount of light as a 60-W incandescent bulb, because the fluorescent bulb is more efficient at converting electrical energy to light energy. The purchase price of a compact fluorescent bulb is much more than that of an incandescent bulb, but the fluorescent bulb costs less over its lifetime because of its greater efficiency, and because it lasts ten times longer than an incandescent bulb. Using the information given below, determine the payback time (in months) for a 15-W compact fluorescent bulb, that is, determine how much time is needed before the extra purchase cost of the fluorescent bulb has been recovered by savings. Assume that the compact fluorescent bulb is "on" for five hours per day, and that the cost of electrical energy is 8.0¢ per kW·h.

Price of 60-W incandescent bulb: $0.50

Price of 15-W compact fluorescent bulb: $20.00

Lifetime of 60-W incandescent bulb: 1000 h

Lifetime of 15-W compact fluorescent bulb: 10,000 h

[Ans. 32 months]

4-10   One estimate of the world's nuclear electrical capacity in the year 2000 is four million megawatts. If all the waste heat from operating these nuclear plants for one year were used to melt ice from the Antarctic ice cap, what would be the increase in the depth of the oceans? Assume that 70% of the surface area of the earth is covered with water, and that in addition to the direct waste heat from the reactors, all the electricity produced by the reactors ends up eventually as waste heat to melt the ice. The radius of the earth is 6400 km.
[Ans. 1 mm]

⑤ | *HEAT TRANSFER*

5.1 MODES OF HEAT TRANSPORT

Energy management is all about energy transfer. Heat must be transferred from a burner to water to run a steam turbine; it must be effectively transferred out of an automobile engine to avoid overheating; we would like to minimize heat loss in our houses in the winter and, of course, energy is transferred from the Sun to the Earth.

Two of the three basic processes by which heat is transferred are easily understood on the premise that thermal energy is simply kinetic energy of the atoms, molecules, or free electrons composing the body. *Conduction* is the flow of heat, without the accompanying bulk movement of matter, from a hot to a cold region. Suppose one end of a copper bar is heated. Copper, being a metal, is composed of atoms arranged in a crystal lattice surrounded by a "sea" of electrons free to move (that is why metals conduct electric current). The thermal energy in the hot end is a result of the kinetic energy of the oscillations of the atoms at their sites, plus the kinetic energy of the randomly moving electrons. This kinetic energy is transferred from atom to atom by the atomic interactions in the crystal lattice or by collisions with the free electrons. In this way the kinetic energy at the hot end is shared with atoms and electrons further along the bar and the heat is thereby "conducted". In the case of metals, the free electrons are the most important factor in the transfer. It is for this reason that good conductors of electricity are usually also good conductors of heat. Insulating materials like wood contain very few free electrons and do not conduct electricity or heat efficiently.

Convection is heat transfer in a fluid (gas or liquid) by actual macroscopic motion of atoms of the hot material into the colder regions. As a fluid in a vessel is heated from below, the lowest layer expands, and therefore its density decreases; as it becomes lighter it will move upwards to the top of the colder denser fluid above it. In older homes, the old-fashioned hot-air heating systems distributed heat entirely in this way; the air was heated in a furnace and, being less, rose up through ducts into the rooms above. Meanwhile, the colder, heavier air near the floor flowed down into return grills and was re-heated in the furnace. In another design based on hot water-filled radiators rather than hot air, the hot radiators heated the air in the rooms by contact, and the air then rose as a convection current, being replaced by cooler air from below.

In many cases natural convection is too slow or has some other drawback. For example, the hot-air heating systems required large diameter pipes that occupied a lot of wall space. In such cases fans or pumps are used to make the process more efficient; such cases may be called "forced convection".

Neither of the above processes can explain the transfer of heat from the Sun to the Earth, since in empty space there can be no conduction or convection. This transfer is by *radiation*; any hot body emits a broad spectrum of electromagnetic waves which carry energy. Radiation will be discussed separately in Chapter 7.

5.2 CONDUCTION

In this section, heat conductance in solid materials is examined quantitatively. Suppose temperature difference $T_1 - T_2 = \Delta T$ is maintained between two parallel faces of area A, of a slab of thickness Δx as shown in Fig. 5-1. The flux J of the heat-flow expressed as energy per unit area per unit time $(J \cdot m^{-2} \cdot s^{-1})$, or power per unit area $(W \cdot m^{-2})$ is given by $Q/(At)$, where Q is the heat transferred in a time interval t across area A. The flux is proportional to the "temperature gradient" $\Delta T/\Delta x$ which drives the flow[1]:

$$J = \frac{Q}{At} = k\frac{\Delta T}{\Delta x} \tag{5-1}$$

---

[1] This is just one example in science of a whole class of phenomena in which the gradient of some potential drives the flow of some measured quantity. Other examples: in electricity, Ohm's Law (V = IR) can also be written as $q/(At) = (1/\rho)(\Delta V/\Delta x)$ where ρ is the resistivity, ΔV is the voltage, and q is the electric charge. In diffusion, $n/(At) = D(\Delta C/\Delta x)$ is Fick's Law where D is the diffusion coefficient, n is the number of moles transferred and ΔC is the concentration difference of the diffusing molecules across the distance Δx.

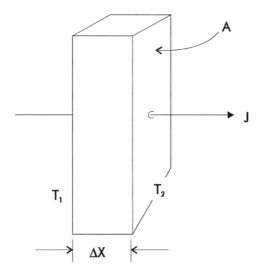

Figure 5-1: Thermal flux (J) due to a temperature difference $T_1 - T_2$.

where k is a constant of proportionality called, in this case, the *thermal conductivity*.

Table 5-1: Thermal Conductivities of Selected Materials

Material	$k \ (W \cdot m^{-1} \cdot {}^{\circ}C^{-1})$
Vacuum	0
Copper	385
Plywood	0.12
Glass	0.8
Brick	1.3
Concrete	1.0
Air (non convecting)	2.4×10^{-2}
Argon (non convecting)	1.6×10^{-2}
Fibreglass	4×10^{-2}

Equation 5-1 can be rewritten to give the heat transferred per unit time, Q/t (i.e., power P):

$$P = \frac{Q}{t} = kA\frac{\Delta T}{\Delta x} \qquad\qquad (5\text{-}2)$$

Equation 5-2 accords with intuition; we certainly expect the heat transferred per unit time, Q/t, to increase with the cross-section area, A, and with the temperature difference, ΔT, and to decrease with an increase in the thickness, Δx. Values of k for a few materials are given in Table 5-1.

Two facts might be noted from the table: Firstly, the conductivity of the metal is about three orders of magnitude greater than for the insulating materials as expected from the arguments given in section 1. Secondly, the conductivity of air is twice as good as that of fibreglass. Why then use fibreglass insulation? The reason is that air, and all gases, are very easily set into convection which greatly increases their ability to transport heat. One way to suppress convection is to divide the gas up into very small cells which, if less than a critical size (~ 1 mm), will not permit convection circulation. This is the function of fibreglass or foam insulation, or even feathers; it allows us to profit from the great insulating property of stagnant air by confining the air in small pockets within the material.

Methods for measuring the thermal conductivity of conductors and insulators are outlined in Problems 5-1 and 5-2.

Example 5-1 *How many joules per day are transferred through a plywood sheet that is 4 ft. by 8 ft (1.22 m × 2.44 m) in size and 5.0 mm thick if the temperature on one side is -10°C and 20°C on the other?*

From Eqn. 5-2, and Table 5-1

$Q = kA(\Delta T/\Delta x)t$

$\quad = (0.12\ W\cdot m^{-1}\cdot{}°C^{-1})(1.22\ m \times 2.44\ m)[20°\text{-}(\text{-}10°)](1/5.0 \times 10^{3}\ m)(24\ hr \times 3600\ s/hr)$

$\quad = 1.9 \times 10^{8}\ J$

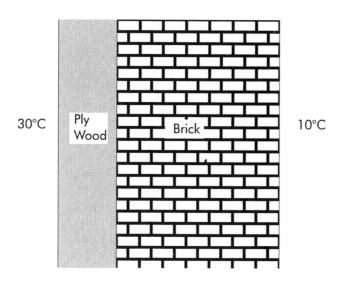

Figure 5-2: Example 5-2.

Example 5-2 *A 2.5 cm thick layer (1") of plywood is glued to a 10 cm thick brick wall (4") as shown in Fig. 5-2; the glue is of negligible thickness. The exposed face of the plywood is maintained at 30°C and that of the brick is 10°C. What power is passed across each square metre of this assembly and what is the temperature, T, of the plywood-brick interface?*

Writing Eqn 5-2 for the plywood and the brick;

$$30 - T = (Q/tA)(\Delta x/k) = (Q/tA)(0.12/2.5 \times 10^{-2})$$

$$\underline{T - 10} = \underline{= (Q/tA)(1.3/10 \times 10^{-2})}$$

adding 20 = $\quad = (Q/tA)(17.8)$

$$\therefore Q/tA = 20/17.8 = 1.1 \ W/m^2$$

Substitute in either of the above expressions for T

$$T = 25°C$$

5.3 THE CONDUCTIVE-CONVECTIVE LAYER

It might seem possible to calculate heat loss through walls, windows etc. by direct substitution into Equation 5-1. However, in addition to the conduction process of Eqn. 5-1 there is also a convection process — one which supplies heat to one side and carries it away from the other. Obviously air will circulate around depending on the temperature gradients, but the thin layer of air right against the surface experiences a frictional drag and cannot follow the flow of the main body. Thus there is a transition air-film called the "conductive-convective layer" across which the heat transfer is dominated by the conduction of the air (near the solid surface) and gradually becomes dominated by convection further away from the solid surface.   For this reason the air actually contributes substantially to the insulating value of a surface; in the case of a window it is the most important part!

Table 5-2: Conduction-convection Parameter in Air at Atmospheric Pressure (ΔT = temperature difference across conductive-convective layer)

	$h(W \cdot m^{-2 \cdot \circ} C^{-1})$
Horizontal surface facing up	$2.5\,(\Delta T)^{1/4}$
Horizontal surface facing down	$1.3\,(\Delta T)^{1/4}$
Vertical surface	$1.8\,(\Delta T)^{1/4}$

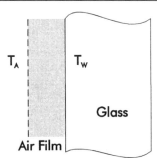

Figure 5-3: The conductive-convective layer of air on a glass window.

As shown in Fig. 5-3 it is assumed that this layer has a temperature drop T_A-T_W across it, where T_A is the temperature of the air in the room and T_W is the temperature of the window surface. (Put your finger on a window in cold weather if you want to be convinced that the inner surface of the window is not at room temperature).

The mathematical description of convection is very complex but fortunately it has been found that, for temperature differences which are not too great, this conductive-convective layer is well described by a simple empirical equation:

$$\frac{Q}{t} = h(T_A - T_W)A \qquad (5\text{-}3)$$

where h is a parameter described in Table 5-2.

The factor $(\Delta T)^{1/4}$ makes the solution of practical problems nonanalytical and slightly awkward, but as will be seen in Example 5-3 a simple iterative method leads quickly to a solution.

■

Example 5-3   *Heat is transferred through a vertical window, glazed with a single pane of glass of dimensions 1.0 m × 1.0 m × 3.0 mm, from a room at 20°C to the outside at 0°C. The thermal conductivity of the glass is 0.80 $W \cdot m^{-1} \cdot °C^{-1}$. Assume that the outside and inside air is perfectly still, so that there is a conductive-convective layer on both sides of the glass as shown in Fig. 5-4. Determine the rate of heat loss through the window, and the temperature difference across the glass.*

Rewriting Eqns. 5-2 and 5-3 to give the temperature drops for the three layers:

$$20 - T_1 = \frac{Q}{t}\frac{1}{A}\frac{1}{h}$$

$$T_1 - T_2 = \frac{Q}{t}\frac{1}{A}\frac{\Delta x}{k}$$

$$T_2 - 0 = \frac{Q}{t}\frac{1}{A}\frac{1}{h}$$

Add: $$20 = \frac{Q}{t}\frac{1}{A}\left\{\frac{2}{h} + \frac{\Delta x}{k}\right\}$$

Re-arranging to solve for Q/t:

$$\frac{Q}{t} = \frac{20A}{\left\{ \frac{2}{h} + \frac{\Delta x}{k} \right\}}$$

Of course Q/t is the same in each equation since, when equilibrium is established, the heat which enters at one side must pass through every interface and be removed at the final surface.

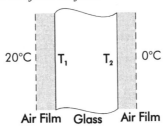

20°C T_1 T_2 0°C

Air Film   Glass   Air Film

Figure 5-4: Example 5-3.

It is necessary to make some reasonable assumptions — or even guesses — about T_1 and T_2, and then iterate the solution to an answer. Let's assume to start with that $T_1 = T_2 = 10°C$ (i.e., assume that the insulating properties of the thin glass are negligible (or $T_1 - T_2 \approx 0$) and that the two air layers are equivalent). Thus, ΔT for each air film is $10°C$, and from Table 5-2:

$$h = 1.8 \, (10)^{1/4} = 3.2 \, W \cdot m^{-2} \cdot °C^{-1}$$

Calculating Q/t with this value of h gives:

$$\frac{Q}{t} = \frac{(20°C)(1.0m^2)}{\left[\dfrac{2}{3.2W \cdot m^{-2} \cdot °C^{-1}} + \dfrac{0.0030m}{0.80W \cdot m^{-1} \cdot °C^{-1}} \right]} = 32 \, W$$

Knowing the value of Q/t, we can now use our earlier expression for $T_1 - T_2$ to determine the temperature change across the glass:

$$T_1 - T_2 = \frac{Q}{t}\frac{1}{A}\frac{\Delta x}{k} = 32W \times \frac{1}{1.0m^2} \times \frac{0.0030m}{0.80W \cdot m^{-1} \cdot {}^\circ C^{-1}} = 0.12\,{}^\circ C$$

instead of zero as we assumed. The calculation could be repeated with ΔT for each air film set equal to $\frac{1}{2}[20 - 0.12]\,{}^\circ C = 9.94\,{}^\circ C$, but it is hardly worth it, since our original guess of $\Delta T = 10\,{}^\circ C$ was very close to this value. ∎

The above example points out several interesting things: The insulating property of a window is almost completely determined by the surface air films; without these films the heat flow would be

$$\frac{Q}{t} = 20A\frac{k}{\Delta x} = 5.3 \times 10^3 \text{ W}$$

or about 170 times larger! The conductivity of the glass is almost irrelevant; if visibility were not a factor it could be a sheet of copper with almost no change in heat loss. The glass only serves to provide two surfaces to form two air films. It is now clear why windows lose so much heat on a windy day. The wind disrupts and effectively removes the outer layer cutting the insulating property of the window in half. It is also clear that the function of storm windows is not to form a "dead air space" as is often stated, but to create two more surfaces to which air layers may adhere. A single glazed storm window cuts heat loss by a factor of two on a still day and three on a windy day over that of an unprotected window.

5.4 THERMAL RESISTANCE

Commercial insulation is usually specified by its "R" value which stands for its "thermal resistance", defined by

$$R = \frac{\Delta x}{k} \tag{5-4}$$

If this is substituted into Eqn. 5-2,

$$\frac{Q}{\Delta t} = \frac{A \, \Delta T}{R} \tag{5-5}$$

This equation can be rewritten as

$$\Delta T = \left[\frac{1}{A}\right]\left[\frac{\Delta Q}{t}\right] R \tag{5-6}$$

which is similar in form to Ohm's Law in electricity, $V = IR$, so the (thermal) resistance is well named.

The numbers usually quoted for the R-value of insulation (e.g. R10) are based on an awkward and archaic set of units: Q in British Thermal Units (Btu), t in hours, A in square feet, Δx in inches, and ΔT in $^\circ$F. The relation between the thermal conductivity based on these units and the SI thermal conductivity is (see Exercise 1);

$$k \, (BTU \cdot hr^{-1} \cdot in \cdot {}^\circ F^{-1} \cdot ft^{-2}) = 6.93 \, k \, (W \cdot m^{-1} \cdot {}^\circ C^{-1}) \tag{5-7}$$

The R-values of several materials are given in Table 5-3 in British units $(Btu^{-1} \cdot ft^{2} \cdot hr \cdot {}^\circ F)$ for 1-inch thicknesses and in SI units $(W^{-1} \cdot m^{2} \cdot {}^\circ C)$ for 1-cm-thick samples.

Table 5-3: Thermal Resistance (R) Values

Substance	R (1 in.) $Btu^{-1} \cdot ft^{2} \cdot hr \cdot {}^\circ F$	RSI (1 cm) $W^{-1} \cdot m^{2} \cdot {}^\circ C$
Air (non-convecting)	6	0.42
Copper	3.7×10^{-4}	2.6×10^{-5}
Plywood	1.25	8.6×10^{-2}
Glass	0.18	1.3×10^{-2}
Brick	0.21	1.4×10^{-2}
Concrete	0.14	1.0×10^{-2}
Fibreglass	3.6	0.25
Polyurethane foam	6.3	0.43

Example 5-4   *Calculate the R-value of 1 inch of non-convecting air (i.e., the first line in Table 5-3).*

First, the value of k for air in British units is found using Table 5-1 and Eqn. 5-7, and then Eqn. 6-4 is used to calculate R:

$$k \quad = 6.93 \times 0.024 = 0.17 \; Btu \cdot hr^{-1} \cdot in \cdot {}^{\circ}F^{-1} \cdot ft^{-2}$$

$$R \quad = \frac{\Delta x}{k} = \frac{1}{0.17} = 6 \quad (as \; shown \; in \; Table \; 5\text{-}3)$$

Of course an R-value of 6 for one inch of air can never be realized in practice because such a thick layer of air would be set into convection and the heat transfer would increase greatly. If the air volume is broken up into small cells by some insulating material, such that convection is not possible, then this high insulating property of air can be utilized. This is the purpose of fibre or foam insulations. As can be seen from Table 5-3, the R-value of such insulations actually approach the ideal value of 6. In some cases it is even exceeded as for polyurethane foam. This is because the gas in the cells is not air but some other gas used in the manufacture of the foam that has a lower thermal conductivity even than air. The same effect operates in some sealed storm windows where the space between the double glazing is filled with the gas Argon; as can be seen from Table 5-1 this increases the insulating R-value by 50% (since $R \propto 1/k$).

The R-value of a conductive-convective layer is given by:

$$R = \frac{1}{h} \tag{5-8}$$

where h is given in Table 5-2.

The R-values are particularly useful since for multilayered structures they add, just as series resistances in an electric circuit.

Let us redo Example 5-3 using the thermal resistances:

Example 5-5   *Re-visit Example 5-3 using thermal resistances.*

Again assume $T_1 = T_2 = 10°C$

$$\therefore\ h = 1.8 \times (10)^{1/4} = 3.2\ W \cdot m^{-2} \cdot {}^{\circ}C^{-1}$$

Air films $: 2 \times (1/3.2\ W \cdot m^{-2} \cdot {}^{\circ}C^{-1})$ $=\ \ 0.625\ W \cdot m^{2} \cdot {}^{\circ}C$

Glass $: \dfrac{3 \times 10^{-3} m/0.80\ W \cdot m^{-1} \cdot {}^{\circ}C^{-1}}{R_{total}}$ $=\ \ \dfrac{0.004\ W \cdot m^{2} \cdot {}^{\circ}C}{0.629\ W \cdot m^{2} \cdot {}^{\circ}C}$

From Eqn. 5-5
$$Q/t = A\Delta T/R = 1\ m^{2}(20\ {}^{\circ}C)/0.629\ W \cdot m^{2} \cdot {}^{\circ}C = 32\ W,\ as\ before.$$

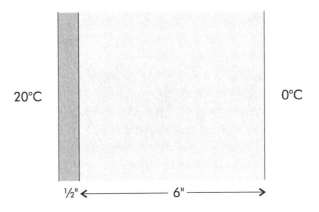

20°C   0°C

½" ←————— 6" —————→

Figure 5-5: Example 5-6.

Example 5-6   *The wall of a room is constructed in two layers. In contact with outside <u>still</u> air at 0.0°C is a 6" (15.2 cm) layer of concrete. A ½" (1.3 cm) layer of plywood is glued in close contact with the inside surface of the concrete. The inside temperature is 20.0°C and the inside air is also still. Find the power transferred across 10 m² of the wall and the temperature at each surface.*

Assume for now that $T_2 = T_3 = T_4$.

Therefore, $T_1 - T_2 = T_4 - T_5 = 10\,{}^{\circ}C.$

$$h = 1.8 \times (10)^{1/4} = 3.2 \ W \cdot m^{-2} \cdot {}^{\circ}C^{-1}$$

Now calculate the R-value for each layer, using R = 1/h for the air films and R = Δx/k for the concrete and wood:

			RSI
Air films:	$2 \times (1/(3.2 \ W \cdot m^{-2} \cdot {}^{\circ}C^{-1}))$	=	$0.62 \ W^{-1} \cdot m^2 \cdot {}^{\circ}C$
Concr.:	$(15.2 \times 10^{-2} \ m)/(1.0 \ W \cdot m^{-1} \cdot {}^{\circ}C^{-1})$	=	$0.15 \ W^{-1} \cdot m^2 \cdot {}^{\circ}C$
Wood:	$(1.3 \times 10^{-2} \ m)/(0.12 \ W \cdot m^{-1} \cdot {}^{\circ}C^{-1})$	=	$\underline{0.11 \ W^{-1} \cdot m^2 \cdot {}^{\circ}C}$
	R (total)	=	$0.88 \ W^{-1} \cdot m^2 \cdot {}^{\circ}C$

Use Eqn. 5-5 to determine Q/Δt:

$$\frac{Q}{t} = \frac{A \ \Delta T}{R} = \frac{(10 \ m^2)(20 \, {}^{\circ}C)}{0.88 \ W^{-1} \cdot m^2 \cdot {}^{\circ}C} = 227 \ W$$

Now that Q/t is known, Eqn. 5-6 can be used to find the ΔT's for the plywood, concrete, and air:

wood:
$$T_2 - T_3 = (1/A)(Q/\Delta t)R = (1/(10 \ m^2))(227 \ W)(0.11 \ W^{-1} \cdot m^2 \cdot {}^{\circ}C)$$
$$= 2.5 \, {}^{\circ}C$$

concrete:
$$T_3 - T_4 = (1/10)(227)(0.15) = 3.5 \, {}^{\circ}C$$

either air layer:
$$T_1 - T_2 = T_4 - T_5 = (1/10)(227)(0.31) = 7.0 \, {}^{\circ}C$$

Notice that the total temperature difference across all the layers (including the two air layers) adds up to 20.0° C, as it should.

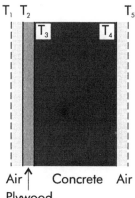

Figure 5-6: Showing the temperatures for Example 5-6.

Now calculate T_4, T_3, and T_2 :

$$T_4 = T_5 + 7.0\,°C = (0 + 7.0)\,°C = 7.0\,°C$$

$$T_3 = T_4 + 3.5\,°C = 10.5\,°C$$

$$T_2 = T_3 + 2.5\,°C = 13.0\,°C$$

A second iteration using $T_1 - T_2 = 7.0\,°C$ gives:

$$Q/t = 213 \text{ W, rounded to } 2.1 \times 10^2 \text{ W}$$

$$T_2 - T_3 = 2.3\,°C$$

$$T_3 - T_4 = 3.3\,°C$$

$$T_1 - T_2 = T_4 - T_5 = 7.2\,°C$$

$$T_3 = 10.5\,°C$$

$$T_2 = 12.8\,°C$$

Notice that this second iteration produces relatively minor changes in the values. ∎

5.5 DEGREE-DAYS

In designing insulation for buildings, one has to remember that outside temperature varies a great deal during the winter. For this reason, units called "heating degree-days" or HDD are used to give a measure of the insulation or heating requirements in various locations. The building industry in metricated countries including Canada use, or are converting to, SI units but in the United States the British system of units is still in universal use. It is assumed that, if the average outside temperature (T_{av}) is over 18°C (65°F), no heat need be supplied to the building. When T_{av} is less than 18°C, then for 1 day

$$\text{No. of Celsius degree-days} = 1 \text{ day} \times (18 - T_{av}) \qquad \textbf{(5-9)}$$

or $\qquad$ $$\text{No. of Fahrenheit degree-days} = 1 \text{ day} \times (65 - T_{av}) \qquad \textbf{(5-10)}$$

The sum of these quantities for every day on which T_{av} is less than 18.3°C (or 65°F) gives the total "degree-days" for the locality. The degree-day map for Canada is shown in Fig. 5-7, and that for the United States in Fig. 5-8.

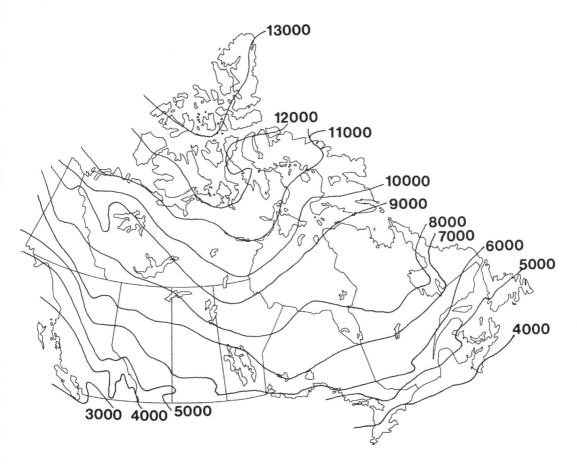

Figure 5-7: Heating Degree Days (HDD) in Canada averaged for one year.

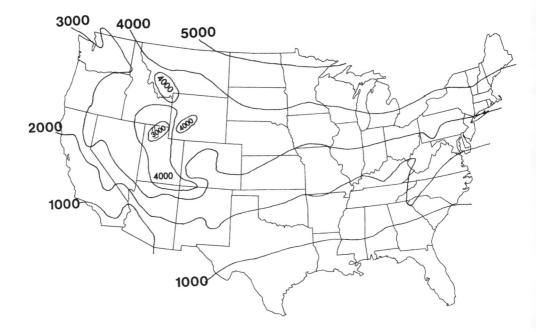

Figure 5-8: Heating Degree Days (HDD) in U.S.A. averaged for one year.

■

Example 5-7 *Suppose we have a house through whose walls, roof and windows we have calculated a heat-loss rate of 4000 W (4000 J/s) on a day when the temperature is 21°C inside and 6°C outside. Then, per degree temperature difference and per day, the loss is given by:*

$$Loss = \left[\frac{4000\,J}{s}\right] \times \left[\frac{3600\,s}{1\,h}\right] \times \left[\frac{24\,h}{1\,day}\right] \div (15°C)$$

= 2.3 × 10⁷ J per day per degree temperature difference

This is then the heat loss per degree-day. In the City of Toronto, the annual heating energy requirement for this house would be about (using Fig. 5-5):

(4000 degree-days) × (2.3 × 10⁷ J/(degree-day)) = 9 × 10¹⁰ J

■

EXERCISES

5-1   Show that: $k(\text{Btu} \cdot \text{hr}^{-1} \cdot \text{in} \cdot {}^\circ\text{F}^{-1} \cdot \text{ft}^{-2}) = 6.93 \ k \ (\text{W} \cdot \text{m}^{-1} \cdot {}^\circ\text{C}^{-1})$

using: 1 Btu = 1054.8 J   1 inch = 2.54 × 10⁻² m

1 ft² = 0.0929 m²   9 F° = 5 C°

[Hint: You will probably discover that $1 \ \text{Btu} \cdot \text{hr}^{-1} \cdot \text{in} \cdot {}^\circ\text{F}^{-1} \cdot \text{ft}^{-2} = 0.144$ $\text{W} \cdot \text{m}^{-1} \cdot {}^\circ\text{C}^{-1}$, and wonder how to get from 0.144 to 6.93. Here is an analogy: You know that 1 cm = 0.01 m; suppose that the length ℓ of some object is measured both in cm and in m. How is the numerical value of ℓ in units of cm related to ℓ in units of m? So, how is the numerical value of k in $\text{Btu} \cdot \text{hr}^{-1} \cdot \text{in} \cdot {}^\circ\text{F}^{-1} \cdot \text{ft}^{-2}$ related to k in $\text{W} \cdot \text{m}^{-1} \cdot {}^\circ\text{C}^{-1}$?]

5-2   Determine the heat energy lost in 1.0 h through a slab of styrofoam that is 3.0 cm thick, and that has a size of 2.0 m × 1.5 m. The temperature difference between the two sides of the styrofoam is 12°C. The thermal conductivity of styrofoam is approximately $0.01 \ \text{W} \cdot \text{m}^{-1} \cdot {}^\circ\text{C}^{-1}$. Neglect the effect of air films on the surface of the styrofoam.
[Ans. 4×10^4 J]

PROBLEMS

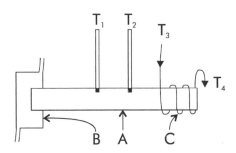

Figure 5-9: Problem 5-1.

5-1 An apparatus to measure the thermal conductivity of copper (called a conductometer) consists of a round copper bar A heated by a steam jacket B at one end and cooled by coils C carrying water at the other. The heat conducted along the copper is removed by the water (specific heat 4186 J· kg^{-1}·°C^{-1}). Two thermometers T$_1$ and T$_2$ are inserted in small holes in the copper 2.0 cm apart. The temperature difference T$_1$ - T$_2$ is 10°C when water flows through the tubing at 50 cm^3 per minute and enters at a temperature of T$_3$ = 10.0°C and leaves at a temperature T$_4$ of 27.3°C. The diameter of the bar is 2.0 cm. Find the thermal conductivity of copper.   [Ans. 3.8 × 10^2 W· m^{-1}· °C^{-1}]

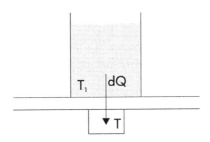

Figure 5-10:   Problem 5-2.

5-2 To measure the thermal conductivity of an insulator, thin samples are usually used as in the method of Fitch shown in Fig. 5-10. A heat reservoir at constant temperature T$_1$ (e.g. a metal pot of boiling water) is placed on the insulating layer (thickness, x, and conductivity, k) which in turn is placed on a metal block (mass, M, cross-section area, A, and specific heat, C) at time t = 0. In a time interval dt an amount of heat dQ flows across the insulator, raising the temperature of the block by an amount dT.
a) Show that:

$$dt = \frac{CMx}{kA} \frac{dT}{T_1 - T}$$

b) Show that a graph of ℓn(T$_1$ - T) vs. t gives a straight line whose slope is -kA/(CMx) from which the conductivity k can be determined.

5-3 The door of a refrigerator is 1.5 m high, 0.80 m wide and 6.0 cm thick. Its thermal conductivity is 0.21 W· m^{-2}· C^{-1}. The inner surface is at $0°C$ and the outer is at $28°C$. What is the heat loss per hour through the door? (neglect convection effects). [Ans. 4.2×10^5 J]

5-4 A beer cooler, made of plastic foam, is in the form of a box. The inside dimensions of the box are 50 cm × 30 cm × 30 cm and the walls are 5.0 cm thick. The plastic foam has a thermal conductivity of 0.040 $W· m^{-2}· C^{-1}$. Ice is placed inside the cooler to keep it at $0°C$ inside. How much ice melts each hour if the outside temperature is $30°C$? [Ans. 0.20 kg]

5-5 Two copper plates 0.50 cm thick have a 0.10 mm sheet of glass sandwiched between them. One copper plate is kept in contact with flowing ice water and the other is in contact with steam. What are the temperatures of the two copper-glass interfaces, and what power is transferred through a 10 cm × 10 cm area?
[Ans. $9°C$, $91°C$, 6.6×10^3 W]

5-6 A home has a concrete basement floor 15 × 15 m square and 10 cm thick. On a winter day, the ground below is at -10.0°C while the basement recreation room is at 20.0°C.
(a) Calculate the rate of loss of heat, remembering to include the air film.
     [Ans. 2.3×10^4 W]
(b) A do-it-yourself enthusiast glues a plywood floor of thickness 1.0 cm onto the concrete. What fraction of the heat loss is eliminated?
     [Ans. 25%]
(c) What is the temperature at the interface of the plywood and concrete?
     [Ans. -2.2°C]

ENVIRONMENTAL EFFECTS OF FOSSIL FUELS

Some newspaper headlines:

"Urban Joggers Risk Lung Damage"
  Guelph [Ontario] *Mercury*, November 13, 1990

"U.S. National Park Beauties Slip Behind Veil of Pollution"
  Toronto *Globe and Mail*, December 9, 1989

"Flames Light Sky for 100 km: Natural Gas Pipeline Explodes"
  Guelph [Ontario] *Mercury*, January 9, 1992

Almost every day newspapers carry stories about air pollution, or "the greenhouse effect", marine oil spills, fires or cave-ins in coal mines, acid rain, or burning oil wells or gas pipelines. All of these serious environmental problems are associated with the use of fossil fuels, and they are so numerous and important that the corresponding materials are split between two chapters of this text:

- Chapter 6 deals with issues related to extracting and transporting fossil fuels, thermal pollution due to the waste heat from burning fossil fuels (especially for production of electricity), and air pollution, including acid rain and smog.
- Chapter 7 considers the greenhouse effect.

6.1 PROBLEMS WITH EXTRACTION AND TRANSPORT OF FOSSIL FUELS

Extraction

In order for fossil fuels to be used, they must first be removed from the ground. In the case of oil and gas wells, the drilling and extraction of the fuels produces local disruptions in ecosystems, although in most areas the impact is rather minor, comparable to that of any small industry. Of course, there are examples of situations where the effects are significant, such as the long-range air and water pollution that resulted from the burning of the oil wells in Kuwait in 1991. As well, in regions such as the Arctic and offshore, spills and blowouts can have major and lasting effects on plant and animal life, and extra care must be taken to minimize the possibility of serious accidents.

Did You Know? *The world's first producing oil field was near Petrolia, Ontario, which brought in the first gusher in 1858. The photograph shows a pump (of Canadian design) still in operation in the Petrolia field.*

In the case of coal, both underground and surface (strip) mining have disruptive consequences. Underground mining can lead to land subsidence, and the risk of fires and explosions is always present. Strip mining produces complete and utter disfigurement of the landscape, unless costly measures are taken to reclaim the land after the coal is removed. In both underground and strip mining, water seepage reacts with sulphur compounds in the coal to produce sulphuric acid, which leaks into rivers and lakes. In addition, mining is an extremely hazardous occupation — in the U.S.A. between 1970 and 1980, an average of 150 people were killed annually in coal-mining accidents. For miners who are not disabled or killed in the mines, there remains the possibility of contracting black lung disease as a result of breathing coal dust for many years. The lungs become damaged to such an extent that breathing can become extremely difficult, and the miners become susceptible to a host of respiratory ailments and problems.

Transport

Once out of the ground, the oil, gas, and coal need to be transported, either to a site where further processing occurs, or to an end-use site. Coal is commonly transported by ship or train, and environmental effects are usually minimal. However, the transport of oil by ship can have disastrous consequences — we have all heard of major spills due to oil tankers running aground and producing local havoc in fragile coastal environments. Although these accidents make major news stories, tankers actually put more oil into the oceans in another way — by flushing ballast water. After delivering oil to a destination, tankers are filled with water as ballast for the return trip. Upon completion of this trip, the water (containing oily residues from the tanks) is discharged into the sea. It has been estimated[1] that ballast water accounts for about 20% of the oil in the oceans, whereas tanker accidents introduce only about 3%. Other major sources of oil[1] reaching the oceans include river runoff (25%), unburned hydrocarbons from the atmosphere (10%), natural seepage (10%), and flushing of ballast water from fuel tanks (not to be confused with oil-cargo tanks) in ships (10%).

Oil and gas are also transported by pipeline, which occasionally have leaks and/or explosions (as indicated in one of the headlines at the beginning of this chapter). Problems such as these are particularly troublesome in environmentally sensitive areas such as the Arctic.

Transporting a fossil fuel requires some energy, of course, and the further the fuel must be moved, the more energy is required. Ships provide the most energy-efficient mode of transport[2], followed closely by pipelines and railroads (Table 6-1). Trucks use more energy, and airplanes considerably more.

---

[1] J.M. Fowler, *Energy and the Environment, Second Ed.* McGraw-Hill, 1984, p. 209 (original source: The Global 2000 Report to the President, *Entering the Twenty-First Century*, A Report Prepared by the Council on Environmental Quality and the Department of State, U.S. Government Printing Office, Washington, D.C., 1981)

[2] Ref. 1, p. 466 (original source: R.F. Hemphill, "Energy Conservation in the Transportation Sector," in *Energy Conservation and Public Policy*, J.C. Sawhill, ed., Prentice-Hall, 1979, p. 92)

Table 6-1: Energy Cost (Range) in Freight Transport

Mode	Energy Cost (Btu/(ton · mile))
Ship	100 - 500
Pipeline	100 - 1300
Railroad	200 - 1000
Truck	1100 - 2000
Airplane	8000 - 25000

6.2 THERMAL POLLUTION

We saw in Chapter 4 that most of the thermal energy produced in a fossil-fuel or nuclear electrical plant is not converted to electricity, but rather is discharged as waste heat to water (or the atmosphere). As more and more electrical generating stations come into service, this thermal pollution becomes an increasing problem.

In nuclear plants, all of the waste heat is removed by cooling water; at a typical efficiency of 30%, the cooling water is removing heat energy 2.3 times the station's electrical rating. In fossil-fuel plants, a significant portion (about 10%) of the available energy goes up the stack as hot gases and is lost to the atmosphere; the demand for cooling water is considerably less than at a nuclear plant both for this reason and because efficiencies are nearer 40%.

In many plants, the cooling water is drawn from a lake, river, or ocean, used once and then returned with its temperature increased by about 10°C. (In Ontario and most other areas of North America and Europe, there are regulations requiring that the rise in temperature in "once-through cooling" be no greater than 10°C.) The temperature increase produces changes in the overall aquatic ecology in the surroundings, encouraging some organisms and discouraging others. Most types of fish are particularly sensitive to temperature changes and can remain healthy only within rather narrow temperature limits, which vary from species to species. Temperature affects fishes' appetites, digestion, growth, behaviour, spawning, and longevity. As well, the warmer water contains less dissolved oxygen, which can affect the viability of aquatic organisms. When an electrical plant is closed down for maintenance or other reasons, fish and other creatures that have thrived in the warmth of the cooling waters are suddenly thrust into a colder environment and might not be able to survive.

Heat energy is not the only pollutant. Upon intake the water is usually treated with chemicals to prevent buildup of slime and to kill various kinds of larvae, etc. After the water is used for cooling, it is discharged along with the chemicals into some larger body of water. (A filter screen prevents fish and debris from entering the cooling system.)

There is increasing concern about the large volumes of water needed to cool electrical plants. Example 6-1 below shows just how much water is needed to remove the waste heat from a typical fossil-fuel fired plant.

Example 6-1   *A coal-burning electrical plant produces 1000 MWe at 40% efficiency. What minimum volume of water is needed per second to remove the waste heat, if the water undergoes a temperature increase of 10° C? (Assume two significant digits in all given data and neglect the heat energy dissipated up the chimney.)*

Consider a time period of 1 s. Since a watt is a joule per second, and the electrical power produced is 1000 MW, then the electrical energy produced in 1 s is 1000 MJ. To determine the waste heat generated in 1 s, first use the efficiency to determine the total energy input to the plant.

$$\text{efficiency} \quad = \quad \frac{\text{useful (electrical) energy out}}{\text{total energy in}}$$

$$\text{Thus, total energy in} \quad = \quad \frac{\text{useful (electrical) energy out}}{\text{efficiency}}$$

$$= \quad \frac{1000 \ MJ}{0.40} \quad = \quad 2500 \ MJ$$

The waste heat energy is just the difference between the total energy and the electrical energy:

$$\text{heat energy} \ = \ (2500 - 1000) \ MJ = 1500 \ MJ$$

131

Now determine the mass of water required by using Eqn. 4-5:

$$Q = mc\Delta T \quad giving \quad m = \frac{Q}{c\Delta T} = \frac{1500 \times 10^6 \ J}{(4186 \ J \cdot kg^{-1} \cdot K^{-1})(10 \ K)} = 3.6 \times 10^4 \ kg$$

Since the density of water is 1.0×10^3 kg/m³, the volume of water required is:

$$volume = \frac{mass}{density} = \frac{3.6 \times 10^4 \ kg}{1.0 \times 10^3 \ kg/m^3} = 36 \ m^3$$

Hence, 36 m³ of water are required every second. ∎

The volume of water determined in the above example is huge! On a daily basis, it works out to 3.1×10^6 m³ or 3.1 billion litres, and this is only for one electrical plant. In Problem 6-2 at the end of this chapter, we see that if once-through cooling were used in all fossil-fuel and nuclear plants in the U.S.A., then the annual cooling water requirement would be 1/16 of the total water available as runoff from precipitation. Of course, some water can be used more than once at different locations along a river, but nonetheless the water requirement for cooling is enormous.

It is fairly obvious that in regions without access to large water bodies, once-through cooling is not practical. Instead, closed-cycle cooling is used in which the same water is used over and over again. There are three basic approaches: cooling ponds, wet cooling towers, and dry cooling towers.

Cooling Ponds

In regions where land costs are low, artificial lakes or ponds can serve to provide cooling water for an electrical plant. These ponds are of such size that the entire volume flows through the plant in a week or two, and the heat received by the water is dissipated mainly by surface evaporation to the atmosphere. In order to provide ample surface for evaporation, the ponds need to be quite large — several square kilometres. Of course a continual water supply is needed to replenish the evaporative loss, but this requirement is much less than for once-through cooling. In addition, the artificial lake has recreational and even sport fishing potential.

Wet Cooling Towers

Figure 6-1: Cooling towers at a nuclear power station in France.

These hyperboloid-shaped towers (Fig. 6-1), having both a height and diameter of 100 to 200 m, are widely used in Europe. The warm cooling water from the condenser in the electrical plant is introduced at the top of the tower (Fig. 6-2); it is either sprayed in as a mist of droplets or enters in bulk to fall on a series of baffles that break it up. In either case the result is that a large surface area of water is exposed to air; the consequent evaporation cools the remaining water prior to recirculation to the condenser. The air in the tower naturally rises as a result of being heated, and is replaced by cooler air from below in this *natural-draft* arrangement. As with cooling ponds, wet cooling towers require makeup water, but only 2% to 4% of that needed for once-through cooling.

The large size of wet cooling towers can be reduced by using fans to force much larger volumes of air into contact with the falling water. In these *forced-draft* cooling towers, the fans use about 4% of the station's energy output.

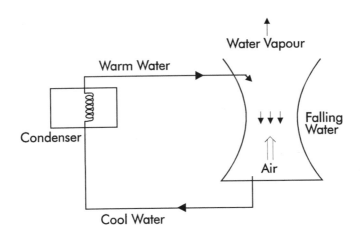

Figure 6-2: Schematic diagram of a wet cooling tower.

Both types of wet cooling tower share a major environmental drawback in that by inducing evaporation, they introduce large amounts of water vapour into the atmosphere. At the very least, the towers are a constant source of clouds and introduce humidity into the local environment. As well, chemicals are added to the water to prevent organic growths and corrosion, and these chemicals are released into the atmosphere.

Dry Cooling Towers

Dry towers do not rely on evaporation, but instead allow the warm cooling water to pass through a large heat exchanger (like a huge car radiator) inside the cooling tower. Air passes up the tower past this exchanger by either natural or forced draft. The environmental problems of introducing moisture and chemicals into the atmosphere are removed, but the cost is two to three times that of a wet tower and up to 8% of the station's energy is needed to operate a forced-draft dry tower. An important point is that a dry tower can be used in a location where there is not sufficient water for a wet tower or cooling pond.

Thermal Enhancement

It should be evident from the foregoing that fossil-fuel and nuclear electrical plants produce large quantities of waste heat. It would seem logical to heat buildings with this energy, or to provide low-temperature thermal energy for industries. In some parts of the world, cogeneration (production of both steam and electricity for industry) and district heating (use of waste heat from electrical production to heat a community) are much more common than in North America. Cogeneration and district heating are discussed in more detail in Chapter 18.

6.3 THE URBAN HEAT-ISLAND

Thermal pollution is not just the waste heat dumped into the environment at a generating station. Most of the energy of fossil fuels consumed by society finds its way eventually as heat energy to the environment; the exception is energy "stored" in manufactured materials. When averaged over the globe, the heat generated by society is a minute fraction of the energy reaching us from the sun; however, in urban areas our energy consumption is highly localized, and as a result the temperature of urban regions is often greater than in the surrounding districts. This increase in temperature is due not only to heat generated in buildings and by motor vehicles, but also to increased absorption of solar energy because of the large quantities of concrete and asphalt. As well, an important cooling mechanism — the evaporation of water — is reduced in cities because rainwater quickly drains away instead of being held by soil and vegetation. In the summer, there is an increasing use of air conditioning in buildings and vehicles; heat energy is extracted from interiors, and expelled to the local atmosphere, producing higher air temperatures. This results in increased use of the air conditioners, which warms the outdoor air even more. As well, air conditioners use electricity, which has thermal pollution associated with its generation.

Many examples of this urban heat-island effect have been documented. In the City of Edmonton, Alberta, the average daily minimum temperature in January is 2.8°C higher downtown than in the surrounding countryside[3]. In July, the average minimum is 2.6°C higher downtown.

6.4 AIR POLLUTION

" Forget six counties overhung with smoke,
Forget the snorting steam and piston stroke,
Forget the spreading of the hideous town;
Think rather of the pack-horse on the down,
And dream of London, small and white and clean,
The clear Thames bordered by the gardens green. "
— William Morris (1834-1896), *Prologue, The Wanderers, 1.1*

The use of fossil fuels — first coal, then oil and gas — has increased the general standard of living dramatically. However, it has also fouled the air so much that in Mexico City in 1991 the air was extremely unhealthy for 192 days out of the year (Figure 6-3). When there were few people on this planet, the rate at which pollutants were sent into the air was low enough that the atmosphere remained relatively clean. Now that there are over five billion of us, air pollution is a serious problem.

Figure 6-3: On most days in Mexico City, a brownish haze hangs in the air.

[3] *The State of Canada's Environment*, Government of Canada, Ottawa, 1991, p. 2-8

Air pollution from fossil fuels can be categorized into six main types, each of which will be addressed in detail in the remainder of this chapter:

- particulate matter (PM)
- sulphur dioxide (SO_2)
- nitrogen oxides (NO_X)
- hydrocarbons (HC)
- carbon monoxide (CO)
- carbon dioxide (CO_2), discussed in Chapter 7 (The Greenhouse Effect).

These substances, released directly from the fossil fuels when burned or processed, are *primary* pollutants. However, the damage they cause is often due to other compounds — *secondary* pollutants — formed as a result of chemical reactions involving the primary pollutants. For example, both SO_2 and NO_X undergo reactions to form acids, which fall to the earth as acid rain, damaging aquatic ecosystems, crops, buildings, etc.

Because air can move quickly and easily over hundreds and thousands of kilometres, pollutants generated in one area can affect locations far away. Many types of air pollution present global problems that require international cooperation for their solution.

Particulate Matter (PM)

Particulate matter consists of a wide variety of particles — dust, smoke, droplets, microscopic bits of metal, tires, etc. — that are small enough to remain airborne for a considerable time. Naturally-generated PM consists of wind-blown dust, soil particles, pollen, smoke from forest fires, sea salt, and volcanic ash. The major human sources are industrial activities such as mining, quarrying, pulp and paper operations, and manufacturing; other sources include fossil-fuel electric power stations, waste incineration, and motor vehicles. Figure 6-4 shows the sources of PM emissions in Canada in 1985: various industrial processes produced about 52%; combustion of fuel in stationary sources such as power plants, industrial plants, homes, and businesses provided about 27%; transportation generated about 7%; and miscellaneous sources such as waste incineration and slash burning of forests produced about 14%. Included in the 27% from stationary fuel-burning is 13% from electric power plants and 8% from fuelwood.

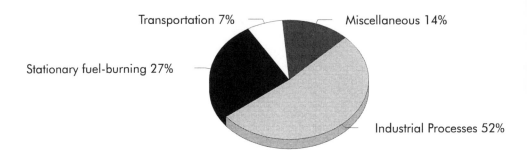

Transportation 7%

Miscellaneous 14%

Stationary fuel-burning 27%

Industrial Processes 52%

Figure 6-4: Sources of PM (Particulate Matter) in Canada in 1985 (*The State of Canada's Environment*, Government of Canada, Ottawa, 1991, pp. 14-6 to 14-7)

Small particles (diameter < about 1 μm) can be inhaled deeply into the respiratory tract and can result in impaired breathing and exacerbation of pre-existing pulmonary problems such as asthma. Particles of this size reduce visibility and scatter sunlight before it reaches the ground. Because of this scattering and because particles can also act as nuclei for formation of raindrops, high concentrations of PM can affect the weather. In addition, particles can carry damaging materials such as sulphuric acid to surfaces, accelerating their deterioration. Particulate matter is an important component of the classical London-type smog, discussed later in this section.

Control of Particulate Matter

Industry and electric power plants produce the majority of PM, and fortunately, there are suitable control devices for these sectors: *electrostatic precipitators* and "*bag houses*." As smoke and dust particles move, they acquire electric charges by rubbing with each other; in electrostatic precipitators, the particles pass through a metal tube containing a central rod at an electric potential of up to 100000 volts relative to the tube. This means that the tube and rod have opposite charges, and each charged particle is attracted to either the tube or rod, where the particle's charge is neutralized. As the particles collect, they agglomerate and become large enough to settle out.

A "bag house" is like a large vacuum cleaner, where a narrow cloth bag (about 15m long) is used to filter out the particles from air blown through the bag by large fans. "Bag houses" are used more in industry than in power plants.

In Canada, air quality is monitored at about 130 stations, which sample air in over 50 Canadian urban centres. The quantity of PM pollutants registered at these stations declined[4] by 44% from 1974 to 1989, and is now at about 60% of the maximum acceptable level. This reduction reflects emission control and cleaner processes in industry, and a trend toward cleaner energy sources such as natural gas. However, from 1985 to 1989 there was no change in measured PM, and a study[5] by the Ontario Ministry of the Environment showed little change in PM in that province from 1981 to 1991. In the U.S.A., where air quality is monitored at over 1600 sites, the measured concentration of PM decreased by 50% from the 1960s to the 1980s.[6] However, there was little reduction in the 1980s (a decrease of only 1% from 1982 to 1989), and in 1989 there were 25 million people in the U.S.A. living in areas where the PM concentration was still above the acceptable level.

Lead

It is nice to read success stories once in a while, and the reduction of particulate lead in the atmosphere is just such a story. Lead is a toxic metal that affects the nervous system, the kidneys, the brain, and the

---

[4] Ref. 3, pp. 2-10 to 2-11

[5] Toronto *Globe and Mail*, April 10, 1991

[6] D.M. Elsom, "Atmospheric Pollution: A Global Problem (2nd Ed.)", Blackwell Publ., Oxford 1992, pp. 219-220 (original source: U.S. Environmental Protection Agency)

cardiovascular system. Children are particularly susceptible to lead poisoning. In 1982, lead additives in motor fuels accounted for almost two-thirds of Canada's lead emissions. These additives were essentially banned in Canada in 1990, and even prior to 1990, there was a decreasing number of vehicles using leaded fuel. The result: from 1974 to 1989, there was a decrease[7] of 93% in the particulate lead measured in Canada's atmosphere. However, leaded gasoline is still available in some countries in the world, and in some states in the U.S.A.

Sulphur Dioxide (SO₂)

All fossil fuels contain some sulphur, although coal contains much more than oil, and oil much more than natural gas. The sulphur content of a fossil fuel is highly variable: coal from western Canada typically contains less than 0.5% sulphur by mass[8], whereas coal from Nova Scotia and New Brunswick has from 2% to over 6%. When a fuel is burned, almost all of the sulphur is released as SO_2, which is a colourless gas with a strong pungent odour. The major sources of SO_2 are coal-burning, especially in electric power plants, and even more important in Canada, the smelting of metallic ores, which usually contain sulphur. Figure 6-5 shows the sources of SO_2 in Canada in 1985. Ore-smelting accounts for most of the 67% from the industrial sector, contributing almost 50% of Canada's SO_2; the 27% from stationary fuel combustion includes 20% from fossil-fuel electric power plants. The biggest single source of SO_2 emissions in North America is the INCO smelter in Sudbury. In many other countries, where there is a heavier reliance on using coal to produce electricity, a larger percentage of the SO_2 comes from power plants. Around the world, more than half of the SO_2 comes from the burning of coal[9].

---

[7] Ref. 3, pp. 2-11 to 2-12

[8] Ref. 3, p. 12-23

[9] W. Fulkerson et al, Sci. Am. **263**, No. 3, p. 130 (Sept. 1990)

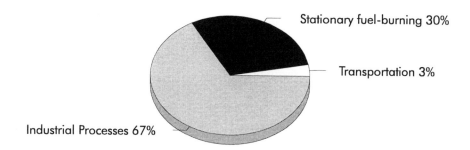

Stationary fuel-burning 30%

Transportation 3%

Industrial Processes 67%

Figure 6-5: Sources of SO_2 (Sulphur Dioxide) in Canada in 1985 (The State of Canada's Environment, Government of Canada, Ottawa, 1991, pp. 14-6 to 14-7)

Sulphur dioxide itself is very toxic to plants, even in very low concentrations (< 0.1 parts per million by volume — ppmv), and can cause respiratory irritation in humans at concentrations of 1 to 2 ppmv. According to Canadian air quality standards, the maximum acceptable concentration (annual average) is 0.023 ppmv. In 1989 the average concentration measured in Canada was about 25% of this maximum, but in some industrial areas the average concentration was above the maximum.[10]

Additional concerns are presented by SO_2 because of the acids that it forms. When SO_2 is released into the atmosphere, some of it oxidizes to give SO_3, which combines with water to form sulphuric acid (H_2SO_4). In addition, some of the SO_2 combines directly with water to form sulphurous acid (H_2SO_3). These acids fall to the earth either as acid rain or by adhering to particulate matter. Further discussion of the effects of these acids is presented later in this chapter.

---

[10] Ref. 3, p. 2-11 to 2-12

Controlling SO$_2$

Reduction of SO$_2$ emissions can be accomplished in a number of ways. The easiest is to switch to a fuel with less sulphur. High-sulphur coal can be replaced with low-sulphur coal, or with oil or gas. However, such a change cannot often be accomplished without cost. It might be necessary to import the low-sulphur fuel over long distances at some expense, and furnaces might need to be retrofitted for a new fuel. There might also be political problems — imagine importing low-sulphur coal into a region rich with high-sulphur coal, and thus putting local coal miners out of work.

It is also possible to clean coal before burning. Some of the sulphur is in the form of iron pyrite (FeS$_2$), which has a higher density than the rest of the coal. If the coal is crushed, and then agitated with water, air, oil, and a small amount of surfactant (a wetting agent), the iron pyrite settles out and can be removed; this process is called *oil flotation*. Cleaning coal in this way is cheaper than removing SO$_2$ from the flue gas, but it does not remove sulphur that is organically bound, and typically removes only up to 30% of the sulphur[11].

A common, but expensive, approach is to use *scrubbers* to remove SO$_2$ by spraying a mixture of water and lime (Ca(OH)$_2$) or limestone (CaCO$_3$) down through the flue gas passing upward in the smokestack. The SO$_2$ combines with the lime or limestone to produce CaSO$_3$, which can be removed as a solid from the water. However, huge holding tanks are required for this removal, and the disposal of the CaSO$_3$ itself presents environmental problems. Scrubbers can remove up to 90% of the sulphur[11], and new scrubbers being developed can remove 97%.

Other alternatives for controlling SO$_2$ that might make significant contributions in the future are fluidized-bed combustion and coal gasification, discussed in Chapter 17.

Although not specifically energy-related, the current efforts to reduce SO$_2$ emissions from smelters focus on:

- chemically capturing some of the SO$_2$ in the emissions and converting it to sulphuric acid to be sold.

---

[11] Ref. 9, p. 130

- removing (as much as possible) non-useful sulphur-containing ore by oil flotation before smelting.

In Canada, SO_2 emissions decreased by almost 45% between 1970 and 1985[12], and the annual average concentration of SO_2 measured at air quality stations decreased from 1974 to 1989 by over 50%, from about 60% of the maximum acceptable level to 25% of this level[13].   This improvement is attributable to a shift away from high-sulphur fuels, and better emission controls in smelters and electric power plants. Emissions of SO_2 will continue to decrease — in March 1985, the Canadian government and seven eastern provinces put in place a program to reduce SO_2 emissions by 50% within a decade. In December 1985, the Ontario Government ordered the four biggest single sources of SO_2 in Ontario to cut acidic emissions by two-thirds by 1994.

In the U.S.A., the use of scrubbers and lower-sulphur coals decreased SO_2 emissions[14] by 33% from 1974 to 1989, even though coal consumption increased by 50%. In October 1990, the U.S. Congress passed a clean-air bill that calls for a two-step reduction of almost 50% in SO_2 emissions by electric power plants by the year 2000.

Nitrogen Oxides (NO_x)

Another important component of air pollution comprises the oxides of nitrogen. Whenever air is heated above 500°C, nitrogen oxide (NO) is formed from nitrogen and oxygen, regardless of the heat source — automobile engines, home furnaces, coal-fired electric plants, etc. In the atmosphere, NO is oxidized to produce nitrogen dioxide (NO_2); whereas NO is colourless, NO_2 is a brownish gas, which absorbs in both the visible and ultraviolet regions of the spectrum.

Once formed, nitrogen dioxide can produce nitric acid (HNO_3) via more than one chemical route, and this acid settles to the earth as acid rain

---

[12] Ref. 3, p. 14-5

[13] Ref. 3, p. 2-12

[14] Ref. 9, p. 130

or on particulate matter. Nitrogen dioxide is also one of the primary pollutants in photochemical smog, discussed later in this chapter.

In Canada, 63% of the NO_X emitted in 1985 came from the transportation sector (Fig. 6-6), primarily motor vehicles. Fossil-fuel burning at stationary sources — electric power plants, industries, homes, and businesses — produced another 30%. Combining the transportation and stationary source sectors, virtually all the NO_X originates with the burning of fossil fuels. About 13% of Canadian emissions come from power plants, but in the U.S.A., the contribution from this source[15] is closer to 50%. Globally, most of the NO_X emissions are produced by burning fossil fuels, with about 30% released by coal-burning[16].

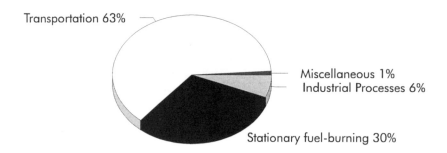

Transportation 63%

Miscellaneous 1%
Industrial Processes 6%

Stationary fuel-burning 30%

Figure 6-6: Sources of NO_X (Nitrogen Oxides) in Canada in 1985 (*The State of Canada's Environment*, Government of Canada, Ottawa, 1991, pp. 14-6 to 14-7)

Controlling NO_X

Since motor vehicles produce huge amounts of NO_X, much attention has been paid to decreasing NO_X emissions from this source. Various generations of catalytic converters have been developed to chemically reduce NO_X in exhaust gases to N_2. The most recent converters remove

---

[15] J.M. Fowler, *Energy and the Environment, Second Ed.*, McGraw-Hill, 1984, p. 164

[16] W. Fulkerson et al, Sci. Am. **263**, No. 3, pp. 129-130 (Sept. 1990)

about 75% of the NO_X from exhaust gases[17], and also have the function of oxidizing carbon monoxide and any unburned hydrocarbons (more about these pollutants later in this chapter).

Emissions of NO_X from coal-burning power plants can be decreased by lowering combustion temperatures through use of new burners that produce a cooler flame, or through injection of steam into the combustion chamber. These alternatives are just now starting to be adopted, and of course have expenses associated with them. For example, it has cost Ontario Hydro $13 million to replace all 320 burners at its largest coal-fired plant with a cooler-burning model[18].

In Canada between 1970 and 1985, NO_X emissions[19] increased by more than 40%, largely because the number of cars and trucks rose by about 70%. Measurements of atmospheric NO_2 (annual mean) at air quality stations decreased by about one-third from 1977 to 1981, but has remained roughly constant since then at about 40% of the maximum acceptable level[20].

In the U.S.A., emissions of NO_X remained relatively constant[21] from 1974 to 1989. As part of the new clean air bill passed in 1990, the U.S. Environmental Protection Agency will be developing, in a two-phase process, performance standards by 1993 and 1997 for NO_X emissions from coal-burning electric power plants.

Acid Rain

One of the pollution problems of which the public is most aware is acid rain. We have already discussed how SO_2 and NO_X produce damaging acids that can fall to the earth in rainwater. Natural rainwater is slightly

---

[17] N. Bunce, *Environmental Chemistry*, Wuerz Publ., Winnipeg, 1990, p. 310   (original source: J. Haggin, *Chem. Eng. News*, May 15, 1989, p. 23)

[18] "Curbing Acid Rain," Ontario Hydro Publication G0096, March 1989

[19] Ref. 3, p. 14-5

[20] Ref. 3, p. 2-12

[21] Ref. 9, p. 130

acidic with a pH of about 5.6, as a result of carbon dioxide dissolving in it to produce carbonic acid (H_2CO_3). However, as a result of SO_2 and NO_X emissions, rain with a pH between 3.5 and 4.5 regularly falls in many areas of the world. In 1974 in Scotland, there was a rainfall having a pH of 2.4 — this is more acidic than vinegar!

The effects of acid rain are many: damage to forests, crops, soils, lakes and rivers, fish, stonework, etc. In the case of forests and crops, it is often difficult to point the finger of guilt solely at acid rain, because there are often other possible causes of damage, such as ground-level ozone. Nonetheless, there is much evidence to indicate that acid rain is the cause of extensive forest declines in Europe and on the higher slopes of mountains in the eastern U.S.A.[22]. In Canada, there has been deterioration in sugar maple groves exposed to acidic deposition and other pollutants. Damage to vegetation and soil depends strongly on the natural ability of soil and bedrock to buffer or neutralize the acid. For instance, the granite bedrock and thin soils of the Canadian Shield have very low buffering capability, and acid rain falling in this region can be especially damaging.

The situation is more clear-cut with respect to the effects on lakes, rivers, and aquatic life. In regions with soils and rock of poor buffering ability, there is no doubt that acid rain has decreased the pH of lakes and rivers. Healthy aquatic ecosystems have pH values above 6. As the pH approaches 6, crustaceans, insects, and some algal and zoo plankton decline or vanish[23]. As pH decreases from 6 to 5, fish populations disappear, with the most highly valued species (trout, salmon, etc.) dying at the higher pH values in this range. At a pH of 5, virtually all fish are dead. In Ontario alone, there are more than 7 000 lakes acidic enough that essentially all the fish have died, and another 12 000 in which other forms of plants and animals have been affected[24]. Many lakes in the northeastern U.S.A. and Scandinavia cannot support fish[25].

---

[22] Ref. 3, p. 24-11

[23] Ref. 3, pp. 24-5 and 24-13

[24] Toronto *Globe and Mail*, January 1990, reporting on a study by the Ontario Ministry of the Environment

[25] J.M. Fowler, *Energy and the Environment, Second Ed.*, McGraw-Hill, 1984, p. 165

Acid rain also attacks buildings, corroding iron and steel, and especially destroying limestone, marble, and concrete. When artistic monuments made of stone are damaged, items of cultural heritage can be irretrievably lost.

There is no evidence that the observed concentrations of acids in rain and snow have had any direct effects on humans. However, people can inhale acidic pollutants in the air and irritation to the respiratory system can result. Humans may also be affected by the leaching of toxic metals into water supplies from soil and bedrock as a result of acid rain.

As a result of controls on acidic emissions put into place in the last few years, the acid levels in precipitation in eastern Canada have decreased somewhat (by as much as 30% in some areas)[26], and will continue to decrease as further controls come into effect in Canada and the U.S.A. However, once all the restrictions are met, it is likely that acidification of some sensitive aquatic systems will remain; perhaps further controls will be mandated. At least some of the credit for the reduction in acid levels in Canada should go to the Canadian Coalition on Acid Rain, which quietly lobbied governments from 1981 until 1991. Having seen acid rain legislation approved in the U.S.A. in 1990, the group disbanded.

Hydrocarbons (HC)

Hydrocarbon pollution is released from any incomplete combustion of fuel and from evaporation of fuels and other organic liquids, and plays an important role in the production of photochemical smog. In Canada in 1985, about 35% of the HC emissions (Fig. 6-7) came from the transportation sector (with automobiles contributing over half of this), 22% from solvent use, 20% from industrial processes (especially in the petroleum, petrochemical, and coal and coke industries), about 10% from fuel combustion in stationary sources such as industrial plants and power stations, and about 13% from miscellaneous sources.

---

[26] Toronto *Globe and Mail*, February 19, 1991, relating information from a report of a federal-provincial acid rain research and monitoring committee

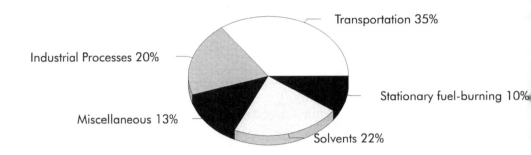

Figure 6-7: Sources of HC (Hydrocarbons) in Canada in 1985 (*The State of Canada's Environment*, Government of Canada, Ottawa, 1991, pp. 14-6 to 14-7)

Controlling HC

Since motor vehicles are an important source of HC emissions, most of the effort in controlling HC pollutants has focused on vehicles. Not surprisingly, California has led the way over the years in setting vehicle emission standards — the serious air pollution problems in the Los Angeles area are well known.

Twenty years ago, about half of the HC from a typical car came from the exhaust, and the other half from "blowby" (gas forced past the piston rings) and evaporation from the carburetor and gas tank. The "blowby" gas is now fed back into the engine, and the evaporation problem is controlled by better seals around the carburetor and tighter gas caps. The HC in the exhaust is now reduced by the catalytic converter, which oxidizes the HC to H_2O and CO_2. In the 1963-69 model years, the HC emission level in the exhaust from a new heavy-duty gasoline-powered vehicle was 10.6 g/km; in the 1987-90 model years, the value was 0.61 g/km.[27]

As a result of emission controls, HC pollutants from transportation sources in Canada[28] dropped about 20% from 1970 to 1985. However,

---

[27] J. Jefferies, Ontario Ministry of the Environment (private communication)

[28] Ref. 3, p. 14-9

emissions from industrial sources increased, giving a net rise of total HC emissions of about 15%. In the U.S.A. total emissions decreased by 12% from 1975 to 1989.[29]

Smog

Travellers arriving by air in Los Angeles often see a brownish blanket over the city — photochemical smog. When the summer Olympics were held in Los Angeles in 1984, organizers worried whether the air would be clean enough for athletes to perform well, and whether events might have to be postponed if the pollution were so bad that it would be unhealthy for athletes to exert themselves.

The modern photochemical smog is not at all the same as the classical smog associated with London, England. Classical smog — literally, smoke and fog — is formed by PM from burning coal, and moisture; another component is SO_2 generated from the coal-burning. Classical smog was common in many industrial regions of the world in the 19th and early 20th centuries, and is still a problem in eastern Europe, the former Soviet Union, and China, where economic development is fuelled by coal. For example, about 75% of Poland's energy comes from coal[30], most of which is burned without emission controls of any kind. Classical smog is usually worst in winter, when a lot of fuel is being burned. In December 1952, there was a "killer smog" in London, with extremely high concentrations of PM and SO_2, resulting in a death rate two to three times normal for several days. About 4 000 deaths were attributed to the smog, with most of those who died being elderly people with respiratory problems. The smog problem in London was finally solved by placing restrictions on coal-burning.

Photochemical smog requires light (as the term "photo" implies), and thus is a problem in summer months in northern or southern regions of the world, and year-round in tropical regions. Also required are NO_x, HC, and warm temperatures (above about 18°C). We will not go into the detailed chemistry here; in brief summary, sunlight drives various

---

[29] Ref. 6, p. 223

[30] W.U. Chandler et al, Sci. Am. **263** No. 3, p. 124 (Sept. 1990)

temperature-dependent chemical reactions involving NO_X and HC to produce secondary pollutants, the most important of which is *ozone* (O_3).

Ozone is a powerful oxidizing agent that damages lungs, and is very injurious to plants. It affects crop production, and is considered by many to be the most serious air pollutant with which we have to deal today. Note that what we are discussing here is *ground-level* ozone. The decline in ozone concentration in the upper atmosphere is a completely different problem, initiated by chlorofluorocarbons (CFCs) and resulting in decreased absorption of harmful ultraviolet radiation from the sun.

Peak one-hour average values of ozone concentration measured in Canada from 1979 to 1989 hovered around the maximum acceptable level[31]. During the very hot summer of 1988, record high ozone levels were recorded in eastern Canada. At one station in North York (a suburb of Toronto), the one-hour maximum acceptable level was exceeded 157 times. Since emissions of NO_X and HC have been increasing in Canada, we can expect the smog problem to become gradually worse in the short term. In 1991 the federal and provincial governments developed a plan to reduce ozone concentrations by the year 2005, by focusing on more effective pollution-control devices on cars and changes in the composition of gasoline, household paints, and dry-cleaning solvents[32].

Ozone is a serious problem in most urban centres in the world, but the city with the worst air quality is probably Mexico City, home to 20 million people. In 1991 the ozone air pollution index passed 200 points on 192 days out of the year[33]. According to the World Health Organization, no one should breathe air containing ozone at a pollution index greater than 100 for more than one hour a year. Along some sidewalks in Mexico City, booths are set up where people can breathe oxygen for a few minutes by inserting money into a machine.

Ozone concentrations could be reduced if people would use mass transit instead of private automobiles. Many North Americans drive to work alone in a car, releasing much more NO_X and HC than if buses, trains, subways, or carpools were used. Unfortunately, many people are

---

[31] Ref. 3, p. 2-13

[32] Toronto *Globe and Mail*, June 14, 1991

[33] Toronto *Globe and Mail*, May 30, 1992

accustomed to the convenience of the single-person vehicle, and are unlikely to switch to mass transit unless it is clearly a faster way to travel. Introduction of special traffic lanes for buses and vehicles with three or more persons, or road tariffs for single-person commuters, might help to solve the problem.

Carbon Monoxide (CO)

The final pollutant to be discussed in this chapter is carbon monoxide, formed whenever a fuel undergoes incomplete combustion. This gas is colourless, odourless, and highly toxic — it binds strongly to hemoglobin in the blood, preventing the hemoglobin from carrying oxygen. High concentrations of CO can be a serious pollution problem in urban centres.

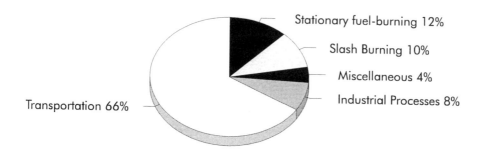

Figure 6-8: Sources of CO (Carbon Monoxide) in Canada in 1985 (*The State of Canada's Environment*, Government of Canada, Ottawa, 1991, pp. 14-6 to 14-7)

Transportation is the most important source of CO in Canada, producing 66% of the emissions in 1985 (Fig. 6-8), with almost all of this contribution coming from road vehicles. Fuel combustion at stationary sources provided 12%, split almost equally between fossil-fuel burning at industrial plants and use of fuelwood; electric power plants contributed only 0.5%. Notice in Fig. 6-8 that slash burning of forests provided 10% of the CO, compared with only 8% from industrial processes.

Globally, the burning of tropical rain forests and savannas produces at least as much CO as does the burning of fossil fuels[34]. This is a strong indication of how rapidly the rain forests are disappearing.

Reducing CO

As is the case with HC and NO_x pollution, control of CO has concentrated on road vehicles. Catalytic converters oxidize the CO to produce CO_2 (which is a serious global pollutant on its own — see Chapter 7.) Modern catalytic converters can eliminate most of the CO from the exhaust; the CO emission from a new heavy-duty gasoline vehicle was 130 g/km for the 1963-69 model years, but only 8.1 g/km for the 1987-90 model years[35].

Measurements of CO in Canada from 1974 to 1989 indicate a decline of 63% in peak eight-hour-average concentrations, to a value that is about 20% of the maximum acceptable level[36]. In 1989, only two air monitoring stations, one in Calgary and the other Regina, recorded peak eight-hour-average concentrations that were greater than half of the maximum acceptable.

6.5 INDOOR AIR POLLUTION

We tend to think of the environment as out-of-doors, but since most people spend a great deal of time inside buildings, air pollution indoors

---

[34] R.E. Newell et al, *Sci. Am.* 261, No.4, p. 82 (Oct. 1989)

[35] J. Jefferies, Ontario Ministry of the Environment (private communication)

[36] Ref. 3, p. 2-12

is important to consider. In recent years, buildings have been sealed more tightly to conserve energy, but this leads to buildup of indoor pollutants. Many modern materials used in building construction, furniture, and consumer products give off dangerous gases such as formaldehyde, especially when new. In addition, there are combustion products such as CO from smoking, NO_X from gas stoves, and PM from tobacco smoke and woodstoves. The radioactive gas radon from the ground and from masonry also presents problems. (The effects of radon are discussed in Chapter 14.) It is quite common for indoor air to contain larger concentrations of these pollutants than outdoor air.

The gas stove is one source of pollution that has been studied in the U.S.A. and Britain[37]. It has been found that levels of NO_X in homes with gas cooking stoves are three to seven times higher than in homes with electric ranges, and that children in these homes have more frequent respiratory ailments. Concentrations of CO and NO_X in a kitchen with a gas stove can even exceed recommended standards for outdoor air quality.

Indoor air pollution can be reduced by sufficient ventilation, but in very hot or cold weather, an expensive heat exchanger might be required to reduce air-conditioning or heating costs.

6.6 COST VERSUS BENEFIT

All of the air pollution problems discussed in this chapter have solutions, and all the solutions require either money, as in the case of scrubbers, or a change of lifestyle, such as using mass transit. How quickly society moves toward the solutions depends on whether people view the benefits of cleaner air to be worth the cost and inconvenience. In the last two decades, North American society has shown a willingness to put up with some expense and inconvenience to improve air quality, and governments have responded with tighter regulations. In some other areas of the world, populations are striving to improve their standard of living; the consequence is that energy consumption and air pollution in these regions are increasing rapidly. For example, in the Middle East and South

---

[37] J.M. Fowler, *Energy and the Environment, Second Ed.*, McGraw-Hill, 1984, p. 190

Asia, energy consumption increased by 6% annually from 1974 to 1989, compared with a world average of only 2.3%.

At a more basic level, the effect of any decrease in per capita production of pollution can always be cancelled by an increase in the number of people. World population continues to grow at a rate of 1.7% annually. Even with this small growth rate, exponential growth can lead to very large numbers in a relatively short time. Exercise 2-1 showed that if the world population were to continue increasing at 1.7% annually, in 1800 years the mass of people would equal the mass of the earth. This bizarre conclusion highlights the fact that world population cannot continue to grow indefinitely; the environmental stresses caused by population and other byproducts of human activities would simply become unmanageable.

6.7 LIFE-CYCLE ANALYSIS

It should now be clear that almost everything we do generates some form of pollution. This simple fact can lead to a general problem. In any given situation, which of two alternatives is the preferable one? A consumer might have to decide between using cloth or disposable diapers, or an electric utility might be faced with building either a nuclear reactor or a hydroelectric dam.

Questions such as these can be addressed by life-cycle analysis, in which the effects of a product or a technology are analyzed in great detail. In the diaper example, the environmental problems generated by the manufacture, use, and disposal of each type of diaper would be determined (as well as possible) and compared. In fact, the 'diaper dilemma' is representative of many environmental conundrums. It is apparent that there are waste problems associated with disposable diapers, but cloth diapers require washing, which generates water pollution and also uses energy (thus producing more pollution). It is not obvious which type of diaper is better for the environment. Other factors are also at work, such as personal convenience and the time available, which can be extremely important for working parents and single parents. With diapers as with other environmental questions, social concerns become intermingled with scientific and environmental concerns.

People often have a general perception that wood products are environmentally friendly because they use a renewable resource, and that plastics are inherently wasteful of oil, a finite resource, and generate landfill problems. A recent paper[38] presents a life-cycle analysis comparing disposable paper cups and disposable polystyrene foam cups. The analysis focuses on the environmental impact of the manufacture of the cups, and on the problems associated with disposal. The paper states that a paper cup requires 36 times as much electricity, 112 times as much steam, and twice as much cooling water as a foam cup. As well, manufacture of a paper cup generates about 580 times the volume of waste water, containing much higher levels of suspended solids and organochlorides. A paper cup releases methane and carbon dioxide (greenhouse gases) when deposited in a landfill site, whereas the polystyrene is essentially inert. The detailed analysis, which is easily readable, leads to a logical conclusion that paper cups are 'more negative' in environmental impact than polystyrene cups. However, life-cycle analyses are seldom simple. The journal in which the paper appeared was flooded with letters[39] claiming that the analysis was faulty on many grounds: it used outdated information on energy use and effluents in the pulp and paper industry, it did not consider the positive function of trees in removing CO_2 from the atmosphere, nor did it discuss the recycling potential of paper cups and the environmental effects of styrene as a pollutant. In sum, this apparently simple question evidently has no simple answer. It is not at all clear which type of cup (paper or polystyrene) is environmentally more innocent. Certainly the analysis itself proved more complicated that one would have suspected.

The plastic-versus-paper debate has also focused on shopping bags. Are plastic bags more "environmentally friendly" than paper bags ... or is it vice-versa? At least one company has decided that plastic is better — Figure 6-9 shows a reproduction of a plastic bag used by one "outdoors" company.

---

[38] M.B. Hocking, *Science* **251**, Feb. 1, 1991, pp. 504-505

[39] H.A. Wells, Jr.; N. McCubbin; R. Cavaney; B. Camo; *Science* **252**, June 7, 1991, pp. 1361-1362

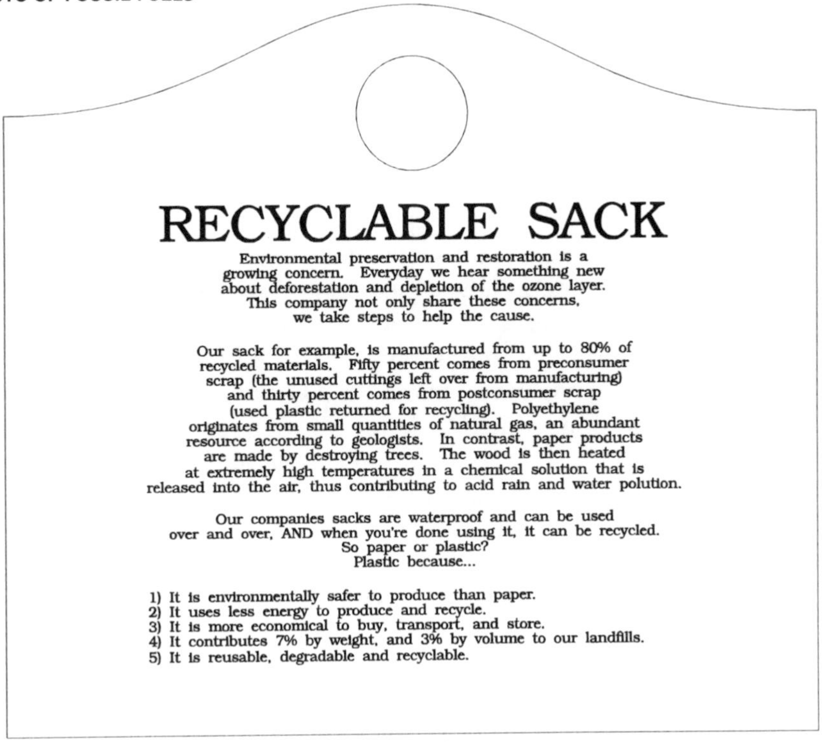

RECYCLABLE SACK

Environmental preservation and restoration is a growing concern. Everyday we hear something new about deforestation and depletion of the ozone layer. This company not only share these concerns, we take steps to help the cause.

Our sack for example, is manufactured from up to 80% of recycled materials. Fifty percent comes from preconsumer scrap (the unused cuttings left over from manufacturing) and thirty percent comes from postconsumer scrap (used plastic returned for recycling). Polyethylene originates from small quantities of natural gas, an abundant resource according to geologists. In contrast, paper products are made by destroying trees. The wood is then heated at extremely high temperatures in a chemical solution that is released into the air, thus contributing to acid rain and water polution.

Our companies sacks are waterproof and can be used over and over, AND when you're done using it, it can be recycled. So paper or plastic? Plastic because...

1) It is environmentally safer to produce than paper.
2) It uses less energy to produce and recycle.
3) It is more economical to buy, transport, and store.
4) It contributes 7% by weight, and 3% by volume to our landfills.
5) It is reusable, degradable and recyclable.

Figure 6-9: A plastic bag used by an "outdoors" company.

EXERCISES

6-1 For a fossil-fuel electrical plant that is operating with an efficiency of 40% (two significant digits), what is the ratio of waste heat to electrical energy produced? [Ans. 1.5]

6-2 For a nuclear electrical plant that is operating with an efficiency of 30% (two significant digits), what is the ratio of waste heat to electrical energy produced? [Ans. 2.3]

6-3 Why is photochemical smog not a problem in Toronto in January?

6-4 Being as specific as possible, what is the major source (> 50%) of
(a) SO_2 in Canada? (b) SO_2 in the world?
(c) NO_X in Canada? (d) CO in Canada?

6-5 For which of the following types of air pollution does transportation provide at least 25% of the emissions in Canada?
(a) PM (b) SO_2 (c) NO_X (d) HC (e) CO

6-6 In a molecule of sulphur dioxide, what is the ratio of the mass of sulphur to the mass of oxygen? Refer to inside bookcovers for molar masses. [Ans. 1:1]

6-7 What mass of sulphur is contained in 15 000 tonnes of $CaSO_3$? Refer to inside bookcovers for molar masses. [Ans. 4000 tonnes]

PROBLEMS

6-1 This problem compares the cooling water requirements of a nuclear plant producing electricity at 30% efficiency and a fossil-fuel electrical plant operating at 40% efficiency. Assume that the electricity production at the two plants are equal. Neglecting the heat that goes up the chimney in the fossil-fuel plant, and assuming two significant digits in given data, determine the following ratio:

$$\frac{\text{cooling water required for nuclear plant}}{\text{cooling water required for fossil-fuel plant}}$$

[Ans. 1.6]

6-2 (a) In the year 1987 the U.S.A. produced 1961 TW· h of electrical energy in conventional (i.e., fossil-fuel) thermal plants and 455 TW·h in nuclear plants. Assuming 30% efficiency for nuclear plants and 40% for conventional thermal plants, determine the (annual) volume of cooling water required to cool these plants in once-through cooling if the cooling water undergoes a temperature increase of $10\,^\circ$C. (Neglect the heat lost up the chimney in conventional plants, and assume two significant digits in given data.) [Ans. 3.4×10^{11} m^3]

(b) The total amount of precipitation in the U.S.A. is estimated to be about 5.6×10^{12} m^3 annually, of which approximately 30% is available as runoff. Confirm the statement made in the text that about 1/16 of the annual runoff would be required for once-through cooling of electrical plants.

6-3 For a typical 1000-MWe fossil-fuel electrical plant that has an efficiency of 40% and is cooled by a wet cooling tower, calculate the amount of water evaporated daily, assuming that the water is at $100\,^\circ$C and

157

that all the waste heat removed from the plant is removed by evaporation. Assume two significant digits in given data.   [Ans. 5.7×10^7 kg]

6-4   In many devices involved in energy industries, a fluid (liquid or gas) flows through a pipe. If the cross-sectional area of a pipe is 2.0 m², and the fluid in it has a speed of 1.5 m/s,
(a)   what is the volume flow rate in m³/s?   (i.e., what volume of fluid passes through any cross-section of the pipe every second?) [Ans. 3.0 m³/s]
(b)   what mass of fluid passes through any cross-section of the pipe every second, if the fluid has a density of 1.0×10^3 kg/m³? [Ans. 3.0×10^3 kg/s]

6-5   An old coal-fired steam-turbine electrical power plant that produces 800 MW (two significant digits) of electrical power is located beside a lake. The overall efficiency of the plant is 34%.
(a)   What is the total thermal power input to the plant? [Ans. 2.4×10^3 MW]
(b)   At what rate is waste heat discharged from the plant (in MW)? [Ans. 1.6×10^3 MW]
(c)   If the lake water used in the condenser is to be raised in temperature by no more than $6.0\,^\circ$C, what minimum volume of water must be made available per second?   [Ans. 62 m³]
(d)   If the lake water enters the condenser through a pipe of diameter 8.0 m, what minimum speed must the water have in the pipe? [Ans. 1.2 m/s)]

6-6   In 1988, 720 million tonnes of coal were burned in the U.S.A. If each tonne of coal had an average sulphur impurity of 1.5% (by mass), how many tonnes of sulphur dioxide could have been put into the air? [Ans. 22 million tonnes]

6-7   A particular coal-fired electric power station uses 7.5×10^9 kg of coal per year to produce 950 MW of electrical power. The coal contains 2.4% sulphur by mass. If a scrubber is used to remove 85% of the sulphur from the exhaust gases in the

Information Tip for Problems 6-6 to 6-8:
Molar masses are provided inside bookcovers.

form of $CaSO_3$, what mass of $CaSO_3$ is produced annually?
[Ans. 5.7×10^8 kg]

6-8 A coal-burning electric power plant produces 850 MWe at an efficiency of 37%. It burns coal of energy content 28 MJ per kg and sulphur content 1.7% by mass. What mass (in tonnes) of SO_2 is produced by this plant in a year? (1 tonne = 1000 kg) [Ans. 8.8×10^4 tonnes]

6-9 The maximum permissible emission rate of carbon monoxide from new automobiles in the 1994-96 model years is 7.88 g/km.
(a) What mass of carbon monoxide is emitted by a new 1994 automobile in its first year of operation, if it travels 11 500 km and its CO-emission rate remains constant at the maximum permitted?
(b)   The CO-emissions of an automobile do not actually remain constant, but increase as a car gets older and has been driven a greater distance. For every 10 000 km that a car travels, the maximum permitted increase in the CO-emission rate is 0.34 g/km. Re-calculate the answer to part (a) of this problem, assuming that the CO-emission rate of the new car begins at 7.88 g/km and then increases linearly at the maximum possible rate.
[Ans. (a) 90.6 kg   (b) 92.9 kg]

ELECTROMAGNETIC RADIATION AND THE GREENHOUSE EFFECT

7.1 The Electromagnetic Spectrum

In Chapter 4 the two heat transfer processes of "conduction" and "convection" were discussed. In this section the third process "radiation" will be introduced. Whereas conduction and convection require the intermediary of matter to transport energy, radiation does not. The energy of the Sun comes to us by the process of radiation through essentially empty space.

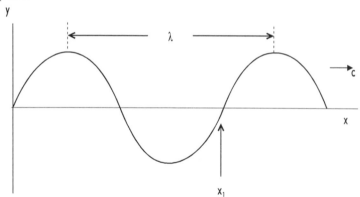

Figure 7-1: An E-M Wave moving in the x direction with speed c.

Radiant energy consists of electromagnetic (E-M) waves of which visible light is only one example. Being waves they are characterized by a wavelength λ and a frequency f. The wavelength is the crest-to-crest distance of the waves (see Fig. 7-1.), and the frequency is the number of full wavelengths that pass any point (such as x_1 in Fig. 7-1) each second while the wave moves at a speed "c". The relationship between these three quantities is

$$c = f \lambda \qquad\qquad (7\text{-}1)$$

The speed of E-M waves is that of light, which in a vacuum is,

$$c = 2.998 \times 10^8 \text{ m} \cdot \text{s}^{-1}$$

If the wavelength is measured in metres then the frequency must have units s^{-1}. This unit is commonly called the "*hertz*", and means "*cycles per second*" and is abbreviated "Hz". Except for radio waves it is very inconvenient to use metres to measure wavelengths. Since the range of wavelengths is so very large, several different subsidiary units have been defined to keep the numbers simple. The total range of wavelengths define the *electromagnetic spectrum* and is described (along with the appropriate conversion factors) in Table 7-1.

Table 7-1: The Electromagnetic Spectrum

Wavelength Range (approx.)		
Type of wave	λ(m)	Conventional units
Radio waves	>0.1	>0.1m
Microwaves	10^{-1}-10^{-4}	100mm - 0.1mm
Infrared	10^{-4}-7×10^{-7}	100 m - 0.7 m
Visible light	7×10^{-7}-4×10^{-7}	700 nm - 400 nm
Ultraviolet	4×10^{-7}-10^{-8}	400 nm - 10 nm
X-Rays	10^{-8}-10^{-11}	10 nm - 0.01 nm
Gamma rays	$<10^{-11}$	<0.01 nm

Note: $1\text{mm} = 10^{-3}\text{m}$, $1\mu m = 10^{-6}\text{m}$, $1\text{nm} = 10^{-9}\text{m}$

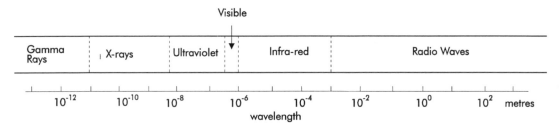

Figure 7-2: The Electromagnetic Spectrum

This very wide range of wave- lengths is shown diagrammatically in Fig. 7-2 where it is evident how narrow is the range of visible wavelengths relative to the whole E-M spectrum.

7.2 Inverse Square Law

It is important to know how the intensity of radiation varies with the distance away from the source of energy. Obviously it decreases as the distance increases; if you stand further from a fireplace you feel a smaller intensity of radiated heat. The quantity *intensity* is defined as the amount of radiant energy passing per unit time through a unit area, or the power per unit area (since power = energy/time).

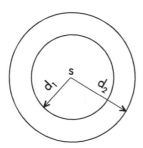

Figure 7-3: Inverse Square Law

In Fig. 7-3 the point s is a source of radiation emitting P joules/second (watts). All of this power passes through the sphere of radius r_1 so that the intensity at a distance r_1 is

$$I_1 = P/4\pi r_1^2$$

The power which goes through this sphere must also pass through the sphere of radius r_2. The intensity at this new distance is

$$I_2 = P/4\pi r_2^2$$

Since the power is the same then

$$I_1/I_2 = (r_2/r_1)^2 \tag{7-2}$$

This equation is the *inverse square law* which says that the intensity of radiation from a <u>point</u> source varies inversely as the square of the distance from the source. It holds for all types of radiation e.g. sound, radioactivity etc. so long as the source can be considered point-like compared to all other dimensions.

Example 7-1 *The "solar constant" is the intensity of energy, supplied by the Sun, arriving at the top of the Earth's atmosphere; it is 1.40×10^3 W/m^2. The distance from the Sun to the Earth is 149.6×10^6 km and to Venus it is 108.2×10^6 km. What is the value of the solar constant on Venus?*

$I_1/I_2 = (r_2/r_1)^2$

$I_v/1.4 \times 10^3 \ W \cdot m^{-2} = (149.6 \ Mkm/108.2 \ Mkm)^2$

$I_v = 2.68 \times 10^3 \ W \cdot m^{-2}$

7.3 Blackbody Radiation

The next thing to consider is how the energy in the E-M radiation is distributed over this range of wavelengths. If the radiation from some hot object is passed through a glass prism, for example, it will be spread out according to wavelength, the shortest wavelengths being deviated the most by the prism and the longest wavelengths the least. The result is a distribution of energy vs. wavelength called the "*blackbody spectrum*"; this is shown in Fig. 7-4 for objects at temperatures of 1500 K, 2000 K, and 2500 K.

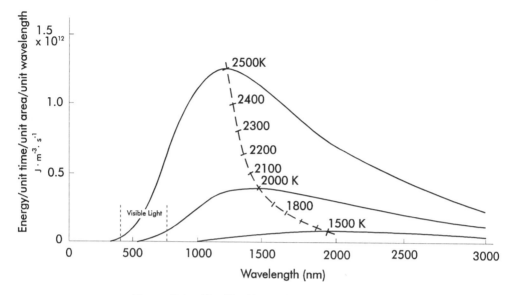

Figure 7-4: The Blackbody Spectrum.

There are several important features of this figure:

1) The area under any curve is the total energy/unit area/unit time, that is the intensity, I(T), radiated by the object at a temperature T.

2) Curves representing different temperatures never cross. This means that as the temperature of an object increases its energy output at every wavelength also increases. The way in which the total intensity increases with temperature is given by *Stefan's Law* which is,

$$I(T) \propto T^4 \qquad\qquad (7\text{-}3)$$

The fourth power of the temperature means that the total energy radiated by objects rises very rapidly as the temperature is increased.

Eqn. 7-3 can be expressed as an equality if a constant of proportionality is introduced. In the resulting equation,

$$I(T) = \sigma T^4 \qquad\qquad (7\text{-}4)$$

the constant σ is Stefan's constant and has the value

$$5.67 \times 10^{-8} \text{ W} \cdot \text{m}^{-2} \cdot \text{K}^{-4}.$$

An object which radiates according to Eqn. 7-4 is called a "perfect blackbody". Most surfaces, however, radiate with less efficiency than the ideal *black* case. This is expressed in the equation by multiplying by another dimensionless parameter ϵ called the "average emissivity"; the meaning of 'average' in this term is discussed later. Thus,

$$I(T) = \epsilon \sigma T^4 \qquad\qquad (7\text{-}5)$$

The value of ϵ lies between 0 and 1. If it is 0 then the object doesn't radiate at all; if $\epsilon = 1$ then the surface is said to be "perfectly black", hence the term "blackbody radiation". Real surfaces have values that vary with wavelength. For example, the metal tungsten, which is used as the filament in incandescent electric light bulbs, has an emissivity of 0.46 at a wavelength of 500 nm (yellow light), but only 0.38 at 1000 nm (infrared) and it is not a perfect blackbody ($\epsilon = 1$) at any wavelength. The average emissivity in Eqn. 7-5 is therefore the average over all wavelengths.

■

Example 7-2 *What is the total intensity I(T) for a perfect blackbody at 2500 K?*

From Eqn. 7-5 with ϵ *= 1 (perfect blackbody),*
$$I(T) = 5.67 \times 10^{-8} \text{ W} \cdot m^{-2} \cdot K^{-4} \times (2500) \text{ K}^4$$
$$= 2.21 \times 10^6 \text{ W/}m^2$$

This is the area under the upper curve in Fig. 7-4.
Can you think of a simple way to verify the order of magnitude of the value from the curve itself? [Hint: think of triangles.] ■

---

3) The wavelength of the peak of each curve moves from long wavelengths at lower temperatures to shorter wavelengths as the

temperature is raised. If λ_m is the wavelength of the peak then it is related to the temperature of the radiating object by *Wien's Law*:

$$\lambda_m(\text{nm}) = \frac{2.8972 \times 10^6 \text{ nm} \cdot \text{K}}{T(K)} \qquad (7\text{-}6)$$

The positions of the peak of the radiation curves are also shown on Fig. 7-4 as a dotted line. At a temperature of about 6000K, as we will see, the peak will appear near 500 nm; this is the temperature of the surface of the Sun and explains why sunlight appears white. It contains all the visible colours approximately equally.

Example 7-3 *What is the wavelength of the peak of the blackbody radiation curve for an object at 2500 K?*

From Wien's law (Eqn. 7-6)

$\lambda_m = 2.8972 \times 10^6 \text{ nm} \cdot K/2500 \text{ K} = 1159 \text{ nm}$

This is the wavelength of the peak of the upper curve in Fig. 7-4.

---

4) Figure 7-4 shows that most of the emission from objects at about 2000 K is in the infrared (heat) region. What visible light they emit is predominantly in the red; i.e. red-hot.

A proper theoretical understanding of the form of this radiation curve had to await the development of the Quantum Theory by Max Planck. The equation for the curve is *Planck's Radiation Equation*:

$$I(\lambda, T) = \frac{2\pi hc^2}{\lambda^5} \frac{1}{e^{hc/\lambda kT} - 1} \qquad (7\text{-}7)$$

In this equation h is *Planck's Constant* and k is *Boltzmann's Constant*. When the well-known values for all constants are substituted, Eqn. 7-7 can be written

$$I(\lambda,T) = \frac{3.746 \times 10^{-16}}{\lambda^5} \frac{1}{e^{1.46 \times 10^{-2}/\lambda T} - 1} \qquad (7\text{-}8)$$

For an object that is not perfectly black then Eqn. 7-7 and 7-8 must be multiplied by the emissivity which is a function of the wavelength. Eqn. 7-8 becomes,

$$I(\lambda,T) = \epsilon(\lambda)\frac{3.746 \times 10^{-16}}{\lambda^5} \frac{1}{e^{1.46 \times 10^{-2}/\lambda T} - 1} \qquad (7\text{-}9)$$

Example 7-4 *What is the intensity per unit wavelength $I(\lambda,T)$ for a perfect blackbody at a temperature of 2500 K at the peak of its radiation curve? The wavelength (from Example 7-3) is 1159 nm.*

From Planck's Radiation Equation (Eqn. 7-9) with $\epsilon(\lambda) = 1$ (perfect blackbody),

$$I(\lambda, T) = \frac{3.746 \times 10^{-16}\ W \cdot m^2}{(1159 \times 10^{-9})^5\ m^5} \frac{1}{e^{1.46 \times 10^{-2}/(1159 \times 10^{-9} \times 2500)} - 1}$$

$$= 1.17 \times 10^{12}\ W/m^3$$

This is the peak value of the upper curve in Fig. 7-4.

7.4 Solar Radiation

When a radiation detector is taken above the Earth's atmosphere by means of a rocket or satellite and the spectrum of the incoming radiation from the Sun is recorded, the upper curve of Fig. 7-5 is obtained. Except for a few features in the wavelength range 200 to 400 nm this curve is that of a blackbody at a temperature of 6000 K. This is how we know that the visible surface of the Sun (the photosphere) has a temperature of 6000 K.

If the radiation detector is operated at the Earth's surface, under the atmosphere, the lower curve of Fig. 7-5 is obtained. The curve retains the shape of a blackbody at 6000 K but is of lower intensity. This loss of solar energy is a result of scattering of sunlight back into space by clouds, dust in the atmosphere, and by the atmospheric molecules themselves. The area of the lower curve is about 65% of the upper curve, so 35% of the incoming radiation is scattered away. Astronomers call the fraction lost in this way the planetary "*albedo*" (α); for the Earth $\alpha = 0.35$.

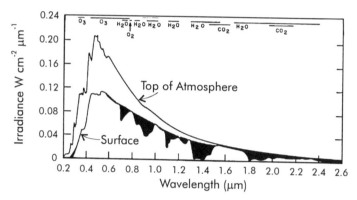

Figure 7-5: Solar Radiation at the Earth.

The Earth, as seen from space, is a very bright planet, that is, its albedo is large. This is caused by the large area of water on the surface and the highly reflective clouds in the atmosphere. For us on the surface this means that there is a smaller fraction (65%) of solar radiation which is actually absorbed. Some other planets are much darker when they are viewed from space, that is their albedos are smaller. For example, the albedo of Mars is 0.16 and the rings of Uranus are as dark as coal dust with an albedo of 0.03.

The lower curve in Fig. 7-5 also has a number of regions in the infrared where the black body curve seems to be eaten away (crosshatched in Fig. 7-5). These are regions where the atmospheric molecules, particularly water (H_2O) and carbon dioxide (CO_2) strongly absorb radiation from the solar spectrum. This absorption has a profound effect on the temperature and climate of the Earth, and is responsible for what is known as the "*Greenhouse Effect*".

7.5 Energy Balance in the Earth's Atmosphere

The Earth and its atmosphere receive energy in the form of solar radiation mostly at short wavelengths (because the Sun has a high temperature) and, as mentioned above, some is reflected or scattered out immediately ($\alpha = 0.35$) and the rest is absorbed. The Earth and the troposphere (lower atmosphere) also radiate energy but do so at long wavelengths since their temperature is much lower. The interchanges of energy in this system are complex and are shown schematically in Fig. 7-6 for an assumed input of 100 units of solar energy.

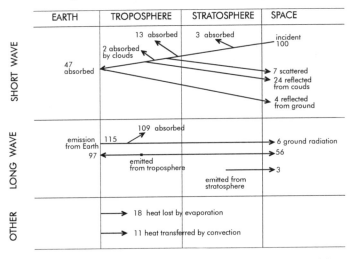

Figure 7-6: Energy Balance in the Earth's Atmosphere. Adapted by R.H. Stinson from *Atmospheric transmission* by Thomas G. Kyle, Pergamon Press, 1991.

Of these 100 units 35 units (7 + 24+ 4) are returned directly to space, accounting for the Earth's albedo ($\alpha = 0.35$). The remaining 65 units are absorbed: 3 by the stratosphere (upper atmosphere), 15 by the troposphere and 47 by the surface of the Earth. The stratosphere re-radiates its 3 units back to space and so need not be considered further.

In addition to the direct solar input, the warm troposphere radiates in all directions and on our energy scale contributes over twice as much energy to the Earth (97 units). The Earth, therefore, receives 144 units (47 + 97). At equilibrium this energy must also be lost by the Earth which occurs in three ways: 29 units (18 + 11) are recycled back to the troposphere by water evaporation and convection, and 115 units are re-radiated for a total of 144 units as expected.

Energy balance must also prevail in the troposphere. It receives 15 units (13 + 2) from the Sun, 109 of the 115 of the Earth's radiation, and 29 from evaporation and convection for a total of 153 units. This is lost as 97 units radiated to the Earth and 56 units radiated to space.

This complicated system exchanging energy back and forth among the Earth, troposphere, stratosphere and space is in thermal equilibrium and each component has its characteristic temperature.

Several important environmental questions follow from this: What would be the effect of disturbing any one of these loops? Of particular concern these days is what would happen if the 109 units of the 115 radiated from the Earth were to increase because of an increased absorbing power of the atmosphere? How would the system adjust? Would any of the other loops be affected?

7.6 The Greenhouse Effect

Imagine that sunlight is falling on a black surface and that all of the energy is absorbed. The temperature of the surface will rise to some value, say T_1, where the rate at which energy is radiated away is equal to the rate at which the Sun provides it; equilibrium is established. The radiation curve of the surface at T_1 is the lower curve in Fig. 7-7. If T_1 is around 300 K then the peak of the radiation curve is far in the infrared region and indeed all of the relevant wavelengths of Fig. 7-7 are in the infrared.

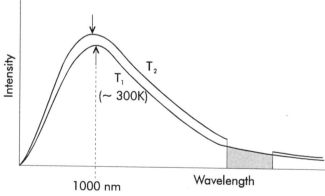

Figure 7-7: The Greenhouse Effect.

171

If a sheet of glass is interposed between the sunlight and the surface then, because the glass absorbs in certain regions of the infrared, the equilibrium is altered. In Fig. 7-7 this absorption is represented simply by the shaded region. The surface can no longer radiate away sufficient energy to maintain equilibrium and so its temperature will begin to rise. The temperature rises to a new value T_2 until the excess energy represented by the region between the two curves is equal to the energy represented by the shaded region. At this stage equilibrium is re-established but at a higher temperature. This is the explanation of the greenhouse effect.

7.7 The Earth's Greenhouse Effect

In the Earth's atmosphere the gases, primarily H_2O and CO_2, provide both absorption and scattering of the incoming solar energy. If the intensity of the solar radiation at the top of the Earth's atmosphere is I_0 (W/m^2), then the total power incident on the atmosphere is that which flows through a circle of radius R, as shown in Fig. 7-8. This power can be written as the product of area and intensity $\pi R^2 I_0$, where R is the radius of the Earth. A fraction α of this is reflected and scattered away and so the power delivered through the atmosphere is $\pi R^2(1-\alpha)$. The power radiated away by the planet at an effective temperature T_e is given by Stefan's Law (Eqn. 7-4) and is radiated from the entire surface of the Earth; it is therefore, $4\pi R^2 \sigma T^4$. If P_{net} is the net power retained by the Earth's surface then

$$P_{net} = \pi R^2(1-\alpha) - 4\pi R^2 \sigma T^4 \qquad (7\text{-}10)$$

and the net power per unit area is

$$I_{net} = (1/4)I_0(1-\alpha) - \sigma T^4 \qquad (7\text{-}11)$$

In a state of equilibrium $I_{net} = 0$. The value of I_0 at the mean distance of the Earth from the Sun is 140×10^3 W/m^2 and with $\alpha = 0.35$ this gives an effective temperature of the Earth's surface of 252 K. The actual mean temperature of the Earth's surface is, however, 15 °C or 288 K. Thus the greenhouse effect accounts for a warming of the surface by 36 K.

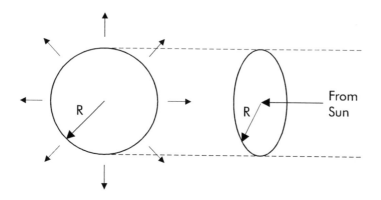

Figure 7-8:   The Energy Balance of the Earth.

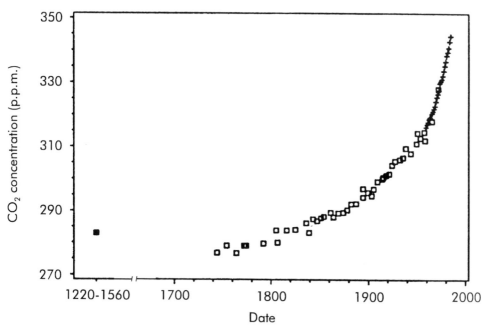

Figure 7-9:   Atmospheric Carbon Dioxide since 1200 AD (H. Friedli et. al. *Nature* <u>324</u> p.237 (1986)).

Actually, as we can see from Fig. 7-6, most of the radiation away from the Earth takes place from the troposphere and not the Earth's surface. The troposphere has a mean temperature of 2°C or 275 K, so the greenhouse warming is nearer 23 K (275-252). The higher estimate is often quoted in popular articles and overestimates the importance of the greenhouse effect.

Any planet with an atmosphere will experience a warming of its surface because of the greenhouse effect. Mars with its thin atmosphere has a small greenhouse effect. By contrast, on Venus, with a very dense atmosphere of CO_2, the greenhouse effect dominates producing temperatures of 700 K on the surface.

Until the 19th century the activities of people had very little effect on the equilibrium of the gases in the atmosphere. The Industrial Revolution, and its rapid development of industry, brought increasing emission of gases into the atmosphere. The evidence that the CO_2 concentration in the atmosphere has increased is undeniable; the effects of its increase are more controversial. Figure 7-9 shows the concentration of CO_2 in the atmosphere from at least the sixteenth century to the present. The most recent figures are from measurements on gases in the contemporary atmosphere, while data representing earlier times come from measurements on air-bubbles trapped in polar ice-caps or in glacial ice.

The concentration of CO_2 remained unchanged at about 280 parts-per-million (ppm) until about 1800 and then started to increase. The increase, while gentle at first, has become very steep in the latter part of the twentieth century and has increased by over 20% from pre-industrial days.

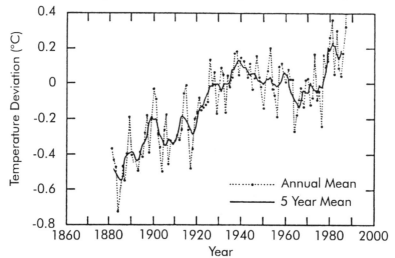

Figure 7-10: Global Temperature Since 1880.

Figure 7-10 shows the mean annual temperature of the Earth since 1880; the solid curve is the five-year average. That the temperature has risen by about one degree in this period is evident; to attribute that rise to the increase in greenhouse gases is much more problematic. It is well known that there are temperature variations of the Earth on many time scales. Ice ages occur on the scale of tens of thousands of years and smaller fluctuations occur on shorter scales. For example, a period of about 70 years ending in 1715 became known as the "Little Ice Age" since it was a period of unaccustomed cold in northern Europe. It is also known that this same 70 years was a period in which there were very few sunspots; in other words, there was little solar *activity*. The appearance of spots on the surface of the Sun generally follows a cycle of approximately 11 years from a maximum number to a minimum and back to a maximum again. The cycle can, however, vary between 10 and 12 years. There is now evidence that when the cycle is short the Earth's mean temperature is higher, and when the cycle is longer the temperature is lower. Since 1880 the cycle has been getting shorter; this is precisely the period in which the temperature has been rising. There are not enough years in this period to assign a cause to the temperature change observed. (Indeed, there may be several contributing causes.) Nevertheless it is quite clear that the temperature changes since 1880 usually correlate closely to the increasing concentration of greenhouse gases, and to sunspot activity.

There are scientists who are reasonably able to argue that the data of Fig. 7-10 are simply evidence of another variation in the Earth's temperature that has nothing to do with the greenhouse effect. Others, just as reasonably, argue that it is evidence of a major disturbance of the environment caused by our activities and refer in particular to the rapid rise in temperature since 1970.

It is notoriously difficult to forecast seasonal or annual temperatures. The problem is that the Earth's atmosphere is so very complicated that detailed calculations and predictions tax the resources of the largest computers. Even if the required calculations could be carried out, some of the interrelated processes in the atmosphere remain unknown or incompletely understood. There are many effects that might serve to decrease the greenhouse effect; these are *negative feedback* effects. For example, it is likely that the oceans of a warmer Earth would evaporate more water and thus there would be more cloud cover. An increased cloud cover would increase the Earth's albedo, α, in Eqn. 7-11; an increase in α would result in a decreased temperature T_e (see problem 7-8). In other words, the greenhouse effect may, to some extent, be self-correcting on a watery planet like the Earth. It is on grounds such as these that some scientists argue against predictions of catastrophic temperature changes in the short term. It is nevertheless true that the majority of atmospheric scientists anticipate temperature rises of between 1 and 5 K in the next 50 years.

7.8 Other Greenhouse Gases

Although the discussion so far has been in terms of CO_2 it is not the only contributor to the greenhouse effect. Other gases of human origin make about an equal contribution. The other important gases are methane (CH_4), chlorofluorocarbons (CFC), ozone (O_3), and nitrous oxide (N_2O). Table 7-2 summarizes their present concentration and rate of increase. (The table also shows the "relative contribution" of each gas to the greenhouse effect — this will be explained later.)

The increase in these gases is almost entirely a result of the activities of humans in the modern industrial society. The increase of CO_2 is not only due to the burning of fossil fuels but is enhanced by massive deforestation, as plants are a major factor in removing CO_2 from the atmosphere.

Methane has many sources: Major changes in agricultural practice, particularly increased growing of rice and increased herds of cattle have resulted in more atmospheric methane. Natural gas is largely methane and its production from drilled gas wells inevitably leads to leakage or incomplete combustion. Finally, methane is copiously produced in landfills.

CFCs found many industrial uses since their discovery, particularly in refrigeration, as a plastic foam expander, and as a driver gas in aerosol sprays. All of these uses result in atmospheric pollution. In addition to the greenhouse effect caused by CFCs in the troposphere, the erosion of the protective layer of ozone by CFCs that diffuse to the stratosphere is a separate and very serious issue. Ozone is produced by sunlight in the stratosphere and is essential to our well-being as it shields us from the harmful effects of the ultraviolet rays of the Sun. It is, however, a pollutant in the lower atmosphere. It is produced by a complex of chemical reactions involving the action of sunlight on the chemical smog that pollutes large urban areas.

Table 7-2: The Greenhouse Gases

Species	Concentration (ppm)	Rate of Increase (% per year)	Greenhouse Contribution (%)
CO_2	350	0.5	60
CH_4	2	1	15
CFC	0.0008	4	12
O_3*	0.01-0.05	0.5	8
N_2O	0.3	0.2	5

* in the troposphere (lower atmosphere)
From: H. Rodhe, *Science* 248 p 1217 (1990)

These gases are not equally effective in producing a greenhouse warming. In addition to the large effect of water, their relative contribution is given in the last column of Table 7-2.

If production of these gases were to stop, the effects would not disappear quickly. The methods by which the gases are removed are various and not at all well understood; this is particularly the case for CO_2. The oceans, plants, and chemical reactions and other unidentified agents remove them at various rates which have been estimated and are given in Table 7-3.

Table 7-3: Decay Time of Greenhouse Gases

Time to decay to 37% (1/e) of initial value	
Species	Time (yr.)
CO_2	120
CH_4	10
CFC	65-120
O_3*	0.1
N_2O	150

* in the troposphere
From: H. Rodhe, *Science* 248 p 1217 (1990)

7.9 Effects of Greenhouse Warming

As explained above, the calculations of future greenhouse warming remain estimates, owing to our imprecise understanding of the processes at work. For the same reason, predictions of climate change are just as imprecise, perhaps more so. However, it seems certain that the global average temperature will rise by 1 to 5 K over the next half-century, and some consequences necessarily follow from this.

If it does nothing else, global warming will cause the ocean levels to rise. This need not be the result of increased precipitation, or melting of the polar ice-caps; the oceans will rise simply from the thermal expansion of water (see problem 7-11). Small though this effect might appear, it may account for a global average increase of 0.2 m in sea level. (Local factors will also play a part in any given locality: changes in ocean circulation, distribution of salinity, etc.) One model calculation gives a rise in the North Atlantic of 0.4 m, but no change (or even a small decrease) in the Ross Sea of Antarctica. Yet a rise of 0.4 m in the North Atlantic would be catastrophic in low-lying areas in, for example, the Eastern Seaboard of the U.S.A., on many Caribbean islands, and perhaps in northwestern Europe, since the North Sea connects to the Atlantic. Elsewhere on our planet, a small rise in sea-levels could be even more catastrophic. Consider for example Bangladesh, which even at present sea-levels can have as much as half its surface covered by water during flood season.

A further consequence of increased temperature will be increased ocean evaporation and the consequent increase of rainfall on the continental land masses. An increased snowfall might even make the polar ice-caps increase in size. An increased rainfall would make the flooding problem in Bangladesh even more severe, and models suggest it may be of even greater magnitude than the ocean rise itself. Paradoxically, increased precipitation and temperature does not necessarily mean increased soil moisture. Shorter winters will trap less moisture and longer summers will increase evaporation so the soil moisture might actually decrease. The large deserts around the Tropics of Cancer and Capricorn, such as the Sahara and the U.S. southwest will increase, and semidesert grasslands now used for grain production like the Canadian prairies will lose productivity.

Some agricultural scientists counter that the increase of CO_2 in the atmosphere will increase the rate of photosynthesis in plants, thus increasing productivity. In addition, they argue, increased plant coverage of the fertile regions of the Earth will increase soil-moisture retention. Thus they believe that, although the producing regions of the Earth may be redistributed, the effects may be globally self-correcting or even positive! Arguments of this type have almost evenly divided agriculturalists and climatologists on the "optimistic" and "pessimistic" sides of the greenhouse debate.

7.10 Solutions to the Greenhouse Effect

From the lifetime data given in Table 7-3 it is obvious that nothing can be done to reverse the trend to greenhouse warming in a short time. Even if production of all greenhouse gases could be stopped immediately, the effects would be felt for a century at least, and longer if irreparable damage to the environment occurs. In addition, although the production takes place under the control of the various nations and their respective political systems, the atmospheric circulation knows no political boundaries. The problem is a global problem in a world that has not been used to solving global problems either peacefully or with dispatch.

Clearly a rapid phase-out of CFCs is necessary for more reasons than global warming. This is one area where some progress has been achieved. The so-called "Montréal Protocol" of 1990 calls for a reduction of CFCs by 50% by the year 2000. This may well be speeded up, however, because of the discovery of an "ozone hole" (i.e., an area of depleted ozone) in the stratosphere over North America and Northern Europe in 1992.

The other greenhouse gases are almost all the result of our society's demand for energy. The utilization of energy cannot be stopped, but methods must be found to minimize gas emission. First, and potentially easiest to address, is the efficiency with which energy is used. Switching from incandescent to fluorescent lighting, insulating houses, lowering winter home temperatures, raising summer air-conditioning temperatures, and building more fuel-efficient cars, are measures which, while variable in the size of their effect, will add together to reduce the rate of increase of greenhouse gases.

To make an absolute reduction of the emission of these gases is more difficult. To do this we must switch from our dependence on fossil fuels to a more diverse mixture of renewable and non-renewable resources. These are discussed in more detail in other chapters of this book.

Finally, the ultimate problem is population; people require energy and energy produces greenhouse gases. While population control is a problem beyond the scope of this book, its solution must involve all of humankind with all its knowledge including the scientific.

Exercises

7-1 While standing on a dock you count 20 wave crests (starting at zero) go by in 80 s. The distance between crests is 4 m. What is the wave speed? [Ans. 1 m/s]

7-2 What is the wavelength of the carrier wave of a radio station broadcasting at 740 kHz? [Ans. 400 m]

7-3 A small point-like source of radioactive material is placed 10 cm below a geiger counter and 900 counts/s is recorded. The counter is raised to 30 cm; how many counts are now recorded in one second? [Ans. 100]

7-4 Two identical lead spheres are hung on thin threads in a vacuum. The temperature of one is 320 K and it radiates 6.0 watts of thermal energy. The second sphere has a temperature of 325 K. How much heat does it radiate? [Ans. 6.4 W]

7-5 At what wavelength is the peak of the radiation curve of the first sphere in Exercise 7-4?
[Ans. 9000 nm]

Problems

7-1 The solar constant at the mean position of the Earth is 1.40×10^3 $W \cdot m^{-2}$; the mean distance from the Earth to the Sun is 149.6×10^6 km. What is the total power developed by the Sun?   [Ans. 4.0×10^{26} W]

7-2   The distance from the Sun to Mars is 2.28×10^8 km. Neglecting the planet's thin atmosphere, what is the solar constant at the surface of Mars?   [Ans. 0.60×10^3 W·m^{-2}]

7-3   The Earth orbits about the Sun in an elliptical orbit with a mean distance of 149.6×10^6 km. The minimum distance (perihelion) is 147.0 $\times 10^6$ km and occurs about Jan. 4. The maximum distance is 152.1×10^6 km and occurs about July 6. The average solar constant is 1.40×10^3 W/m^2; what is it on Jan. 4? What is the total percent change between Jan. 4 and July 6? Does this account for the seasons? [Ans. 1450 W/m^2, +7%]

7-4   Using the result of problem 1 calculate the surface temperature of the Sun. The radius of the Sun is 6.98×10^5 km and Stefan's constant is 5.67×10^{-8} in SI units.   [Ans. 5800 K]

7-5   Using the solar constant given in problem 1, calculate the mean total power intercepted by the Earth. The radius of the Earth is 6400 km. [Ans. 1.8×10^{17} W]

7-6   Use the result of problem 5 to calculate the surface temperature of the Earth in the absence of the atmosphere and assuming that the surface is totally black, that is, the albedo is zero.   [Ans. 279 K]

7-7   Repeat problem 6 assuming the correct albedo for the Earth of 0.35.   [Ans. 254 K]

7-8   Show that an increase of the Earth's albedo by a small amount $d\alpha$ will produce a decrease in the surface temperature given by
$$dT = -71 \, d\alpha/(1 - \alpha)^{3/4}$$
If the albedo increases by 1%, what is the change in temperature? [Ans. -1 K]

7-9   What is the ratio of the intensity per unit wavelength of a perfect blackbody at a wavelength of 500 nm to that at 1000 nm if the temperature is 3000 K? [Ans. 0.25]

7-10 Repeat question 7-9 for the case of tungsten (see Sec. 7.3).
  [Ans. 0.30]

7-11 The oceans of the world cover 3/4 of the earth's surface and have an average depth of 3.8 km. If the temperature of the oceans were to rise by 1 K how much would the sea level rise on average due to thermal expansion alone. Assume that the surface area of the oceans would not change. The thermal expansion of a volume V_0 of liquid to a volume V due to a temperature change T is given by $V = V_0 (1 + \beta T)$ where β is the volume expansion coefficient. The value of β for water is $0.2 \times 10^{-3} \, °K^{-1}$.
[Ans. 0.8 m]

⓼ | *ELECTRICITY*

8.1 Electric Charge

Electric charge is a fundamental property of matter. One of the reasons that electric science developed later than mechanics was that most matter in bulk displays no effect of electric charge; the electric effects are subtle and difficult to evoke. This is of course due to the fact that electric charge exists in two forms called "positive" (+) and "negative" (-), and matter normally contains equal amounts of both, producing a null effect. In addition we know that "like" charges (i.e. +/+ or -/-) repel each other, whereas "unlike" charges (+/-) attract.

Extensive experimentation in the 17th and 18th centuries elucidated the fundamental laws of the forces produced by stationary and moving charges on other stationary and moving charges, and the magnetic effects resulting from moving charges. Much of this fundamental electrical science will not be needed here as the primary concern is the production of electric current and potential, the flow of current through conducting materials, and its use as a carrier of useful energy.

In the 20th century, as the structure of the atom became better understood, it was realized that the electric neutrality of bulk matter was a result of several facts:

1. The unit of electric charge, both positive and negative, is quantized. This means that charge does not exist in arbitrary amounts but occurs in integral multiples of some smallest finite amount called the "fundamental unit of charge". To the greatest precision of our measurement capabilities, the absolute value of the fundamental units of positive and negative charges are equal. In the SI system of units this charge has a magnitude of

$$e = 1.602 \times 10^{-19} C \qquad \text{(8-1)}$$

where C stands for "coulombs".[1]

2. Ordinary matter is electrically neutral (equal numbers of positive and negative charges) right down to the level of the individual atoms.

3. The negative electric charge resides on the atomic particle called the "electron", and the positive charge on the "proton".

4. In the Rutherford model of the atom the massive proton resides in the atomic nucleus and the much lighter electron in the space around the nucleus. The electrons are much more weakly bound to the atom than are the protons.

8.2 Electric Current and Conductors

As a result of number 4 above, most electric currents (and all that we will deal with) are a result of the movement of electrons

> **Did you know?** The terms positive and negative for the two types of charge were first applied by Benjamin Franklin (1706-1790).

(i.e. negative electric charge) from one place to another. The binding of the electrons to the atoms does, however, vary over a wide range. In some atoms the electrons are very tightly bound and so do not easily move through the material; these materials are called "*insulators*". Other materials, particularly the metals, have a few outer-orbit electrons so weakly bound that, in the bulk material, they are virtually free and so can move to make an electric current; these materials are called "*conductors*".

It is one of the unfortunate circumstances in physics that the mobility of the negative charges was realized long after the definitions and conventions had been firmly established. Unfortunately they were established assuming that it was the positive charges that moved. As a

---

[1] For the definition of the coulomb (C) consult any elementary physics textbook.

result we maintain the fiction that the *electric current* is the flow of positive charge in some direction in a conductor when we know very well that it is really a flow of electrons in the opposite direction. This fiction will plague students to the end of time!

8.3 Electric Potential Energy and Potential

Figure 8-1:   Electric Field and Force.

Since an electric charge is able to exert a force on another electric charge, the first charge (say +Q in Fig. 8-1) is surrounded by an "*electric field*"[2] (**E**) which acts on the second charge (+q) to produce the force **F**. The field and the force are in the same direction for two positive charges. The electric field of Q is defined by

$$\mathbf{E} = \frac{F}{q} \qquad (8\text{-}2)$$

Notice that the units of the field are "newtons per coulomb".

---

[2] When boldface type is used, as in **E** and **F** the symbol stands for the vector quantity i.e. the magnitude as well as the direction. E and F in ordinary type stand for the magnitude of the vector alone.

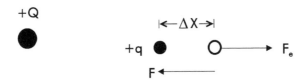

Figure 8-2: Electric Potential Energy.

In Fig. 8-2 the charge q is acted on by the electric field to produce the electric force F_e. Imagine that an external force F, essentially equal in magnitude to F_e, is applied in the opposite direction and the charge is moved a distance Δx. The amount of work done by this force on the charge is ΔW where

$$\Delta W = \mathbf{F} \cdot \Delta x = qE \cdot \Delta x . \qquad (8\text{-}3)$$

Since an amount of work ΔW has been done on the charge q its *electric potential energy* has been increased by the same amount ΔU, i.e. $\Delta W = \Delta U$.[3]

A useful new quantity, which is related to the potential energy, is called the *"electric potential"* or *"potential difference"* (ΔV); it is the *electric potential energy per unit charge* or

$$\Delta V = \frac{\Delta U}{q} . \qquad (8\text{-}4)$$

The units of this new quantity are joules per coulomb, termed "volt" (V)[4]. Notice that the units of the electric field can be volts per metre ($V \cdot m^{-1}$) as can be seen by combining Eqn. 8-3 and 8-4.

---

[3] Recall in mechanics: If a mass "m" is lifted vertically a distance "h" in the earth's gravitational field, the force exerted is F = mg and the work done by the upward applied force is Fh = mgh. The increase in *gravitational potential energy* is also mgh.

[4] Unfortunately in this case the symbol for the quantity (electric potential) and the unit (volt) is the same i.e. V.

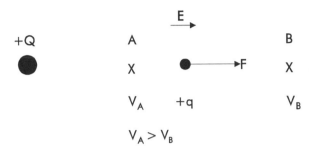

Figure 8-3:   Electric Potential for the case $V_A > V_B$.

In Fig. 8-3 a charge +Q creates an electric field E around it; the direction of an electric field is always away from a positive charge. The small test charge +q experiences a force in the direction of the field. Left to itself it would move from the point A to point B. We say, therefore, that the potential at A is higher than at point B, that is, it would move from the point of higher potential to the point of lower potential. Thus the situation is analogous to that of a mass in a gravitational field.

The analogy (between a gravitational field and an electric field) breaks down when one of the charges has a negative sign. For example if Q is <0 then the direction of the field would be reversed and the potential of A would be lower than that of B. In that case the charge +q would move from B to A, which is as expected since "unlike charges attract". Negative charges move from positions of low potential to positions of high potential. Specific examples are left to the exercises at the end of the chapter.

Example 8-1   *Two metal conductors are charged such that the potential of the first is 100 V higher than that of the second, i.e. $V_1 = V_2 + 100$. A small object carrying a charge $q = 5 \times 10^{-19}$ C is released at conductor 1 and moves to conductor 2. What is the change in KE of the object?*

$\Delta U = q\Delta V, \quad \therefore \Delta U = (-100 \ J/C) \cdot (5 \times 10^{-19} \ C) = -5 \times 10^{-17} \ J$
$\Delta KE = -\Delta U \quad \therefore \Delta KE = 5 \times 10^{-17} J$

Example 8-2   *(The "Electron-Volt") The "electron volt" or "eV" is a unit of energy which is very useful for systems on the atomic scale. It is the energy expended when a charge of one electron is moved through a potential of one volt. What is 1 eV in joules?*

Using Eqn. 8-4
$$\Delta U = \Delta V \cdot q = (1\ V)(1.602 \times 10^{-19}\ C) = 1.602 \times 10^{-19}\ J$$

8.4 Electric Current and Resistance

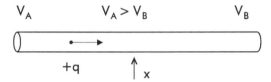

Figure 8-4: Electric Current.

It is important to remember the comment about the "conventional" current made in Section 8.2; even though an electric current is in fact a flow of electrons, we imagine that it is a flow of positive charge in the opposite direction.   If an electric conductor such as a copper wire as shown in Fig. 8-4 has one end at a positive potential with respect to the other end, that is, $V_A > V_B$ the charges will flow from A to B creating a current of charge in the conductor for as long as the potential difference is maintained. We define the *electric current* (I) at the point x as the total amount of charge (Q) that passes the point per unit time.

$$I = \frac{Q}{t} \tag{8-5}$$

The units of I are "coulombs per second" or "*amperes*", A.

Of course, the situation shown in Fig. 8-4 cannot last for long; eventually enough charge will flow from A to B to eliminate the potential difference and the current will stop.   <u>Useful</u> currents must flow in <u>continuous</u> circuits with some device that has the ability to remove the charge at B and reinsert it at A thus maintaining the potential difference, much as a pump removes water from the low pressure end of a closed pipe system and inserts it again at the high pressure end maintaining the current of liquid.

There are several devices that perform this function of providing a source of potential difference; the most common is the dry battery found in flashlights and other devices, the automobile battery, and a large number of rotating machines generally called "dynamos".

It seems logical that in a given conductor a higher potential difference should produce a larger current, but the mathematical relation between I and V is not obvious. This relation was investigated experimentally by George Simon Ohm (1789-1854) in 1826. He found that for most conductors the current was proportional to the potential difference, i.e. I $\propto$ V. Introducing a proportionality constant R this can be written as

$$V = IR \qquad\qquad (8\text{-}6)$$

This simple, but important, equation is known as "Ohm's Law". The quantity "R" is called the "*resistance*" and has the units volts/amperes which we rename the "ohm" with the symbol Ω.

Example 8-3   *A battery with potential difference 1.5 V causes a current of 0.3 mA to flow through a conductor. What is the resistance of the conductor?*

$V = IR$ $\therefore R = V/I = 1.5V/0.3 \times 10^{-3}A = 5 \times 10^3 \ \Omega$

Most common conductors, like metals, obey Ohm's Law and are called "Ohmic conductors". Some materials, however, do not have a linear relationship between potential difference and current, but a more complicated one; they are called "non-ohmic" conductors. Some of these materials, like semiconductors, are very important and will be discussed later in connection with solar cells; for now our attention will be restricted to ohmic conductors.

8.5 Resistivity

Figure 8-5 shows a conductor of length "ℓ" and uniform cross sectional area "A" connected to a battery of voltage "V" producing a current "I"; what does the current depend on? Clearly, from Ohm's Law, it depends on V. It is reasonable that it varies inversely with the length "ℓ"; if the length were doubled the resistance would be doubled and the current would be halved. It is also reasonable that it depends on "A"; if the area were doubled the resistance would surely be halved (See section 8.6 below). Finally it will depend on some intrinsic resistive property of the material itself; if copper were changed to iron and then to silver etc. the current would differ in each case. This property of the material is called the "*resistivity*" ρ. From these facts we can write

$$I = \frac{AV}{\rho\ell} \tag{8-7}$$

Table 8-1: Resistivities of Materials at Room Temperature

Material	$\rho\ (\Omega \cdot m)$
Copper	1.72×10^{-8}
Silver	1.47×10^{-8}
Steel (10% Ni)	29×10^{-8}
Tungsten	5.51×10^{-8}
Carbon	3.5×10^{-5}
Silicon	2.6×10^{3}
Wood	4×10^{11}
Fused Quartz	5×10^{13}

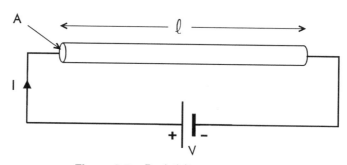

Figure 8-5: Resistivity.

The resistance is

$$R = \frac{\rho \ell}{A} \qquad (8\text{-}8)$$

The units of resistivity are ohm-metres ($\Omega\cdot$m), and values for some common materials are given in Table 8-1.

Example 8-4: *What is the resistance of a conductor made of copper 1.0 cm in diameter and 1.0 km in length?*

Using Eqn. 8-8,

$R = (1.72 \times 10^{-8}\ \Omega\cdot m) \times (1000\ m)/[\pi \times (0.50 \times 10^{-2})^2\ m^2] = 0.22\ \Omega$

8.6 Simple Electric Circuits and Resistance Combination Rules

The simplest possible electric circuit is shown in Fig. 8-6 where a source of potential difference V (perhaps a battery) is connected by conductors of negligible resistance to some device whose resistance is R. A current I flows from the positive terminal of the battery through the resistance to the negative terminal. Since all the resistance of the circuit is assumed to be in the device, then all the heat will be developed there. The figure also shows the standard symbols used for batteries and resistors.

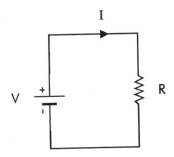

Figure 8-6: Simple Electric Circuit.

There are devices capable of measuring both voltage and current which are called "voltmeters" and "ammeters". The former are assumed to have infinite resistance and so can be connected *across* any part of a circuit without affecting it in any way; today's modern instruments approach that ideal very closely. Ammeters are assumed to have zero resistance so they can be placed *in* the circuit to measure the current again without disturbing the circuit. In Fig. 8-7 is shown the circuit of Fig. 8-6 with a voltmeter and ammeter added. We say that the ammeter is measuring the current "through", and the voltmeter is measuring the voltage "across", the resistor.

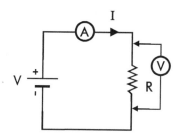

Figure 8-7: Electric Circuit with Meters.

Now let us suppose, as shown in Fig. 8-8 that the resistor R is in reality two resistors R_1 and R_2 connected in series; what is the relationship between R, R_1 and R_2? In other words, "What is the equivalent resistance of the two resistors in series"? Clearly the current I goes through both resistors 1 and 2 and the voltages across them V_1 and V_2 must add up to the original voltage V. Therefore, using Eqn. 8-6,

$$V = IR = V_1 + V_2 = IR_1 + IR_2$$

From this it follows that

$$R = R_1 + R_2$$

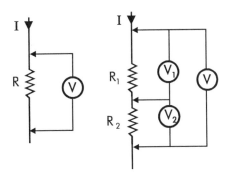

Figure 8-8: Resistors in Series.

This can be generalized to any number of resistors in series:

When resistors are connected in series the total resistance is the sum of the individual resistances.

Alternatively, the resistor R might be made up of two resistances connected in parallel as shown in Fig. 8-9. Now the voltage across each resistor is V but a current I_1 goes through R_1 and I_2 through R_2. The two currents must, however, add up to the original current. Thus

$$I = \frac{V}{R} = I_1 + I_2 = \frac{V}{R_1} + \frac{V}{R_2}$$

from which it follows that

$$\frac{1}{R} = \frac{1}{R_1} + \frac{1}{R_2}$$

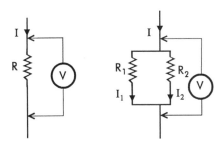

Figure 8-9: Resistors in Parallel.

Again this can be generalized to any number of resistors in parallel:

When resistors are connected in parallel the reciprocal of the total resistance is equal to the sum of the reciprocals of the resistances taken individually.

Example 8-5 *Figure 8-10 shows a circuit consisting of three resistors connected in series and parallel across a battery. What is the current through each resistor and what is the voltage across each resistor?*

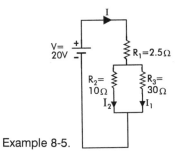

Figure 8-10:   Example 8-5.

If the resistance of R_2 and R_3 in parallel is R' then

$1/R' = 1/R_2 + 1/R_3 = 1/10 \ \Omega + 1/30 \ \Omega = 4/30 \ \Omega$

$\therefore \ R' = 30 \ \Omega/4 = 7.5 \ \Omega$

(The two resistances in parallel are equivalent to a single resistance of 7.5 Ω)
The total resistance $R = R_1 + R' = 2.5 \ \Omega + 7.5 \ \Omega = 10 \ \Omega$

The current $I = V/R = 20 \ V/10 \ \Omega = 2.0 \ A$.
The voltage V_1 across R_1 is: $V_1 = IR_1 = 2.0 \ A \times 2.5 \ \Omega = 5.0 \ V$.

The voltage V' across the two parallel resistors can be obtained in two ways:
i) $V' = IR' = 2.0 \ A \times 7.5 \ \Omega = 15 \ V$, OR
ii)   The total voltage is 20 V; 5 V appears across R_1 so the rest, or 15 V must appear across the other resistors.

The current through $R_2 = I_2 = V'/R_2 = 15 \ V/10 \ \Omega = 1.5 \ A$.

The current I_3 through R_3 can now be obtained in two ways:
i) $I_3 = V'/R_3 = 15 \ V/30 \ \Omega = 0.50 \ A$, OR
ii)   The total current is 2.0 A; 1.5 A goes through R_2 so the remainder, or 0.5 A, must go through R_3.

You will perhaps note from Example 8-5 two useful points that might almost be called "circuit rules" that help in solving problems:

1. Current is not lost but is conserved; if a current I enters a device the same current must leave.

2. The voltage of the source (e.g. a battery) is equal to the sum of the "voltage drops" in the various elements of the external circuit.

8.7 Electric Power

Just as water flowing in a continuous loop, driven by a pump, encounters resistance to the flow due to friction, so electric charges flowing in a closed loop driven by a battery also encounter resistance to the flow. Also, the energy lost due to friction in the flowing water appears as heat and that energy must be provided by the pump. In exactly the same way, the resistance to an electric current creates heat from energy provided by the battery. If a potential difference V causes a total charge Q to flow in a conductor for some time interval t then from Eqn. 8-4, the work done W is given by

$$\Delta W = VQ \ .$$

Divide both sides of the equation by the time t to give

$$\frac{\Delta W}{t} = \frac{VQ}{t} \ .$$

But Q/t is, from Eqn. 8-5, just the current I, and W/t is the power P so

$$P = IV \ . \tag{8-9}$$

It is left as an exercise to show that the units of the product VI are indeed watts as they must be.

Example 8-6 *What power is provided by the battery of Example 8-3?*

$P=VI$ $\therefore P=1.5 \ V \times 0.3 \times 10^{-3} \ A = 4.5 \times 10^{-3} \ W = 4.5 \ mW$

Using Eqn. 8-6 and 8-9 the power can be expressed in other ways for ohmic conductors:

$$P = \frac{V^2}{R} = I^2R \qquad\qquad (8\text{-}10)$$

What is happening at the atomic level is that the electrons are accelerated by the electric force, but before they can travel very far they collide with atoms setting them into vibratory motion; such motion of the atoms of a solid is the basis of "heat". The slowed-down electrons are again accelerated by the electric force and again suffer an energy loss by collisions. Thus the electrical energy which is the kinetic energy of the moving electrons can be converted to heat by letting a current flow through a material that offers resistance to the passage of current.

This is all that happens in a toaster or in any electric heating element. The process is very efficient; all the energy is converted into heat. Obviously the 100% efficient conversion of electricity into heat is extremely convenient in the case of toasters. We should remember though that the initial generation was accomplished from fossil fuel or uranium at only 30-40% efficiency.

Example 8-7 *The power used by a toaster is typically 1000 W; if the voltage applied is 100 V, then what is the current drawn by the toaster and how much energy does it require to make a piece of toast if the toasting time is 2.0 minutes?*

From Ohm's Law the current drawn is 1000 W/100 V = 10 A.
If it takes 2 minutes to make toast then the energy used is 1000 J/s × 120 s =
1.2 × 10⁵ J.

For comparison with Example 8-7, a typical 100 W lightbulb operated at

100 V draws 1 A.

In toasters, irons, stoves, kettles and other devices that operate through resistive heating, the wire is often not bare since at high temperatures it can react chemically with the air.   It is supported or embedded in a material (often a ceramic) that is solid and chemically inert at the temperature encountered.

Sometimes the heating effect can cause problems and is wasteful. Electrical wiring in buildings and electrical transmission lines across the country are made of copper because its low resistivity affords a low resistance (see Example 8-4).   If the electrical energy is to be used in a device, one does not want it being wasted as heat in the wiring. Although the resistivity of copper is very small, the length of transmission lines is very great between generating sites and cities.   The resistance of transmission lines can contribute significantly to the further loss of efficiency in electrical distribution systems, as shown in Example 8-8.

Example 8-8   *A transmission line is made from copper of diameter 1.0 cm.   The distance from the generating station to the user is 160 km.   The station develops 10 MW of power at a voltage of 40 kV as shown in Fig. 8-11. What is the current in the transmission line and what power is lost in heating the line?*

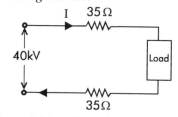

Figure 8-11:   Example 8-8.

For a complete circuit there must be a transmission line going in both directions between generator and load.

Using Example 8-4 the resistance of 160 km of line is

$$R = 0.22 \ \Omega \cdot km^{-1} \times 160 \ km = 35 \ \Omega$$

Since there must be a complete circuit, two such lines are needed, therefore the resistance of the total transmission line is 70 Ω.

Since both the power developed by the station, and the voltage across it, are known the current through the station can be calculated and thereby the current through the whole circuit.

The current is $I = P/V = (10 \times 10^6 \ W)/(4 \times 10^4 \ V) = 250 \ A$

The power loss in the lines is

$I^2R = 250^2 \ A^2 \times 70 \ \Omega = 4.3 \times 10^6 \ W$ or 43% of the output! ∎

How could the performance of the transmission lines in Example 8-6 be improved without making them thicker and heavier? Since the power loss varies with the square of the current, increasing the voltage and thereby decreasing the current will reduce the power loss markedly. If the voltage is increased to 400 kV the loss falls to 0.0043 MW; you should verify this result for yourself. (This shows why electricity is transmitted at high voltage using tall unsightly towers.)

8.8 Chemical Sources of Voltage

There are many uses of direct current (DC) electricity on a small scale, for example, electronic devices are replete with DC power supplies or small batteries. Most large scale energy systems rely on alternating current (AC) which is discussed in Chapter 9. DC is used on a large scale in a few applications, the most important of which is in electric traction. Most electric trains, trolleys and undergrounds use DC generators and motors. There is one other class of DC electric supply which is important in large scale power and that is the chemical cell or battery, familiar to all of us in the flashlight cell and the automobile storage battery.

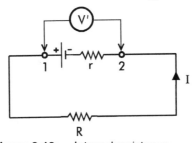

Figure 8-12:   Internal resistance.

All sources of voltage, whether they are generators or batteries, are not perfect but have an *internal resistance*. This is illustrated in Fig. 8-12 where a battery of voltage V and internal resistance r is connected to an external resistance R passing a current I. The actual terminals of the battery are labelled 1 and 2 and a voltmeter could be connected across them. The circuit current flows through the internal resistance producing a voltage drop Ir so that the voltmeter does not read V but V′=V-Ir. The more current that is drawn from the battery, the greater is Ir and the less is the effective voltage V′ of the battery. If a battery is to deliver a useful current and maintain an almost constant effective voltage then the internal resistance must be kept as small as possible. When batteries go "dead" what has happened is that the internal resistance has increased to the point where a small current produces a large voltage drop and so large currents are impossible.

Batteries and storage cells are of interest to us since they are a means of storing chemical energy which they liberate in the form of electrical energy. When two dissimilar metals are immersed in a common conducting medium called the "electrolyte", chemical processes produce a potential difference between them.

1. The Simple Voltaic Cell

This cell shown in Fig. 8-13 consists of copper and zinc plates immersed in dilute sulphuric acid. In the electrolyte, hydrogen ions H^+ and sulphate ions $SO_4^=$ exist separately. Zinc dissolves in the acid much more easily than does the copper; the zinc goes into solution as Zn^{++} ions which react immediately with the $SO_4^=$ ions to form neutral zinc sulphate $ZnSO_4$; hydrogen ions in solution thus find themselves without negative partners. Meanwhile the constant loss of Zn^{++} ions leaves the zinc plate with a net negative charge on it. The same process occurs at the copper plate but much more slowly so that the zinc plate becomes negatively charged with respect to the copper. When the battery is not in use, it comes to an equilibrium where the further dissolving of zinc is prevented because the negative charge prevents the loss of further positive zinc ions. The voltage developed by this cell is 1.1 V.

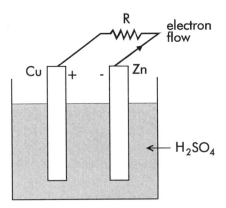

Figure 8-13:   Voltaic Cell.

If the two terminals are connected by a wire, the excess electrons will flow to the positive copper plate through the wire (conventional current from copper to zinc) where they neutralize H^+ ions causing hydrogen gas to bubble up. The removal of electrons from the zinc plate allows more ions to be dissolved replacing the hydrogen ions that have escaped in the gas. Overall, zinc is replacing hydrogen in the solution. Eventually the battery becomes dead when the sulphuric acid is saturated with zinc sulphate.

2. The Dry Battery

Dry cells, such as are used in transistor radios and flashlights, have zinc (the outer casing) and carbon (the central terminal) electrodes and an ammonium chloride paste electrolyte. They also have a finite lifetime as a result of chemical saturation and increasing internal resistance. (There are also combinations of electrodes that operate with an alkaline electrolyte.)

3. The Lead-Acid Storage Cell

A storage cell differs from a battery in that, when it runs down, it can be returned to its initial state by passing an electric current through it in the opposite direction from an external source and reversing the chemical reactions; it is "rechargeable". The best known example is the automobile battery.

The lead-acid cell consists of a negative lead and a positive lead oxide electrode in dilute sulphuric acid. The electric potential of this cell is 2 V;

three in series provide a 6 V "battery" and six produce 12 V. Lead from the lead plate forms a lead sulphate coating right on the plate. The electrons released in this reaction travel through the external circuit and combine with $SO_4^=$ ions to coat the positive plate with lead sulphate. When both plates are covered with lead sulphate so that no further reactions can take place, the cell is "dead". Passing a current through the cell in the other direction redissolves the sulphate and "recharges" the battery. Clearly if the cell is to provide a lot of charge and last a long time the plates must have a large surface area.

The lead-acid cell is not the only rechargeable cell; there are many others of which the "nickel-iron" and the "nickel-cadmium" or NiCad are the most common.

8.9 Capacity of Storage Cells

The longer a cell can supply current the better, and this is determined by the amount of materials used as well as the type. The objective is to store as much energy in the cell as possible. Storage cells contain a lot of metal and can be quite heavy, as anyone knows who has lifted a car battery. Lighter batteries are very desirable but difficult to achieve. (Satellites don't use lead-acid cells, for example, because they are simply too heavy.) A more meaningful characteristic than energy capacity is *specific energy capacity*, that is, the energy stored per kilogram of weight, usually measured in W· hr/kg.

Of course there is another consideration, and that is the rate of delivery of the stored energy. In many applications — an electric car for example — a cell that stores large amounts of energy in a small weight would be useless if that energy can only be taken out very slowly. In other words, we also want a large power per unit weight, usually expressed as W/kg. How are these two related? Here is a case where the internal resistance is very important. If the energy is extracted very rapidly, i.e. the current I is large, then we will sacrifice much of the stored energy as heat I^2r in the internal resistance. If we draw a small current we will not waste so much energy, but we will not be able to develop as much power. Figure 8-14 shows a "Ragone plot", which represents the general characteristics of a cell. It shows that the useful energy is large at small power, and vice versa.

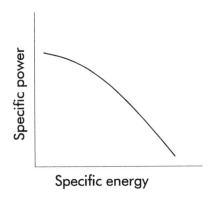

Figure 8-14:   Ragone Plot.

In an automobile we require large bursts of power for acceleration and large specific energy if we wish to make long trips on one charge. The Ragone plot shows that these two requirements are incompatible, a fundamental fact that has plagued the development of the electric car and limited it to slow-moving vehicles like milk vans and indoor tractors.

There is a lot of research in progress on storage cells but progress is slow. Figure 8-15 is a Ragone plot prepared by General Motors in 1975 describing various types of storage cells; the situation has changed very little in the subsequent years. A study of the plot will show that storage cells fall far short of the internal combustion engine in performance. Consider as an example a 1000 kg electric car with a range of 80 km on a charge in which the batteries account for 30-35% of the vehicle weight. The required specific energy is about 40 $W \cdot h/kg$ and the mean specific power is about 16 W/kg, with peaks to 80 W/kg. The Ragone plot shows that lead-acid (Pb/PbO_2) is totally unable to satisfy the peak power requirement, although it is suitable for steady driving (i.e. milk vans etc.). Only the Zn, Na, and Li based batteries qualify but they are all, at present, in the development stage and are currently expensive and unreliable.

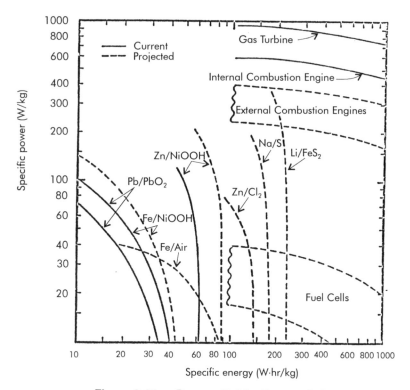

Figure 8-15:   Ragone Plot for Storage Cells.

8.10 Fuel Cells

A fuel cell is similar to a storage cell in that its function is to convert chemical energy directly into electrical energy. This is done at a fixed temperature so the device is not a heat engine and thus is not subject to the Carnot limitation on efficiency; as with a storage cell, the efficiency of energy conversion can be very high. The fuel cell differs from the storage cell in that it is supplied with a constant flow of two chemicals, one being the fuel and the other oxygen or air. Instead of burning the fuel with the oxygen to produce heat, the two chemicals react so as to release electrons directly, which flow as a current around an external circuit.

An example is the hydrogen-oxygen fuel cell illustrated in Figure 8-16. The H_2 and O_2 flow at high pressure into two porous electrodes immersed in a liquid electrolyte solution such as potassium hydroxide (KOH).

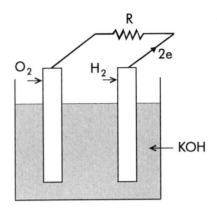

Figure 8-16:   The Hydrogen-oxygen Fuel Cell.

The H_2 is fed to the anode (+) and the O_2 to the cathode (-). The anode has a platinum-surface coating which acts as a catalyst to convert the hydrogen into H^+ ions, thereby releasing electrons to the external circuit. The H^+ ions combine with the OH^- ions in the electrolyte to form water. The reaction at the anode is

$$H_2 + 2OH^- \rightarrow 2H_2O + 2e^- . \qquad (8\text{-}11)$$

At the cathode the electrons combine with oxygen and water in the reaction

$$2e^- + O_2 + H_2O \rightarrow 2OH^- + O . \qquad (8\text{-}12)$$

Thus OH^- ions are continuously formed at one electrode and destroyed at the other, maintaining equilibrium in the electrolyte. Combining Eqns. 8-11 and 8-12 the total reaction is

$$2H + O \rightarrow H_2O . \qquad (8\text{-}13)$$

The KOH is not affected and the waste product is water which must be continuously removed. This constant-flow method of operation eliminates the need for recharging as in the case of storage cells.

The voltage output is determined by the two chemicals used; for H_2-O_2 this is 1.229 V which reflects the molecular binding energy released when the reaction of Eqn. 8-13 proceeds. This cell can produce 100-200 mA of current per cm^2 of electrode surface. There is no aquatic thermal pollution since no condensers are used, and no gaseous or particulate air pollution since there is no burning. Efficiencies are of the order of 40%. Hydrogen is not the only possible fuel nor is oxygen the only oxidant. Of course a fuel cell that uses the oxygen directly from air is very attractive from the point of view of cost.

Thus far the fuel cell's use has been limited. It has nonetheless found a niche in the space programme wherein the rocket fuels (hydrogen and oxygen) are already on board and the waste product (water) is essential as well.

Exercises

8-1   An electric charge of 2.0×10^{-15} C is placed in the vicinity of another electric charge where it experiences a force of 5.0×10^{-13} N. What is the electric field strength at the site of the 1st charge?   [Ans. 250 N$\cdot$C^{-1}]

8-2   The charge in exercise 8-1 moves a distance of 1.0 mm in the direction of the field. Assuming that the field is constant over that distance how much work is done on the charge?   [Ans. 5.0×10^{-16} J]

8-3   If the charge in exercises 8-1 and 8-2 was initially at rest and was carried by a dust mote of mass$=5$ μg, what is its speed after moving the 1 mm?   [Ans. 0.4 mm/s.]

8-4   Through what potential difference has the charge of exercise 8-2 moved?   [Ans. 0.25 V]

8-5   Show that the units of electric field can be either N$\cdot$C^{-1} or V$\cdot$m^{-1}.

8-6   In a chemical process an average current of 25 A was maintained for 3 hours. What electric charge was transferred?   [Ans. 27×10^4 C]

8-7   If the chemical process in exercise 8-6 was an electroplating process and a singly charged metal ion was involved, how many moles of the metal were plated?   [Ans. 2.8 moles]

8-8   A battery of voltage 1.1 V is connected across a resistor of 2.2 Ω; what current is drawn?   [Ans. 0.50 A]

8-9 What power is dissipated in the resistor of exercise 8-8?
[Ans. 0.55 W]

8-10 A copper bar 0.50 cm. in diameter is 5.0 m long; what is its resistance? [Ans. $4.4 \times 10^{-3} \, \Omega$]

8-11 Two resistors each of 0.50 Ω are connected in series; what is the total resistance? [Ans. 1.0 Ω]

8-12 The resistors of exercise 8-11 are connected in parallel; what is the resultant resistance? [Ans. 0.25 Ω]

Problems

8-1 A battery of internal resistance 2.0 Ω is used to drive current through a device which is equivalent to a 20 Ω resistive load. What fraction of the battery's power is dissipated in the load? [Ans. 91%]

8-2 A 100 V power supply contains a 10 A fuse. What is the maximum power rating of any device run from this supply? [Ans. 1 kW]

8-3 The resistors in the network in Fig. 8-17 are:
$R_1 = 6.0 \, \Omega$ $R_4 = 7.5 \, \Omega$
$R_2 = 3.0 \, \Omega$ $R_5 = 5.0 \, \Omega$
$R_3 = 15 \, \Omega$ $R_6 = 4.0 \, \Omega$
If the current I is to be 14 A, what is the voltage of the battery that must be connected across a and b? What is the voltage across R_1 and R_2? What is the voltage across the three resistor section? [Ans. 119 V, 28 V, 35 V]

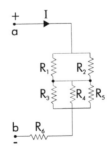

Figure 8-17:   Problem 8-3.

8-4 An electric kettle of 7.0 Ω resistance containing 500 cm^3 of water at 10°C is connected to the 110 V supply. If 10% of the electrical energy is wasted by imperfect insulation, calculate how long it takes to convert all the water into steam. (Latent heat of vaporization of water = 2.26 × 10^6 J/kg). [Ans. 850 s]

8-5 Many cheap electric stoves have three settings on their switch (besides OFF) labelled LOW, MED and HIGH. This is actually achieved by making the heating element with three resistive coils each of equal resistance R. The switch then arranges these resistors i) in series, ii) with two in parallel connected in series to the remaining one, and iii) all three in parallel.

a) Which combination gives LOW, MED, and HIGH?
b) What is the total element resistance in each case? [Ans. i)3R, ii) 3R/2, iii) R/3]
c) If the HIGH setting is 1000 W what is the power of LOW and MED? Is it really a good set of values for LOW, MED, and HIGH? [Ans. P_H = 1000 W, P_M = 222 W, P_L = 111 W]

8-6 A factory is supplied with 100 kW of power from a generating station some distance away.   The supply and return lines each have 5.0 Ω resistance. The factory receives the power at 4000 V. What power does the generator actually have to supply? If the factory took its power at 2000 V what power would have to be generated? [Ans. 106 kW, 125 kW]

8-7 A copper transmission line, 100 miles long and 1.0 cm. in diameter is replaced by one of twice the diameter. At the same time new users are connected and the current drawn increases by 50%. What is the fractional change in the heating losses in the transmission line? [Ans. 44% decrease]

8-8 A transmission line is made of a core of copper surrounded by a concentric circular sheath of steel to give it strength. The copper core is 1.0 cm in diameter and the tightly fitting sheath has an outer diameter of 2.0 cm. What is the resistance of 1.0 km of this transmission line? What is the ratio of the current carried by the core to that carried by the sheath? What is the ratio of the power lost in the core to that lost in the sheath? [Ans. 0.19 Ω, 5.6, 5.6]

8-9   100 MW of electrical power is delivered at 200 kV to a distribution centre over a transmission line which consumes 10% of the total power developed by the generating station in heat losses.

a.   What is the value of the voltage fed to the transmission line at the generator end? [Ans. 220 kV]

b.   How much power would be delivered to the same load if the resistance of the transmission line were halved but the rest of the system were unaltered? [Ans. 110 MW]

9 | *MAGNETISM AND A.C. ELECTRICITY*

In Chapter 8 some properties of direct current (DC) electricity were examined and several important relationships were derived. Most commercial electricity, however, is distributed not as direct current but as an alternating current (AC). That is, the current alternates in direction, flowing first one way and then in the reverse direction many times each second.

The widespread use of AC (rather than DC) for consumer use is for a technological reason; electric current is most easily generated, manipulated, and distributed on the large scale as an alternating rather than as a direct current. Since AC electrical machinery utilizes the magnetic effects of an electric current, in this chapter it seems logical to explore magnetism first.

9.1 Basic Magnetism

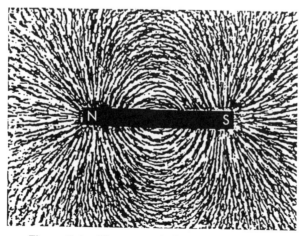

Figure 9-1: The magnetic field of a bar magnet.

The magnetic effects of natural magnets such as "lodestone" have been known for centuries. It is the effect which causes a compass needle or filings of iron to align themselves along invisible "lines of force" which seem to exist in the "magnetic field" surrounding a natural magnet.[1] Figure 9-1 is a sketch of iron filings sprinkled on a sheet of paper placed over a bar magnet. The lines of force as delineated by

> **Did you know?** The technique of rendering magnetic field lines visible by means of iron filings was described by William Gilbert (1544-1603) and "magnetic figures" were apparently known to the Roman philosopher Lucretius (95BC-55BC)

the filings seem to point at two locations (or at least small regions) near the ends of the magnets called the "poles". They are designated "north" (N) and "south" (S) according to which pole of the Earth they point to if the magnet is suspended and allowed to swing freely about its centre.

A further observation about the poles is that unlike (N/S) poles attract and like (N/N, S/S) poles repel each other, much like positive and negative charges. A basic difference between electric and magnetic fields is that whereas electric field lines originate on positive charges and terminate on negative charges, the magnetic field lines occur only in closed loops. In Fig. 9-1 the lines must be imagined to extend from the N to the S pole and return to close the loop within the magnet itself as shown in Fig. 9-2.

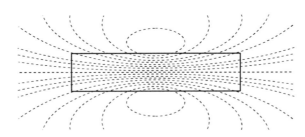

Figure 9-2: The magnetic field inside a bar magnet.

---

[1] Much investigation of the magnetic phenomena occurred before these terms were adopted. They were used extensively, if not actually coined, by Michael Faraday (1791-1867).

9.2 Electromagnetism

For a long time electricity and magnetism were thought to be unrelated separate phenomena until, in 1820, H. C. Oersted (1777-1851) discovered that a compass needle placed near a current-carrying wire was deflected. With this discovery of *electromagnetism* it was realized that magnetic fields are a result of moving charges. In the case of the wire the moving charges are the flowing electrons; in the case of a permanent magnet they are the electrons orbiting in their individual atoms. (*It is important not to confuse electric and magnetic fields*: electric fields exist between electric charges whether stationary or moving; magnetic fields accompany only moving charges.)

Oersted's investigations showed that the magnetic field which accompanied a current-carrying wire was in the form of circular lines about the wire as shown in Fig. 9-3. The direction of the field can be determined from the following *right-hand thread rule*:

> *To advance a right-hand threaded screw (the ordinary kind) in the direction of the conventional current, the screw driver would have to be rotated in the direction of the magnetic field.*

Examine Fig. 9-3 in light of this rule.

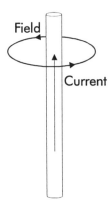

Figure 9-3: The magnetic field about a current-carrying wire.

Figure 9-4 shows a uniform magnetic field established between the poles of two large magnets. If a charge +q is injected at a velocity v, at angle θ, into this field of strength B (defined below) it will experience a force which is at right angles to both the magnetic field lines and the direction of the velocity. This means that in Fig. 9-4 the force is directed either into, or out of, the page. Whether it is into or out of the page is determined by another *right-hand thread rule* shown in Fig 9-5.

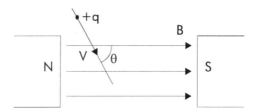

Figure 9-4: The interaction of a moving charge with a magnetic field.

A screwdriver turned in the direction of vector v rotated into vector B would advance a right-hand threaded screw in the direction of the force.

In the situation shown in Fig. 9-4 the force would be out of the page.

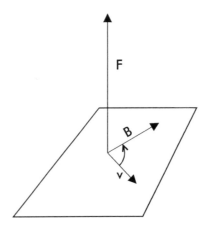

Figure 9-5: The direction of the electromagnetic force.

This rule, like all rules in electricity, has been given for a positive charge. Of course, if the charge is negative the force will be in the opposite direction.

Experiments show that the force experienced by the charge is proportional to the magnitude of the charge, the speed, and the sine of the angle θ. From these observations *magnetic field strength* **B** is given by

$$F = qvBsin\theta \qquad\qquad (9\text{-}1)$$

In Eqn. 9-1 when the force is in newtons, the charge in coulombs, and the speed in m/s, the magnetic field strength has units $N \cdot s \cdot C^{-1} \cdot m^{-1}$; this unit is called the "tesla"[2] (T). A magnetic field has a strength of 1 T if it exerts a force of 1 N on a 1 C charge that is moving at $1\ m \cdot s^{-1}$ at right angles to it.

Example 9-1 *A He^{2+} ion travels at $45°$ to a magnetic field of 0.80 T with a speed of 4.0×10^5 m/s. Find the magnitude of the force on the ion.*

The charge on a He^{2+} ion is twice the elementary charge or

$2 \times 1.602 \times 10^{-19}\ C$

From Eqn. 9-1, $F = qvBsin\theta$

$= 2(1.602 \times 10^{-19}C)(4.0 \times 10^5 m/s)(0.80T)sin45° = 7.2 \times 10^{-14}\ N.$

9.3 Electromagnetic Induction

Oersted's discovery that an electric current produced a magnetic field raised the question as to whether or not a magnetic field would produce an electric current or, more precisely, an electric potential which could, in

---

[2] Named after Nicola Tesla (1875-1943) a brilliant but eccentric electrical engineer who invented the induction motor and the transformer.

turn, drive a current in a circuit. This question was first seriously investigated by Michael Faraday who, after lengthy investigation, discovered *electromagnetic induction*.

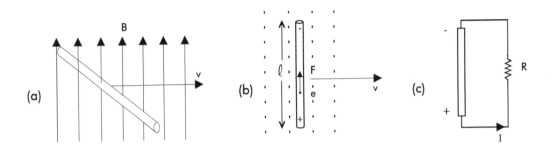

Figure 9-6: A Conductor Moving in a Magnetic Field.

In Fig. 9-6a a conductor of length ℓ, such as a wire, is being pulled by an externally applied force, at a speed v through a magnetic field of strength B; for simplicity we choose v, ℓ, and B to be at right angles to each other.

In Fig. 9-6b the situation is viewed from above, where the dots represent the points of the arrows representing B which are pointing out of the page. Imagine a free electron in the wire being carried along with the wire. According to the right-hand thread rule it will experience a force F in the direction shown. Since the electron, like all the others, is free to move in the wire, it will move toward one end making it negatively charged with respect to the other end. When sufficient charge has been separated, the electric attractive force will counterbalance the magnetic force and the process will come into equilibrium so long as the wire continues to move in the magnetic field. A voltage has been developed along the wire, called the "induced voltage". Balancing the electric and magnetic forces of Eqn.

8-2 and 9-1 ($\theta=90°$),

$$eE = evB$$

or

$$E = vB$$

But also, $E = V/\ell$ (See Eqn. 8-4 and following discussion) so

$$V = vB\ell \qquad (9\text{-}2)$$

Thus the motion of the wire through the field induces a voltage across it. The wire can be considered as a voltage source for some external circuit containing a resistance and so the moving wire would act like the battery in a circuit, driving current, I, through the external resistance, R as shown in Fig. 9-6c; the current would continue to flow so long as the wire continued to move. The current is given by

$$I = V/R = vB\ell/R \qquad (9\text{-}3)$$

In a time dt the wire moves a distance vdt tracing out an area $dA=v\ell dt$ as shown in Fig. 9-7 and so from Eqn. 9-2

$$dA/dt = v\ell \quad \text{and}$$

$$V = B(dA/dt) = (d/dt)(BA) = (d/dt)\phi \qquad (9\text{-}4)$$

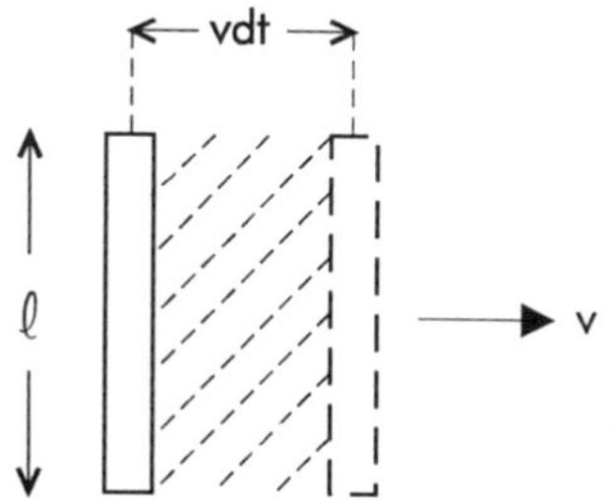

Figure 9-7: Magnetic Flux $d\phi=BdA$.

The quantity ϕ = BA is called the "*magnetic flux*"[3]. Eqn. 9-4 is Faraday's law of electromagnetic induction which states that

The voltage induced in a closed circuit is equal to the time rate of change of the magnetic flux.

Obviously the flux in a circuit can change in two ways: The circuit can move in the magnetic field, changing the area through which the magnetic field acts, or the magnetic field itself could change in a circuit of fixed size. Of course in some instances both might happen. The operation of a generator is an example of the former and a transformer is an example of the latter.

If the closed circuit is a coil of N turns, then the magnetic flux through the coil is just N times the flux through one coil or

$$\phi = \text{NBA}. \tag{9-5}$$

Example 9-2 *A circular loop of wire 5.0 cm in diameter is placed between the poles of a magnet perpendicular to the lines of a uniform magnetic field of strength 0.20 T. The leads from the loop are fed through a tube and are connected to a resistor of 393 Ω. A steady pull on the leads causes the loop to shrink until its area is halved after a time of 2.0 s. What is the current in the resistor during this process?*

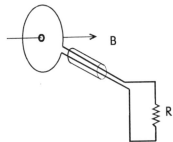

Figure 9-8: Example 9-2.

---

[3] The word "flux" is often encountered in science where some physical quantity flows or passes through some area. For example, a "sound flux" would be the total sound energy falling on some area; a "fluid flux" is the flow from a pipe of a given cross-section area.

The initial flux through the loop is, from Eqn. 9-5,

$\phi = NBA = 1(0.2T)\pi(2.5 \times 10^{-2}m)^2 = 3.93 \times 10^{-4} \; T \cdot m^2$

The final flux is half this amount since the area is reduced to one half. Therefore the change in the flux is $\Delta\phi = 3.93 \times 10^{-4}/2 = 1.96 \times 10^{-4} \; T \cdot m^2$.

From Eqn. 9-4, $V = d\phi/dt = 1.96 \times 10^{-4} T \cdot m^2/2 \; s = 9.81 \times 10^{-5} V$

From Ohm's Law $I = V/R = 9.81 \times 10^{-5} V/393\Omega = 2.5 \times 10^{-7} A = 25 \mu A$

9.4 The Generation of AC Power

Figure 9-9 shows a single-loop coil of area A rotating with an angular frequency[4] ω in a magnetic field of strength B. As the coil rotates, the flux through the coil varies from 0, when the coil sides lay in the direction of the field to a maximum value of $\phi_{max} = BA$ when the plane of the coil is perpendicular to the field. At some arbitrary angle $\theta = \omega t$, as shown in the figure, the area presented by the coils to the field is $A\sin\omega t$ and so, from Eqn. 9-4 the induced voltage is

$$V = (d/dt)(AB\sin\omega t)$$

$$V = \omega AB\cos\omega t \qquad\qquad (9\text{-}6)$$

Figure 9-9: Rotating Coil in a Magnetic Field.

---

[4] The angular frequency is the angle swept out per unit time measured in "radians per second".

If the coil has N turns then the induced voltage appears N times so

$$V = \omega ABN\cos\omega t = V_p\cos\omega t \qquad (9\text{-}7)$$

where $V_p = \omega ABN$ is the peak value of the induced voltage. This is an alternating voltage and is entirely different from the direct voltage produced by cells. In one complete rotation of the coil, the voltage acts in one direction round the loop for one half of the cycle and in the reverse direction for the other half as shown in Fig. 9-10. There are $f=2\pi\omega$ complete oscillations of the voltage per second, where f is the *frequency* of the voltage measured in *hertz (Hz)*.

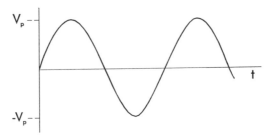

Figure 9-10: Alternating Voltage from a Rotating Coil.

If the ends of the wire making the coil are led out of the region of the field to an external circuit of resistance R, then a current will flow such that

$$I = I_p\cos\omega t \qquad (9\text{-}8)$$

where $I_p = V_p/R$. The current flows first in one direction and then in the reverse direction, following the voltage.

The basic elements of an AC generator are shown in Figure 9-11. The coil is rotated by a turbine (or any other mechanical engine) in the magnetic field of a pair of magnets. In practice, in a generator of high power output, the magnets would be electromagnets . The leads from the coil are connected to sliding contacts (brushes) on a "slip-ring commutator" which applies the AC voltage to the external circuit.

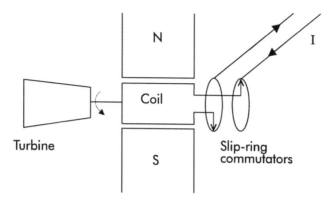

Figure 9-11:   Basic AC Generator.

Electrical generators are very efficient, converting well over 90% of the turbine's rotational energy into electrical energy. The limitation on the overall efficiency of a fuel-heat-turbine-generator system is the thermodynamic limitation of the engine explained in Chapter 3. (This reduces the overall efficiency — for the conversion of chemical energy into electric energy — to 30 - 40%.)

In large scale power installations this machinery is very large. At the Robert Moses Power Plant at Niagara Falls, each turbine has a mass of 200 tonnes and is rotated by the impact of falling water. Each generator contains a rotor 10 m in diameter on a shaft of diameter 1.2 m and mass 55 tonnes. One complete generator has a mass over 10,000 tonnes of which 65% is rotating. The electrical power output of one generator is 150 kW at 13,800 V.

9.5 Power in AC Circuits

From Eqn. 8-9 the power at any instant t is given by

$$P(t) = V(t)I(t) = V_p I_p \cos^2 \omega t$$

What is of more interest to us is the average value of this power P_{av}. Common AC power in North America has a frequency of 60 Hz and most electrical applications involve times much longer than 1/60 of a second. Since the average of $\cos^2$ or $\sin^2$ is 1/2,

$$P_{Av} = (1/2)V_p I_p \qquad (9\text{-}9)$$

A useful voltage (or current) is termed the "*root mean square (rms)*" voltage (or current). To determine this quantity, first square the voltage, next calculate its average (mean) voltage during one complete oscillation, and finally take the square root. The square of the voltage is

$$[V(t)]^2 = V_p \cos^2 \omega t$$

Since the average of $\cos^2$ is 1/2 then

$$[V(t)]^2_{av} = (1/2)V_p \qquad (9\text{-}10)$$

and taking the square root

$$V_{rms} = V_p / \sqrt{2} \qquad (9\text{-}11)$$

Similarly

$$I_{rms} = I_p / \sqrt{2} \qquad (9\text{-}12)$$

Rewriting Eqn. 9-9 using 9-11 and 9-12

$$P_{av} = V_{rms} I_{rms} \qquad (9\text{-}13)$$

which is identical in form to Eqn. 8-9. Thus, the equations developed for DC electricity can be applied to the AC situation if the rms values of the voltage and current are used. When AC voltmeters and ammeters are used their scales are usually calibrated in rms values.

9.6 Transmission of AC Power

The same considerations of transmission-line power-loss as discussed for DC power in Chapter 8 apply as well to AC power. Because the power-loss heating of the transmission lines depends on the square of the current, it is important that the current be kept as low as possible. The only way this can be done is to raise the transmission-line voltage as high as possible. Modern high-power transmission lines can be operated at voltages as high as 750 kV. The generators that produce the electric power, however, have much lower output voltages of the order of 10 to 30 kV because of size and magnetic field limitations. The higher transmission voltages are achieved by means of the *voltage transformer*.

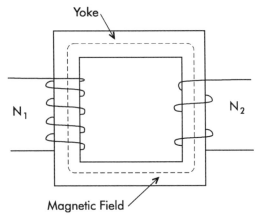

Figure 9-12:   The transformer.

A simple transformer is depicted in Fig. 9-12; it consists of an iron yoke, in the form of a closed loop, on which two coils (of N_1 and N_2) turns have been wound. If an AC source of voltage $V_1(t)$ is connected to coil 1, a field $B(t)$ is created which has a flux $\phi(t)$. This flux is almost entirely inside the iron yoke and is related to the voltage V_1 by Eqn. 9-4:

$$V_1 = N_1(d\phi/dt) \qquad\qquad (9\text{-}14)$$

Virtually the entire flux is trapped inside the iron and so is carried through the second coil inducing a voltage in it given also by Eqn. 9-4:

$$V_2 = N_2(d\phi/dt) \qquad\qquad (9\text{-}15)$$

Dividing Eqn. 9-14 by 9-15

$$V_1/V_2 = N_1/N_2 \qquad\qquad (9\text{-}16)$$

The ratio of the "primary" to "secondary" voltage on the transformer coils is the ratio of the number of turns on the primary and secondary coils.

Example 9-3   *A turbine generator produces AC power at a voltage of 20 kV. It is desired to transmit this at a voltage of 200 kV. If the transformer has 500 turns on its primary coil, how many turns must there be on the secondary coil?*

From Eqn. 9-15

200/20 = N_1/500

N_1 = 10 × 500 = 5000 turns

The transformer described in Example 9-3 is a "step-up" transformer. At a power sub-station the transmission line voltage is reduced to a few thousand volts with a "step-down" transformer and further reduced by the pole transformer near your home to the familiar 120 V_{rms}. Resistive heating losses in the transformer require that they be cooled efficiently; larger units are usually filled with oil for this purpose.

A typical pole transformer that serves a street of homes is shown in Fig. 9-12. The 8000 V high voltage connection to the primary is from the well insulated line at the top of the pole; the other side of the primary is grounded. The secondary consists of two 120 V high-current leads and a grounded line which emerge from the side of the container and are connected to the local distribution.

Example 9-4   *A 2400 V to 120 V suburban step-down transformer, of output 10 kW operates with 90% efficiency. What is the turns-ratio, the power in the primary, and the primary and secondary currents?*

From Eqn. 9-16 N_1/N_2 = 120/2400 = 1/20

Secondary power = 10 kW

Primary power = 10 kW/.9 = 11.1 kW

Primary current = I_1 = P_1/V_1 = 11100 W/2400 V = 4.6 A

Secondary current = I_2 = P_2/V_2 = 10000 W/120 V = 83 A

A small amount of current delivered to the transformer is stepped up to supply enough current for several homes. ∎

Figure 9-13:   A stepdown pole transformer.

Exercises

9-1 If a horizontal wire carries a DC current from east to west, and is located in a magnetic field that is vertically downward, what is the direction of the magnetic force on the (moving charges in the) wire? [Ans. south]

9-2 In the figure 9-14 a charge is moving with a velocity v in a magnetic field B. The non-zero components of v and B are given by the subscripts. Find the direction of the resulting Force in each case.
[Ans. a) y, b) -y, c) -z, d) -y]

223

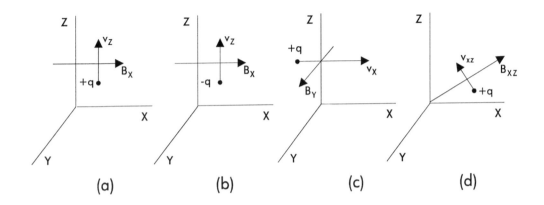

Figure 9-14: Exercise 9-2.

9-3 A wire of length 0.20 m moves at a speed of 15 m/s at right angles to a magnetic field of strength 0.004 T. The wire is also at right angles to the field. What is the induced voltage in the wire? [Ans. 0.012 V]

9-4 Consider a rotating rectangular generator coil as shown in the figure 9-15. When the area is perpendicular to the field B, as in figure 9-15a, consider the direction of the force on a negative charge in each of the four sides of the coil by answering the following questions:

a) What is the value of the force on a charge in any portion of the coil in figure 9-15a? [Ans. Zero]
b) What is the instantaneous current in the wire? [Ans. Zero]
c) What is the direction of the force on a negative charge in portions 1 and 3 in figure 9-15b? Does this force contribute to a current in the coil? [Ans. 1) toward 4, 3) toward 2, Yes]
d) What is the direction of the force on a negative charge in portions 2 and 4 in figure 9-15b? Do they contribute to the current flowing in the coil? [Ans. Perpendicular to the length of the wire, No]
e) In what direction is the (conventional) current in the coil in figure 9-15b? [Ans. 1→2→3→4]

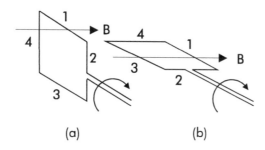

(a)   (b)

Figure 9-15:   Exercise 9-4.

9-5 A 900 W (2 sig. digits) toaster is plugged into a standard 120 V_{rms} outlet.
   a) What is the rms current in the toaster? [Ans. 7.5 A]
   b) What peak current does this correspond to? [Ans. 10.6 A]
   c) What is the resistance of the heating element in the toaster?
      [Ans. 16 Ω]

9-6 An electric stove element operates at 220 V and has a power of 2000 W. Which of the following fuses will blow if put in the protective circuit of this element: 1 A, 5A, 10 A, 15 A, 20 A? [Ans. 1 and 5 A]

9-7 What is the resistance of the element in Exercise 9-6? [Ans. 24 Ω]

9-8 If V_{rms} is 120 V, what is the peak voltage? [Ans 170 V]

9-9 A step-down transformer in a TV set reduces the 120 V_{rms} line supply to 6.3 V_{rms} for the tube filaments. If the primary has 400 turns, how many has the secondary? [Ans. 21]

Problems

9-1 A horizontal wire 0.75 m long is falling at a speed of 4 m/s perpendicular to a magnetic field of strength 2.0 T. The field is directed from north to south. What is the induced voltage of the wire? Which end of the wire is positive? [Ans. 6.0 V, East]

9-2 A copper bar 1.0 cm long is dropped from rest at a height of 3.0 m and falls in a horizontal orientation through the gap of a magnet which produces a uniform field of 0.2 T. What voltage appears across the ends of the bar when it is in the gap? [Ans. 15 mV]

9-3 In the figure 9-16 a metal rod rests on two conducting rails completing a circuit which includes the resistor of 3Ω. The circuit is perpendicular to a magnetic field, of strength 0.15 T, which acts into the figure as shown. The rod is pulled at a constant speed of 2.0 m/s. What current flows through the resistor? In what direction does the current flow? [Ans. 0.050 A, counter clockwise]

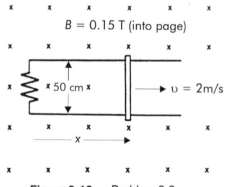

Figure 9-16: Problem 9-3.

9-4 An electric generator has a 20-turn coil of area 0.04 m² rotating uniformly at 60 rev/s in a magnetic field of 1 T. Calculate the peak value of the induced voltage and the power dissipated in a 100Ω resistive load. [Ans. 302 V, 455 W]

9-5 A square coil 5.0 cm on a side and containing 828 turns spins at 3600 rpm in a uniform field of 0.20 T. What is the peak power developed if this coil is connected to a resistor of 55 Ω? [Ans. 440 W]

9-6 What is the rms voltage, current and power in problem 9-5?
[Ans. 110 V, 2.0 A, 220 W]

9-7 A neighbourhood pole transformer reduces the voltage from 44,000 V (2 sig. fig.) to 120 V for household distribution. If the current flowing in the primary coil is 5.5 A, what is the current flowing in the secondary coil? [Ans. 2.0×10^3 A]

COMMERCIAL
ELECTRICITY

We are all accustomed to flicking a switch and having a light turn on, essentially instantaneously. The electrical energy used by the light has not been stored anywhere — it is being generated simultaneously. How do electrical utilities ensure that there is always enough energy — and not too much — for everyone?

This chapter is concerned with various aspects of the commercial generation and distribution of electricity. What are the advantages and disadvantages of hydroelectric energy? How do utility companies deal with changing electrical demand? Are electromagnetic fields from power lines and appliances dangerous to health?

10.1 MEETING ELECTRICAL DEMAND

Figure 10-1: Electrical generators at the Hoover Dam hydroelectric plant on the Nevada-Arizona border. Each of these massive generators can produce about 100MW of electrical power.

Power companies normally have a number of electrical generators (Fig. 10-1) which can be used, some of them virtually all the time, while some are active only during periods of peak demand. Electrical demand is continuously monitored, and the number of generators in operation is adjusted to meet the demand. Figure 10-2 illustrates how a typical company might use various generators to meet demand on a daily basis. There is a baseload which must be generated no matter what the time of day. This load is met by large reliable generators which use the least expensive fuel — hydroelectric generators, nuclear plants, or perhaps coal-fired generators. The hydroelectric stations used to provide baseload power might be a mix of run-of-river stations and dam installations. (As the name implies, a run-of-river station uses water which is actively flowing down a river; a hydroelectric dam creates a lake from which water is extracted when needed.) The intermediate load is met by smaller generators which can be varied easily in power output; steam turbines using coal, oil, or gas as a fuel are often used for this purpose. The peak demand is met by bringing more turbines into service at hydroelectric dams, and by using gas turbines (sometimes called combustion turbine units). In a gas turbine, natural gas is burned and the hot exhaust gases themselves turn a turbine without the use of steam. Gas turbines have the advantage that they are virtually "instant-on, instant-off" and so can accommodate rapid changes in demand.

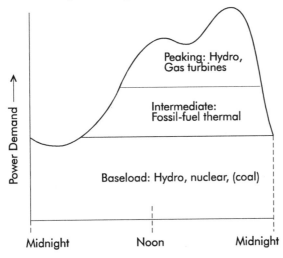

Figure 10-2: Typical generation-stacking-orderfor an electrical utility.

Of course, it takes some minutes to put another generator into service or to increase the output of a running generator (for example, by burning fuel at a faster rate). How are fluctuations in demand on a smaller time scale handled? Suppose that there is a sudden increase in power demand. There is a large amount of energy stored as rotational kinetic energy in the rotors of the generators themselves, and so short-time surges in demand can be taken from this energy, resulting in a slightly slower rate of rotation of the generators. The generators are designed to rotate with a frequency of 60 Hz, but this can decrease to as low as 59.5 Hz as energy is withdrawn, or increase to as high as 60.5 Hz if power demand suddenly drops. Changes in frequency are kept no larger than this because of the use of synchronous electrical motors in industry. These motors run exactly in step with the applied current, and their efficiency decreases drastically if the frequency changes very much. In particular, the efficiency of the synchronous motors which lubricate and cool the generators themselves would be adversely affected by appreciable changes in frequency.

Utilities need to be prepared for the shutdown of any given generator for maintenance or repair, and must have spare generating capacity in reserve. For example, Ontario Hydro has 33 GW of maximum generating capacity to meet a peak demand of about 23 GW. However, occasionally a large number of generators might be out of commission because of mechanical or other problems, and a utility is then forced to import energy from neighbouring utilities (at higher cost). In addition, power can be cut off to large industrial users who have elected to pay a lower price for using interruptible electricity. In extreme cases, a utility can deliberately reduce the voltage, producing a "brownout," or curtail power to users in a "blackout."

Long-term Planning and Demand-side Management

It takes ten to twenty years for a new power plant to be approved and constructed; hence, electric utility companies are continuously projecting demand for many years in the future and planning for new plants. In recent years, many utilities have realized that it is cheaper, more efficient, and more environmentally sound to save a megawatt than to generate a megawatt, and have embarked on measures to reduce both total energy consumption and peak power demand. This aspect of a utility's operation is referred to as *demand-side management*.

To reduce total consumption of energy, utilities can offer advice and assistance on more efficient energy use. They may, for instance, advise large industrial and commercial clients on ways to reduce energy needs while maintaining production or business volume, or they may offer cash incentives to consumers for using high-efficiency motors or low-wattage lighting.

It is also helpful to decrease the peak power demand, which determines how many generators a power company needs to have available. Reducing future peak demand means that construction of new generators can be postponed. To accomplish this, a power company can shift load from peak times to off-peak times by changing the pricing structure to offer lower rates (especially to large industrial users) during the night when total demand is low. As well, in an effort to avoid brown-outs during peak periods, some power utilities offer lower rates to customers who are willing to have their power interrupted on short notice.

10.2 TRANSMISSION TECHNOLOGIES

Once electricity has been generated, it must be transmitted to the consumer. As both the number of consumers and the per-capita consumption of electricity increase, utilities are presented with the problem of delivering an increasing amount of electrical power.

We have already covered the basic physics of this topic in Chapters 8 and 9. Recall that the power P produced by a generator is the product of the voltage V and current I: $P = VI$. Some power is lost as heat in the transmission lines themselves, and is usually expressed as I^2R, where R is the resistance of the lines: $P_{LOST} = I^2R$. These ohmic losses are not insignificant — typically 10% of the power transmitted. The losses can be decreased by increasing the voltage fed to the lines by the step-up transformer at the generator, thus decreasing the current. Typical voltages for long-distance transmission of electricity are now in the range of several hundred kilovolts; the highest voltage used in Canada[1] is 735 kV. Overhead lines are held high in the air by unsightly towers (Fig. 10-3) to avoid arcing (sparking) from the high-voltage lines to the ground.

---

[1] *Electric Power in Canada 1990*, Energy, Mines & Resources Canada, Ottawa, 1991, p. 68

Figure 10-3: Typical high-voltage transmission towers. (Photo courtesy of Ontario Hydro)

To illustrate this point with a simple example, imagine that an urban area doubles its demands for power; if the power is supplied at the same voltage, then the current must double. The I^2R heating loss in the cables therefore quadruples. What options present themselves if this loss is to be avoided? There are four chief alternatives.

> **Did You Know?** In Nova Scotia in 1989, about 100 of the estimated 250 active osprey nests were on transmission-line structures.
> (*Source:* The State of Canada's Environment, Government of Canada, Ottawa, 1992, p. 12-26)

(a) Duplicate the entire transmission line using the same voltage.

In terms of land use, this is neither a desirable nor even, in many cases, a feasible solution. In scenic areas the impact on the landscape is adverse, while in many urban areas the necessary land is simply not available or is extremely expensive. Even when land is available, many people do not want a transmission line located near their property. (This is one example of the NIMBY syndrome: "not in my back yard.")

(b) Increase the voltage applied to the line.

If we double the voltage, then the doubled power can be transmitted with the same current and thus the same I^2R losses. Increasing the voltage introduces limitations from both technology and land use. The higher the voltage on an object, the more efficiently it must be insulated to prevent arcing. Thus, higher transmission voltages require improvements in the insulating materials used to support the cables at the transmission towers. In addition, the main insulator between a cable and the ground is air, and the only way to increase the insulating power of air is to provide more of it by using higher and larger towers, thus increasing the land required for right-of-way. There is increasing opposition to demands for land, for the reasons mentioned in (a). (However, the land required for one ultra-high-voltage line is less than that needed for two lines at half the voltage.)

(c) Reduce the resistance of the line.

Almost all wiring is made of copper, and copper is the most-conductive material available at low cost. Thus, any scheme to reduce the resistance of the line would require a fresh strategy. One method which could prove useful would be simply to cool the line, since resistance falls with temperature (Problem 10-2). Another approach might be available through the phenomenon of *superconductivity*; this is the complete loss of electrical resistance in certain materials when cooled to extremely low temperatures. Superconducting metals at liquid-helium temperature (4 K) have been used successfully in other applications, such as in the magnets of the huge atom-smashing machines used in nuclear research. In 1986, Alex Müller and Georg Bednorz, working at the IBM Research Laboratory in Zurich, Switzerland, discovered that an oxide of barium, lanthanum, and copper exhibited superconductivity at a much higher temperature, about 35 K. Continued research has produced superconductors that operate at temperatures greater than 120 K, well above the temperature of liquid nitrogen (77 K). Since liquid nitrogen can be produced cheaply and easily, it would be inexpensive to keep superconductors at a temperature of 77 K or higher. However, it remains for these superconductors to be manufactured in the form of wires capable of carrying a high enough current to be of use to the electrical industry.

A superconducting line would have to be kept cold by liquid nitrogen, but the cost of this and necessary heat insulation would be more than compensated for by the ability to pass immense quantities of power. If it becomes possible in the distant future to develop a *room-temperature* superconductor, it would have tremendous applications for long-distance transmission of electricity.

For superconducting lines, the resistance to direct current is zero, but, unfortunately, for alternating current there is a small problem. Alternating current gives rise to a continually varying magnetic field, which exerts a force on electrons in the material around the cable; the electrons move as a result of this force, heating the material. To remove this heat, additional refrigeration would be necessary.

(d)   Use underground cables.

Land demands can be greatly reduced by laying cables underground, but there are disadvantages. Underground lines cost about 5 to 15 times as much as overhead lines[2], insulating the wires is more difficult, and the soil is not as good as air in carrying away the I^2R heat that is generated, thus producing the possibility of overheated cables (which degrade the insulation). One improvement being investigated is to cool the cables with oil or gas circulated in a pipe surrounding the cables.

10.3 HEALTH EFFECTS OF ELECTRIC AND MAGNETIC FIELDS

There has been much concern expressed in the news media during recent years about the possible health effects of the electric and magnetic fields produced by transmission lines, house wiring, and common electrical appliances. These fields oscillate with a frequency of 60 Hz in North America (50 Hz in Europe and some other parts of the world). Because the frequency is so low, the energy carried in 60-Hz fields is much too small to break molecular bonds. However, there is some evidence to suggest that these fields can have some biological effects.

Three general types of studies have been carried out: laboratory studies that expose single cells, groups of cells, or organs to fields; laboratory studies that expose animals or humans to fields; and epidemiological

---

[2] J.M. Fowler, *Energy and the Environment, Second Ed.*, McGraw-Hill, 1984, p. 141

studies of various human populations to investigate possible connections between exposure to fields and various diseases. Some experiments have demonstrated that under certain circumstances, fields can interact with cell membranes to produce changes in the rate at which the cell makes hormones, enzymes, and other proteins. The details of the processes which are occurring are not understood. Many of the results are rather unusual — there is evidence that weak fields sometimes produce larger effects than strong fields, and other effects appear only for pulsed fields with special pulse shapes. It will take considerable research and ingenuity to sort out all the complicated information and explain it. People exposed to strong fields under laboratory conditions can experience changes in heart rate and reaction time, and studies of people who use electric blankets report changes in the level of the hormone melatonin, which is important in our circadian rhythm (daily biological cycle). However, it is not at all clear what (if any) the health effects of these changes might be. Indeed, it is still not apparent that there is a direct cause-effect relationship at all.

The two epidemiological studies which have received the most attention involve childhood leukemia in Denver, Colorado. These studies indicate that the incidence of leukemia is greater by about a factor of two in children who live near major electrical transmission lines. (For comparison[3], smoking increases the risk of lung cancer by about a factor of 10.) However, because of the small number of people in the Denver studies, the statistical uncertainty is very large, and the effect could be much smaller or even nonexistent. Even if the statistical uncertainty were smaller, the studies do not necessarily mean that the fields around the lines are a cause of leukemia. Major transmission lines are usually built near busy roads, and such areas typically have more air pollution and noise. Other factors might also be important, such as diet and various socio-economic considerations. Nonetheless, in 1990 the U.S. Environmental Protection Agency concluded, in a somewhat controversial report, that there is a significant link between exposure to extremely-low-frequency electromagnetic radiation and the occurrence of human cancer[4]. However, a truly definitive statement on whether 60-Hz electric and

---

[3] M. Shepard, *EPRI Journal*, Oct.-Nov. 1987, p. 7

[4] S. Shulman, *Nature* 345, June 7, 1990, p. 463

magnetic fields play any role in causing or promoting leukemia or other cancers will have to wait for further research.

Table 10-1 presents data on electric and magnetic fields due to transmission lines and common appliances[5]. The magnitudes of the electric fields are given in kilovolts per metre (kV/m), and the magnetic fields in microtesla (μT). Individuals wishing to reduce exposure to such fields should obviously avoid close contact with any kind of electrical appliance and minimize time spent near high-voltage transmission lines.

Table 10-1:   Typical Electric and Magnetic Field Strengths

Source	Location	Electric Field (kV/m)	Magnetic Field (μT)
500-kV transmission line	Ground level underneath	5	100
Low-voltage residential distribution line	Ground level underneath	0.05	5
Electric blanket	Very close to source	1	30
Hair dryer	Very close to source	1	1000
Toaster	Several cm from source	0.01	5

[5] M.G. Morgan, *Electric and Magnetic Fields from 60 Hertz Electric Power: What do we know about possible health risks?*, Dept. of Engineering and Public Policy, Carnegie Mellon University, 1989, p. 3; and K.E. Donnelly, Ontario Hydro (personal communication)

10.4 HYDROELECTRICITY AND THE ENVIRONMENT

The three major sources of electrical energy are fossil fuels, nuclear energy, and hydro (water) energy. The problems associated with the use of fossil fuels have been discussed in Chapters 6 and 7, and nuclear energy is addressed in Chapters 11 to 15. The present section shifts the focus to hydroelectricity.

Hydroelectricity has many positive features: it is renewable, highly efficient, and has no fuel costs, no combustion products, and no radioactive waste. Once constructed, a hydroelectric plant has low operating costs, and hydroelectric generators can be started quickly to meet peak demands for electrical power.

However, there are a number of disadvantages. To generate large amounts of reliable hydroelectric energy, dams must be constructed at great expense and huge areas of land flooded. For example, the five reservoirs of the existing La Grande Phase I portion of the James Bay hydroelectric project in Quebec cover 11 400 km^2, of which 9 700 km^2 are flooded land[6]. There are many obvious environmental problems associated with a dam and flooding: sedimentation, erosion, spawning difficulties for fish, and large changes to ecosystems. In addition, the toxic metal mercury is released from soil and vegetation in areas flooded by new dams, and makes its way, for example, into fish, building up to extremely high concentrations. Flooding also means that some people will probably have to be moved to new locations; such displacement has produced serious distress for native peoples affected by the James Bay project. Concern has recently been expressed about methane (a greenhouse gas) released by rotting flooded vegetation. Finally, with a dam there is always the spectre of a catastrophic dam failure, with loss of human life.

Remaining Hydroelectric Resources

Most of the best sources of hydroelectric energy have already been tapped. In Canada, the generating capacity of hydroelectric plants in operation and under construction is about 60 000 MW. The total remaining potential is about 190 000 MW, but when future sites are

---

[6] *The State of Canada's Environment*, Government of Canada, Ottawa, 1991, p. 12-26

removed from consideration because of technical, economic, or environmental constraints, the remaining potential is only about 40 000 MW.[7]   Development of some of these sites would likely encounter resistance from various interest groups.

Worldwide, it has been estimated that about 17% of the technically feasible potential has been exploited[8]. With increasing awareness of the negative aspects of hydroelectric power, many new sites that could have been developed without public opposition 20 or 30 years ago would face greater hurdles today. It is likely that much of the increase in hydroelectric energy production in the near future will come from improving current hydroelectric installations by adding or upgrading generators, converting non-hydro dams to power-producing dams, and developing small run-of-river projects.

PROBLEMS

10-1 Suppose that electrical power is distributed from a power plant through a 350-kV transmission line; the power lost to heat in the line is 14% of the total power supplied. The plant is going to be doubled in size, and a selection must be made between duplicating the 350-kV line, or using a single 700-kV line. This single line would have the same resistance, and would pass the same current, as one of the 350-kV systems. What percentage of the total power supplied would be lost in the lines for (a) the two 350-kV line system? (b) the single 700-kV line system?
[Ans. (a) 14% (b) 7%]

10-2 The resistivity of copper at 77 K (the boiling temperature of liquid nitrogen) is 13% of the value at 295 K (22°C). Compare two current-carrying copper transmission lines of the same length and diameter, one at 77 K and the other at 295 K. If the thermal losses are the same in the two lines, what is the ratio of the current in the cold line to that in the warmer line? [Ans. 2.8]

---

[7] Ref. 1, p. 54

[8] Ref. 2, p. 403 (original source: D. Deudney and C. Flavin, *Renewable Energy: The Power to Choose*, Worldwatch Institute, W.W. Norton, New York, 1983, p. 168)

THE ATOMIC NUCLEUS AND RADIOACTIVITY

This chapter is concerned with the structure and masses of the atomic nuclei, and the results of rearrangements of the nuclear structure which give rise to the phenomenon of *radioactivity*. These topics are essential to an understanding of the processes of fission and fusion, which in turn are central to current debates on energy supply.

11.1 Structure of Nuclei

The Rutherford model of the atom pictures a very small dense atomic nucleus which contains virtually all the mass of the atom and all of its positive electrical charge. Orbiting around this nucleus in a volume about 10^{24} times greater are a number of negatively charged electrons, equal in number to the number of positive charges in the nucleus, so that the atom as a whole is electrically neutral. These orbiting electrons determine the chemical nature of the atom (a neutral atom with 6 electrons is always carbon for example) and in our discussion we have no further interest in them. This chapter will focus on the stability of nuclei and the radioactive decay of those which are unstable.

The nucleus is composed of two types of particles of almost equal mass: the proton (p) and the neutron (n). The total number of n and p in the nucleus is the *mass number* (A) of the nucleus. The proton has a single positive electrical charge and the neutron has no charge. Some basic properties of these particles are given in Table 11-1.

Table 11-1: Properties of Proton, Neutron, Electron

Particle	Mass	Charge
Proton	1.6726×10^{-27} kg	e
Neutron	1.6750×10^{-27} kg	0
Electron	9.1×10^{-31} kg	-e
	$(e = 1.6 \times 10^{-19}$ coulomb)	

A chemical element can, however, exist with different nuclear masses; for this to be the case it must be the number of neutrons which varies since the number of protons is fixed. For example, the element carbon exists with mass numbers 10, 11, 12, 13 and 14. Since all carbons have 6 p, then these forms of carbon must have 4, 5, 6, 7, and 8 n in their nuclei. These are called the "*isotopes*" of carbon.

The notation we use to specify a particular isotope of element X is

$$_Z^A X_N \tag{11-1}$$

where Z is the *Atomic Number*, that is, the number of p in the nucleus (also the number of orbiting electrons in the neutral atom), N is the number of n in the nucleus, and A as defined previously is the mass number.[1] Clearly A = Z + N and so one of the numbers is redundant; if a number is to be omitted it is usually N and our symbol becomes,

$$_Z^A X \tag{11-2}$$

The isotopes of carbon mentioned above are $_6^{10}C$, $_6^{11}C$, $_6^{12}C$, $_6^{13}C$, and $_6^{14}C$.

11.2 The Stability of Nuclei

There are about 100 different elements but many elements have more than one stable isotope; there are about 300 stable isotopes. (For example, oxygen has only one stable isotope $_8^{16}O$ whereas tin has ten.) Not every imaginable mixture of p and n will form a stable nucleus. Almost all stable nuclei have a number of protons (Z) which is less than the number of neutrons (N). This stability can be understood on the basis of simple electrostatics; if there are too many protons the mutual Coulomb electrical repulsion of the positive charges overcomes the forces holding the nucleus together. In a very few cases a nucleus is stable with a number of neutrons equal to, or even less than the number of protons. These few cases occur only for the very lightest elements: $_2^3He$ has N (1) less than Z (2); for $_2^4He$ the numbers are equal.

---

[1] Notice that to 2 or 3 significant figures the mass number, A, is the same as the molar mass. The mass number of ^{16}O is 16; its molar mass is 15.99491 g/mol.

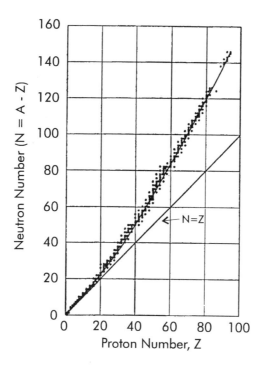

Figure 11-1:   N-Z Plot for Stable Nuclei.

For all heavier nuclei N is slightly greater than Z but not by a large amount; nuclei also are unstable if they have too many neutrons. Figure 11-1 is a plot of N vs. Z showing the narrow band of nuclear stability. The band lies mostly above the line of N = Z.

Theories of the structure of the nucleus are much more complicated than for the atom. In the latter case the structure is dominated by the Coulomb force between the orbiting electrons and the nucleus, which on the atomic scale is just a point. In the nucleus itself there is no such simplicity; fortunately the details of nuclear structure are not required for a discussion of energetics. It is sufficient to know that any nucleus which has too few or too many n relative to p will be unstable and that the nucleus will change in some way to redress the imbalance. The method by which it does so is to emit various particles, in a process called "*radioactivity*".

11.3 Radioactivity

It might be expected that a nucleus with an excess of some particle might simply emit the requisite number of those particles and so produce a new stable nucleus called the "*daughter*" nucleus. Because of the internal structure of the nucleus this almost never happens; neutron-emitting or proton-emitting nuclei are very rare. The unstable nucleus achieves stability by emitting two other types of particles, sometimes in a series of transformations. In the early days of nuclear physics these two particles were unidentified and were simply labelled alpha (α) and beta (β). It was also recognized that there was another radiation which often accompanied α and β and it was labelled gamma (γ). Very quickly these radiations were identified; their properties are given in Table 11-2. Note that the β radiation has two different forms depending on the charge of the emitted particle.

Table 11-2: Radioactive Emissions

Particle		Identity
alpha	(α)	Nucleus of helium atom $^4_2$He
beta	(β)	β^-, Ordinary electron $^0_{-1}$e
		β^+, Positive electron or positron $^0_1$e
gamma	(γ)	Electromagnetic wave of very short wavelength

Alpha emission

Some nuclei — particularly those at the high-mass end of the periodic table — achieve stability by the emission of a tightly bound cluster of particles which constitute the nucleus of normal helium, $^4_2$He. Two examples of practical importance are $^{238}_{92}$U and $^{234}_{94}$Pu. The decay scheme of the former is

$$^{238}_{92}U \rightarrow\ ^{234}_{90}Th\ +\ ^{4}_{2}He \tag{11-3}$$

These transformations are subject to certain conservation laws which determine the balance of the two sides of the equation. First, electrical charge must be conserved; this is the same as the atomic number (subscript). It therefore follows that the atomic numbers on both sides of the equation must add up to the same value ($92 = 90 + 2$).

Secondly, the mass number on the nuclear scale is conserved, this means that the superscripts on each side of the equation must add up to the same value ($238 = 234 + 4$). This is actually a loose way of formulating the correct rule which is called "the law of conservation of baryon number".

The α-particles are emitted with well-defined energies, typically a few MeV, and because of their large mass and charge, they interact strongly with matter. As a result they are easily shielded, being effectively stopped by a sheet of paper. Alpha emitters tend to have extremely long lifetimes.

Beta emission

By far the predominant method of radioactive adjustment for unstable nuclei is by β emission. For example, ^{3}H undergoes radioactive decay by emitting a β^{-} particle. A transformation equation describing this process is written,

$$^{3}_{1}H \rightarrow\ ^{3}_{2}He\ +\ ^{0}_{-1}\beta \tag{11-4}$$

Note that the rules for the conservation of atomic number (charge) and atomic mass number still hold. It must be noted that the mass of the electron is negligible on the scale of nuclear particle masses. Accordingly it is assigned a mass number of zero. In fact the mass of the electron is only 1/1840 of the proton's mass.

The transformation equation for ^{14}C is

$$^{14}_{6}C \rightarrow\ ^{14}_{7}N\ +\ ^{0}_{-1}e \tag{11-5}$$

Note that in the process a nucleus of carbon has been transformed into one of nitrogen; no further transformations will take place in this case as this isotope is stable.

An example of a nucleus which emits a positron is $^{11}_{6}$C; its decay is given by

$$^{11}_{6}C \rightarrow \, ^{11}_{5}B + \, ^{0}_{1}e \qquad (11\text{-}6)$$

Again the daughter boron nucleus is stable. On occasions when the daughter of a β decay is not stable, one or more subsequent decays will occur until stability is achieved. Notice again that the rules for balancing nuclear equations hold in Eqns. 11-5 and 11-6.

Beta particles from radioactive nuclei have speeds close to that of light, and kinetic energies of the order of one MeV. They travel for about one metre in air or a few mm in water or human tissue before coming to rest. In the process of coming to rest in tissue they can do much damage as will be seen in Chapter 15. It is rather easy to shield a β-emitter; a plastic sheet one cm thick affords complete protection. If, however, β-emitting materials are ingested via food, air, or water, the betas can cause considerable damage.

A given radioisotope may emit betas of varying energies. What is observed is a continuous spectrum of energies from zero to some maximum. Since the initial and final nuclear masses are fixed the β energy should also be fixed. This requirement can be met if there is another particle released along with the β to share the energy. For this reason physicists assumed that such a particle existed and it was subsequently found. It is a particle without charge, and if it has a mass it is very small. The particle, called the "*neutrino*", has a velocity essentially equal to that of light and is very difficult to stop or detect; a neutrino will go right through the planet Earth unaffected. We will not consider neutrinos further as they are almost irrelevant to terrestrial problems of energy generation.

Gamma Emission

The γ rays are very short wavelength electromagnetic waves. They are essentially high-energy X-rays. After the emission of an α or β particle, the daughter nucleus, in most cases, is left with excess energy in an "excited" state. This excess energy is emitted, as a γ ray, very shortly after the primary event (10^{-14}s), and permits the nuclear particles to readjust into their lowest energy (ground) state. This is very similar to the readjustment of orbital electrons in excited atoms, where low energy electromagnetic waves are emitted as X-rays or light.

Gamma rays are very penetrating, having energies around one MeV. Typically several centimetres of lead are required to diminish them to an acceptable level and form an effective shield.

Only a very few radioactive isotopes occur naturally in substantial quantities. The reason is that their lifetime must be very long to have survived since their formation in whatever cosmological event was involved, e.g. the formation of the universe itself (20×10^9 yr) or the formation of the solar system (5×10^9 yr). Examples are $^{235}_{92}U$ (α emitter) and $^{40}_{19}K$ (β emitter). A few unstable nuclei are produced continuously by the action of cosmic rays in the atmosphere, but the quantities are minuscule. Examples are the production of $^{14}_6C$ (so important in *carbon dating* in archaeology), and ^3_1H or *tritium*, the radioactive form of hydrogen found in trace quantities in water. Most of the exposure to radiation we experience comes from a small number of natural radioactive isotopes, (See Chap. 15).

With nuclear reactors and high-energy particle accelerators we have the technology to transmute stable nuclei into radioactive ones by adding or removing neutrons or protons. For example, the isotope ^{60}Co, widely used in cancer treatment, is manufactured by exposing the natural stable ^{59}Co to neutrons in a nuclear reactor. The resulting ^{60}Co is long-lived; each radioactive nucleus decays by emitting a β particle followed by two γ rays. The highly penetrating nature of these gammas enables them to reach and destroy deep-seated tumors.

11.4 Radioactive Series

Some very heavy nuclei are so very far from nuclear stability that they require many radioactive events to occur before they achieve stability. This results in a *radioactive series*. Such series usually begin with a long-lived parent whose slow rate of decay determines how many each of the subsequent species are found downstream in the various daughter nuclei.

An example of such a series is that which begins with $^{238}_{92}U$ and ends with $^{206}_{82}Pb$. The series with its emissions and lifetimes is given in Fig. 11-2.

Several other series are known as well: One begins with $^{235}_{92}U$ and ends with $^{207}_{82}Pb$, and another begins with $^{232}_{90}Th$ and ends with $^{208}_{82}Pb$.

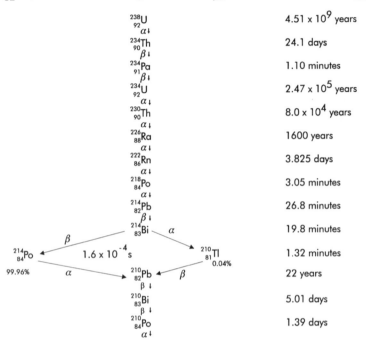

$^{238}_{92}U$	4.51×10^9 years
$^{234}_{90}Th$	24.1 days
$^{234}_{91}Pa$	1.10 minutes
$^{234}_{92}U$	2.47×10^5 years
$^{230}_{90}Th$	8.0×10^4 years
$^{226}_{88}Ra$	1600 years
$^{222}_{86}Rn$	3.825 days
$^{218}_{84}Po$	3.05 minutes
$^{214}_{82}Pb$	26.8 minutes
$^{214}_{83}Bi$	19.8 minutes
$^{210}_{81}Tl$	1.32 minutes
$^{210}_{82}Pb$	22 years
$^{210}_{83}Bi$	5.01 days
$^{210}_{84}Po$	1.39 days

Figure 11-2:   The Uranium Series.

11.5 Radioactive Decay and Half-life

Suppose that we start with a number N_0 of nuclei at time zero. The time at which a given nucleus decays is entirely random so we can only look at the average behaviour of a large number of nuclei. Let λ be the probability that in unit time a given nucleus will decay; this is called the "*decay constant*". If after a time t the number of nuclei remaining is N, then in the next short time dt the number decaying will be proportional to both N and dt, therefore

$$dN = -\lambda N \, dt \qquad\qquad (11\text{-}7)$$

The minus sign expresses the fact that the number N can only decrease as t increases. Eqn. 11-7 in the form,

$$dN/N = -\lambda \, dt \qquad\qquad (11\text{-}8)$$

has the well known solution

$$N = N_0 \, e^{-\lambda t} \qquad\qquad (11\text{-}9)$$

or, alternatively

$$\ln(N/N_0) = -\lambda t \qquad\qquad (11\text{-}9a)$$

The decrease takes place in an exponential manner with time. The behaviour described by Eqn. 11-9 is illustrated in Fig. 11-3a and that of Eqn. 11-9a in Fig. 11-3b.

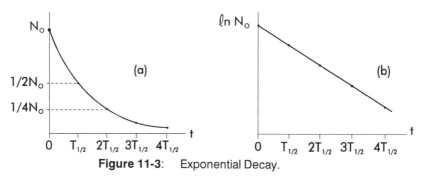

Figure 11-3:   Exponential Decay.

A characteristic of exponential decay is that it can be characterized by a unique time called the *"half-life"* ($T_{1/2}$), that is, the time for any given starting number N_0 to decrease to $N_0/2$. Substituting $N = N_0/2$ into equation 9 or 9a gives

$$T_{1/2} = 0.693/\lambda \qquad\qquad (11\text{-}10)$$

The half-life is clearly illustrated in Fig. 11-3a. Half-lives are usually specified for radioisotopes in preference to decay constants, since it immediately conveys the important information of how long the isotope will survive. For example, ^{239}Pu, a potent carcinogen, has a half-life of

24,400 years.

Example 11-1 How long will it take for a stored radioactive ^{239}Pu waste to decay to 1% of its present level?

$\lambda = 0.693/24,400 \ yr^{-1} = 2.84 \times 10^{-5} \ yr^{-1}$

$N = 0.010 \ N_0$

$0.010 = e^{-\lambda t}$

$t = -(ln \ 0.010)/\lambda = -(ln \ 0.010)/2.84 \times 10^{-5} \ yr^{-1} = 162,000 \ yr$

11.6 Effective Half-Life

If a radioactive species is ingested by a living organism the effective half-life of the species in the organism can be significantly altered by the biological activities of the organism. Although the isotope is decaying with a physical half-life of $_pT_{1/2}$ (decay constant $= \lambda_p$), the organism may be eliminating the isotope in some manner; animals excrete, perspire, exhale etc. The rate of elimination is often proportional to the amount present, so the amount present in the organism would also decay exponentially with a biological half-life $_bT_{1/2}$ (decay constant $= \lambda_b$). The total decay is given by the product of the two decay exponentials

$$e^{-\lambda_p t} \cdot e^{-\lambda_b t} = e^{-(\lambda_p + \lambda_b)t} = e^{-\lambda_e t}$$

where λ_e is the *effective decay constant* and

$$N = N_0 e^{-(\lambda_p + \lambda_b)t} = N_0 e^{-\lambda_e t}$$

Therefore $\qquad \lambda_p + \lambda_b = \lambda_e \qquad\qquad$ (11-11)

From Eqn. 11-11 it follows that

$$1/_pT_{1/2} + 1/_bT_{1/2} = 1/_eT_{1/2} \qquad\qquad \text{(11-12)}$$

Example 11-2: *Iodine ingested by humans it is eliminated in such a manner that one-half of the body's iodine content is excreted every 4.0 days. Radioactive ^{131}I with a physical half-life of 8.1 days is administered to a patient. When will only 1% of the isotope remain in the patient's body?*

$1/_eT_{1/2} = 1/_pT_{1/2} + 1/_bT_{1/2} = 1/8.1 \ d + 1/4 \ d = 0.37 \ d^{-1}$

$\therefore \ _eT_{1/2} = 2.7 \ d$

$\lambda_e = 0.693/_eT_{1/2} = 0.693/2.7 \ d = 0.26 \ d^{-1}$

$N/N_0 = e^{-0.26t} = 0.010$

$ln(0.010) = -0.26t$

$t = 18 \ d$

11.7 Activity

Activity is a term which refers to the number of radioactive nuclei that disintegrate per second and might be considered a measure of the *strength* of the sample. It is clear that the activity, A, will depend on both the number, N, of nuclei present and the half-life; the shorter the half-life the faster the nuclei decay and the greater the activity. Using Eqn. 11-7

$$A = |dN/dt| = \lambda N \qquad (11\text{-}13)$$

Since N decays exponentially then so also will A. It follows that

$$A = A_0 e^{-\lambda t} \qquad (11\text{-}14)$$

Thus the amount of radiation, α, β, or γ, emitted per second, falls off exponentially. The current unit of activity is called the "*becquerel*" and is defined as one disintegration per second; the abbreviation is "Bq". Another, older unit which is gradually losing currency is the Curie (Ci) equal to 3.7×10^{10} Bq.

Example 11-3 *A 3.7×10^{14} Bq (10 kCi) source of ^{60}Co is used for cancer treatment. Each disintegrating nucleus emits two γ rays, one of energy 1.17 MeV and one of 1.33 MeV. What is the mass of ^{60}Co present in the source and how much energy is emitted per second in the form of γ rays? The half-life of ^{60}Co is 5.3 years.*

$\lambda = 0.693/(5.3 \ yr \times 365 \ day/yr \times 24 \ hr/day \times 3600 \ s/hr) = 4.31 \times 10^{-9} \ s^{-1}$

Since $A = \lambda N$ the number of radioactive nuclei is

$A/\lambda = 3.7 \times 10^{14} \ s^{-1}/4.31 \times 10^{-9} \ s^{-1}$
$\qquad = 8.58 \times 10^{22} \ atoms$

Since 6.02×10^{23} atoms of ^{60}Co have a mass of 60 g or 0.060 kg, 8.58×10^{22} atoms have a mass of 8.58(0.06/60.2) = 0.0086 kg.

The energy per second = $3.7 \times 10^{14} \ s^{-1}(1.17+1.33)$ MeV
$\qquad = 9.25 \times 10^{14} \ MeV/s$
$\qquad = 9.25 \times 10^{14} \ MeV/s \times 1.6 \times 10^{-13} \ J/MeV$
$\qquad = 1.5 \times 10^{2} \ J/s = 1.5 \times 10^{2} \ W$

11.8 Some Important Radioactive Isotopes

Although there are hundreds of artificially produced radioactive isotopes, there are a few which, for various environmental reasons, merit special consideration:

1. Cobalt-60, ^{60}Co, This isotope which has been referred to earlier has a half-life of 5.3 years. Each disintegrating nucleus emits a β particle, whose kinetic energy is in the range 0 - 0.3 MeV, followed by two γ rays of 1.17 and 1.33 MeV (see Example 11-3 above). The time delay between these three successive emissions are each about 10^{-12} s, i.e. utterly negligible on our time scale so we regard them as simultaneous.

   The isotope is produced by bombarding natural cobalt with neutrons in a nuclear reactor. It finds widespread use in the radiation treatment of deep tumors, a technology pioneered by Atomic Energy of Canada Limited.

2. Strontium-90, ^{90}Sr is a principal component of fallout from atmospheric tests of nuclear bombs. It has a half-life of 28 years and its betas have a mean energy of 0.2 MeV. The daughter nucleus ^{90}Y has a 64 hour half-life, emitting betas of mean energy 0.8 MeV. Strontium is particularly important as its chemistry is similar to that of calcium and so it becomes incorporated in the skeleton of animals which ingest it. Concern about it has been decreasing with the worldwide ban on atmospheric nuclear-bomb testing.

3. Cesium-137, ^{137}Cs, is an important waste product of fission reactors which has some chemical similarity to potassium. It decays mainly via beta decay of mean energy 0.2 MeV, with a half-life of 30 years. (The β is followed by a 0.66 MeV γ.)

4. Plutonium-239, ^{239}Pu, is an α emitter with an 24,400 year half-life; it is formed from uranium in nuclear reactors. Of more importance than its radioactivity is the extreme toxicity of plutonium.

5. Carbon-14, ^{14}C, is a β emitter of very low energy (0.05 MeV) and half-life 5700 years. It is formed by the interaction of cosmic ray neutrons with ^{14}N in the atmosphere. The atmospheric ^{14}C then enters the plant-animal cycle, so it is present in all biological material. When the organism dies, no more ^{14}C enters and the amount already present declines through radioactive decay as time progresses. By measuring the remaining ^{14}C content the age of the specimen can be determined.

11.9 Nuclear Reactions

It was mentioned earlier that reactors and accelerators can transmute stable nuclides into radioactive ones, of which the production of ^{60}Co is an example; these transmutations are called "*nuclear reactions*". Reactors, as we shall see in a later chapter, produce copious amounts of neutrons; being uncharged they can easily penetrate within a stable nucleus, and if captured there they increase the mass number A by one unit without affecting Z. Our example was

$$\ce{^{59}_{27}Co + ^{1}_{0}n} \rightarrow \ce{^{60}_{27}Co} \tag{11-15}$$

Notice that the rules for balancing these equations are still obeyed.

In the case of bombardment of a stable nuclide by charged particles, e.g. protons, the bombarding particle must acquire sufficient kinetic energy to enable it to overcome the electrostatic repulsion of the positively charged target nucleus. This can be done by accelerating it to a high speed in a cyclotron or similar accelerating machine. In charged particle reactions a new nuclide is again produced but frequently an additional particle is also emitted. For example, when a ^{26}Mg target is bombarded with protons of high energy, neutrons are observed coming off; the reaction is

$$\ce{^{26}_{12}Mg + ^{1}_{1}p} \rightarrow \ce{^{26}_{13}Al + ^{1}_{0}n} \tag{11-16}$$

The particular interest in specific types of nuclear reactions will become clear as fission and fusion, two energy sources of great importance, are discussed. As an example, neutron-induced reactions are important in creating some of the hazardous radioactivity that builds up in the materials used in nuclear reactors.

An important concept in the physics of nuclear reactions is that of the "*cross section*". You might imagine that a reaction would automatically take place if a particle (p, n, etc.) collided with a target nucleus. If this were automatically so, then one could calculate the number of reactions. Consider a flux of N particles per second per square metre striking a thin target at right angles, as shown in Fig. 11-4. The target material has a mass of m grams. It contains, in its 1 m^2 of area, a number of nuclei n given by

$$n = N_a \, m/A \tag{11-17}$$

where A is the molar mass of the target material and N_a is Avogadro's number. Each of these nuclei has a cross sectional area σ m^2 and so the total area presented to the incoming beam is $n\sigma$ m^2. It follows that the number of particles, ΔN, intercepted from the beam is given by

$$\Delta N/N = n\sigma/1 = N_a \, m \, \sigma/A \tag{11-18}$$

The number of reactions that occur per second is

$$\Delta N = n\sigma N = N_a \, m\sigma N/A \tag{11-19}$$

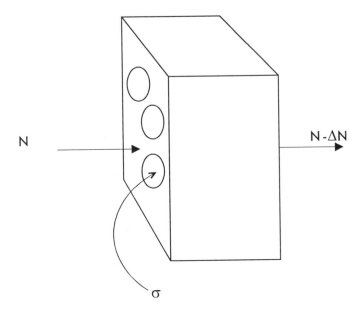

N

$N - \Delta N$

σ

Figure 11-4:   Absorption cross-section.

When a reaction is studied experimentally it is found that a different number of reactions occur than is predicted by Eqn. 11-19. Obviously a collision does not automatically ensure that a reaction will take place. The probability of occurrence of a reaction is dependent on details of the internal structure of the target nucleus and on the energy of the incident particle. It is still possible to describe nuclear reactions by the simple form of Eqn. 11-19 if we use a more flexible definition for the quantity σ. Rather than being the "geometrical cross section" of the nucleus, this is a *reaction cross section*. Thus the nucleus presents an effective area to the beam other than its real area; σ has a large value if the reaction is a highly probable one and a low value if it is less probable.

Values of σ must be determined experimentally by bombarding thin sections of material with particle beams at various energies. (Equation 11-19 permits a calculation of the cross-section.) Cross sections are measured in units of area; in this case m^2 is very inconvenient since the areas are so very small. A unit called the "barn" is used. The word "barn" was used as a code in the atomic bomb project during World War II. It is thought to have been suggested by the common phrase that something is 'as big as a barn', since a cross section of 10^{-28} m^2 is indeed very large relative to the cross sectional area of a nucleus.

$$1 \text{ barn} = 10^{-28} \text{ m}^2 \qquad \text{(11-20)}$$

Table 11-3 shows a selection of a few neutron reaction cross-sections.

Table 11-3: Neutron Absorption Cross-Sections

Isotope	Cross-section in barns
$^{1}_{1}H$	0.332
$^{2}_{1}H$	5.2×10^{-4}
$^{59}_{27}Co$	37
$^{113}_{48}Cd$	2×10^{4}
$^{115}_{49}In$	198
$^{135}_{54}Xe$	2.64×10^{6}
$^{157}_{64}Gd$	2.54×10^{5}
$^{197}_{79}Au$	99
$^{235}_{92}U$	582
$^{238}_{92}U$	2720

Most of the entries in this table have been chosen because of their relevance in Chap. 13, but it is useful to note a few things at this stage:

1) Heavy hydrogen $^{2}_{1}H$ is a much poorer absorber of neutrons than is ordinary hydrogen $^{1}_{1}H$ by a factor of over 600. This is very relevant to the design of the Canadian CANDU power reactor which uses heavy water instead of ordinary water (see Sec. 13.4).

2) The relatively light metal cadmium (Cd) is often used as a shield

against neutrons as its isotope $^{113}_{48}Cd$, which constitutes 12% of the natural material, has such a very high reaction cross-section for neutrons.

3) To make measurements of the strength of neutron radiations, the soft metal indium is often used in the form of thin foils. The isotope $^{115}_{49}In$ constitutes 96% of the natural material and has a rather high reaction cross-section for neutrons. The resulting isotope $^{116}_{49}In$ is radioactive with a half-life of 54 min. It transforms according to

$$^{116}_{49}In \rightarrow {}^{116}_{50}Sn + {}^{0}_{-1}e$$

A measure of the induced radioactivity of the indium is thus a measure of the neutron flux to which it was exposed.

4) The isotope $^{135}_{54}Xe$ is produced continuously in copious quantities in nuclear reactors; it is radioactive with a half-life of 9.2 hr. This isotope is of great importance to the operation of nuclear reactors because of its enormous cross-section for neutrons. (It becomes a major source for the loss of neutrons in the reactor.)

5) Natural uranium consists of two isotopes ^{238}U (99.3% of the total) and ^{235}U (0.7%). The absorption of a neutron by ^{235}U leads to fission as discussed in the next chapter. The isotope ^{238}U, which is the most common, absorbs neutrons without fission and with a rather large cross-section.

$$^{238}_{92}U \rightarrow {}^{239}_{92}U + {}^{1}_{0}n$$

The further fate of ^{239}U will be explored in Chap. 13.

Example 11-4 *A 1 g piece of cobalt is bombarded by neutrons in a reactor for one hour. The neutron intensity is 10^{17} per s per m^2. How many nuclei are transmuted to ^{60}Co? Express the resulting radioactivity in Bq and Ci units. You may ignore the fact that some of the radioactive nuclei decay during the first hour since the half-life of ^{60}Co is 5.3 years.*

The number of nuclei, $n = (1/59)$ mole $\times 6 \times 10^{23}$ atoms/mole $= 10^{22}$

The value of σ for ^{59}Co, from Table 11-3 is 37 barns

$\therefore$ The area presented to the neutron beam $= n\sigma$

$= 10^{22} \times 37 \times 10^{-28}\ m^2$

$= 3.7 \times 10^{-5}\ m^2$

The number of neutrons intercepted is given by Eqn. 11-19

$\Delta N = n\sigma N = 3.7 \times 10^{-5}\ m^2 \times 10^{17} m^{-2}\ s^{-1} = 3.7 \times 10^{12}\ s^{-1}$

The number activated per hour $= 3600s \times 3.7 \times 10^{12}\ s^{-1}$

$= 1.3 \times 10^{16}$

From Eqn. 11-3, the activity $= \lambda \times \Delta N$

$\lambda = 0.693/(5.3 \times 365 \times 24 \times 3600)\ s = 4.1 \times 10^{-9}\ s^{-1}$

Activity $= 4.1 \times 10^{-9}\ s^{-1} \times 1.3 \times 10^{16} = 5.5 \times 10^7\ Bq = 1.5m\ Ci.$ ∎

Calculations such as those in Example 11-4 can also be used to determine the buildup of radioactivity in the structural materials of a nuclear reactor. In practice the calculation is more complicated, as several other factors have been ignored, in particular the fact that as the radioactive species build up they in turn start to decay; this omission is more serious for short half-lives. These calculational details will not be a part of the present analysis.

EXERCISES

11-1 Write nuclear reaction equations for the following processes:
   a) The production of ^{14}C from ^{14}N in the atmosphere by interaction with cosmic-ray neutrons.
   b) The decay of ^{226}Ra to ^{222}Rn by α emission.
   c) The decay of ^{14}C by β emission.
   d) The production of tritium (^3_1H) in the atmosphere from normal hydrogen by high energy protons from the Sun.
   e) The decay of tritium to ^3_2He.
   f) The β decay of $^{40}_{19}K$.

11-2 Alpha particles emitted by nuclei generally have energies in the range 4 - 9 MeV. Calculate the corresponding speeds in m/s. [Ans. 1.4×10^7 m/s at 4 MeV]

11-3 A simplified sketch of a deuterium atom is shown below. Open circles are protons; filled circles are neutrons; dots are electrons in their orbits. Make similar sketches for: $^3_2$He, $^4_2$He, $^6_3$Li.

Figure 11-5:   Exercise 11-3.

11-4 How many protons and neutrons are in the nuclei of $^{11}_5$B, $^{234}_{90}$Ac, $^{235}_{92}$U, $^{238}_{92}$U? [Ans. n-6, 144, 143, 146]

11-5 The decay constant of ^{232}Th is 5.0×10^{-11} yr^{-1}. What is its half-life? On the basis of this value for $T_{1/2}$ do you expect to find ^{232}Th in the Earth's rocks? Why?   [Ans. 1.4×10^{10} yr; yes]

11-6 The radioactive isotope ^{131}I has a half-life of 8 days. A sample is prepared on July 1, and placed in front of a Geiger counter where 20,000 counts are recorded in each second. It is left in place until July 25; how many counts per second are recorded on that day? [Ans. 2500]

11-7 The stable nucleus $^{115}_{49}$In absorbs a neutron to form a radioactive nucleus. Write the equation of the reaction.

PROBLEMS

11-1 The radioactive isotope ^{22}Na has a half-life of 2.602 y and ^{35}S has a half-life of 87.0 days. They each emit positive β particles or positrons. A sample of each is prepared on June 1 and when placed on the window of a Geiger counter the ^{22}Na gives 5000 counts per second and the ^{35}S 20,000 counts per second. When will the two samples have the same activity, and what will the count rate be at that time?
[Ans. Dec. 9, 4350 counts per second]

11-2 Assume that when the earth was created there was the same amount of ^{235}U and ^{238}U; the present ratio of ^{235}U to ^{238}U is 0.7%. The half-life of ^{235}U is 8.8×10^8 yr. and for ^{238}U is 4.5×10^9 yr. What is the age of the earth? [Ans. 8×10^9 yr; this is about twice too large]

11-3 The biological activity of some organs can be investigated using radioactive isotopes. For example, iodine is important in the action of the thyroid. A subject is injected with a small amount of radioactive ^{131}I and after a few days a geiger counter is placed near the thyroid and 5500 counts per second are recorded. Six days later the counter records 1200 counts per second. What is the biological half-life for iodine in humans? The radioactive half-life of ^{131}I is 194 hours. [Ans. 99 hr.]

11-4 It has been estimated that by the year 2000, the U.S.A. alone will have to store 10^{21} Bq (27000 megacuries) of radioactive waste. One problem is the heat generated, which may be sufficient to damage containers. Calculate the total heat generated per second by this waste if, on average, each disintegration yields about 1 MeV in the total kinetic energies of the emitted α, β and γ rays. How does this compare with the power output of a large power station? [Ans. 160 MW, less]

11-5 Radon is a radioactive, α-emitting gas which constitutes a hazard in uranium mines. Its half-life is 3.8 days, but it is continuously generated by the decay of a long-lived radioactive parent, radium. How many α particles are emitted in 1 minute from 5 cm^3 of radon at room temperature and standard atmospheric pressure? For radon the density of the gas at 20 $°$C and 1 atm. is 9.7 kg/m^3; A = 222. [Ans. 1.7×10^{16}]

11-6 Suppose that a sample of radioactive waste consists of equal activities of ^{89}Sr, ^{137}Cs, and ^{106}Rh. Each of these isotopes decays into a stable daughter nucleus. After one year, what fraction of the original activity of the total sample remains, and which isotope(s) would be <u>primarily</u> responsible for this residual activity? The half-lives are: ^{89}Sr-54 days, ^{137}Cs-30 yr, ^{106}Rh-30 s. [Ans. Cs]

11-7 ^{90}Sr emits β-particles with a maximum energy of 0.546 MeV with a half-life of 28.1 yr. Its daughter is ^{90}Y which also emits β-particles with a maximum energy of energy of 2.280 MeV and a half-life of 64.2 hr. After a few days equilibrium is established and there are the same number of Y and Sr nuclei decaying each second; it is just as if the Sr were emitting both β-particles. About 60% of the energy is taken away by the neutrinos and is lost leaving 40 % in the β-particles.
   a) What is the activity of 100 mg of ^{90}Sr after it has achieved equilibrium? [Ans. 1.05×10^{12} Bq]
   b) What is the power developed by this source? [Ans. 0.19 W]

11-8 The old unit of radioactive activity, the curie (Ci) was defined as the activity (number of disintegrations per second) from one gram of radium ^{226}Ra. The half-life of ^{226}Ra is 1628 years. How many disintegrations per second are there in 1.0 Ci of radioactive material? [Ans. 3.6×10^{10}]

11-9 The neutron flux in a reactor is 10^{17} neutrons/(s· m^2) in the moderator and cooling water.
   a) If the reactor uses ordinary water as coolant and moderator, how many heavy hydrogen nuclei are produced in 1.0 year by neutron capture on $^1_1$H in each kilogram of water? [Ans. 7×10^{21} nuclei]
   b) If the reactor uses heavy water (D$_2$O), how much tritium ($^3_1$H) is produced in one kilogram of water in one year? (Since the half-life of tritium is 12.3 yr., the decay of the tritium can be neglected. [Ans. 10^{19} nuclei]
   c) Tritium is radioactive with a half-life of 12.3 years. What is the activity of one kilogram of the reactor water after one year? [Ans. 1.8×10^{10} Bq]

12 ENERGY FROM THE NUCLEUS - FISSION

Useful — and also destructive — energy can be obtained from some atomic nuclei. This nuclear energy can be obtained from very heavy nuclei by the process of nuclear fission and from very light nuclei by the process of nuclear fusion. In this chapter we will discuss the basic science for the realization of useful nuclear energy through fission and its applications.

12.1 Mass and Energy Units

In dealing with the masses of individual nuclei we must use very small numbers. The atomic mass scale is based on the definition that the atomic mass of ^{12}C is exactly 12 atomic mass units (u). Since 12×10^{-3} kg of ^{12}C contains exactly Avogadro's number of atoms then it follows that 1 atomic mass unit is given by

$$1 \text{ u} = 12 \times 10^{-3} \text{ kg}/12 \times N_a = 1.66 \times 10^{-27} \text{ kg} \quad \textbf{(12-1)}$$

Einstein's famous equation,

$$E = mc^2 \quad \textbf{(12-2)}$$

where c is the speed of light (2.998×10^8 m/s), tells us that mass and energy are interconvertible and so the mass of Eqn. 12-1 could just as well be expressed in energy units. Then

$$1 \text{ u} = 1.66 \times 10^{-27} \text{ kg} \times (2.998 \times 10^8 \text{ m/s})^2 = 1.49 \times 10^{-10} \text{ J} \quad \textbf{(12-3)}$$

Usually, however, a different energy unit is used in nuclear physics; the electron volt (eV, see Example 8-2) defined as

$$1 \text{ eV} = 1.602 \times 10^{-19} \text{ J} \quad \textbf{(12-4)}$$

Using Eqns. 12-3 and 12-4 we find that

$$1 \text{ u} = 931.5 \times 10^6 \text{ eV} = 931.5 \text{ MeV} \qquad \text{(12-5)}$$

MeV is the usual unit in which nuclear masses and energies are measured.

12.2 Binding Energy

Since we know the mass of the proton ($m_p = 1.0078252u$) and of the neutron ($m_n = 1.0086652u$) very accurately, and the number of each in every nucleus, we ought to be able to predict the mass of every nucleus. For example, the mass of $^4_2$He should be the mass of 2 protons and 2 neutrons or,

$$2 \times 1.0078252 + 2 \times 1.0086652 = 4.0329808u \qquad \text{(12-6)}$$

In fact, the mass of the $^4_2$He nucleus is 4.002603u, a number that is slightly smaller than expected. Expressed in energy units, Eqn. 12-6 is

$$2 \times 938.79 + 2 \times 939.57 = 3756.7 \text{ MeV} \qquad \text{(12-7)}$$

whereas the actual mass of $^4_2$He is 4.002603u $\times$ 931.5MeV/u = 3728.4 MeV. This is a difference of 28.3 MeV.

In general for a nucleus with Z protons and N neutrons the mass (or energy) should be $Zm_p + Nm_n$. However, in every case experiment has demonstrated that the actual nuclear mass is less, i.e.

$$Zm_p + Nm_n > m_{nucleus} \qquad \text{(12-8)}$$

This excess is called the "*mass excess*" and is a result of the *nuclear binding energy*. It is the extra energy we would have to provide to the nucleus to overcome the nuclear forces and break it up into its constituent parts. In the case of $^4_2$He there is a total binding energy of 28.3 MeV. It is usual to express this as the "binding energy per nucleon"; since $^4_2$He has 4 nucleons (2 p + 2 n), its binding energy per nucleon is (3756.7-3728.4)/4 = 7.07 MeV.

As we will see below, except for the lightest nuclei, the binding energy per nucleon is approximately constant throughout the periodic table. This immediately tells us something about the force that binds the nucleons together. Suppose that the nucleon-nucleon force had a long range, so that a given nucleon bonded to every other nucleon in the nucleus. In this case its binding energy would simply increase proportionately to the nucleon number A; but this is not what we observe. The approximate constancy of the binding energy per nucleon indicates that a given nucleon bonds to only a very small number of close neighbours; the addition of further nucleons to make a heavier nucleus has little further effect on the one we are considering.

The molecules in a water drop provide an analogy; each molecule binds only to its closest neighbours. Consider a drop of mass m containing N molecules. The binding energy is the *heat of vaporization*, Q, which is the energy needed to dissociate (evaporate) the drop into its constituent molecules. The average binding energy per molecule is Q/N. For a drop which is twice as large — that is, double the number of water molecules — the energy (heat) needed to dissociate the drop also doubles. The binding energy per molecule has not changed.

This comparison gives rise to the *Liquid Drop Model* of the nucleus from which we are able to predict and understand many nuclear properties including fission and fusion.

In Fig. 12-1 the average binding energy per nucleon is plotted vs. A (mass number) for all the stable elements of the periodic table. Except for some large variations for the light elements, the curve has low values for light and heavy elements, with a maximum in between; the maximum actually occurs for ^{56}Fe. The average binding energy per nucleon for a few important nuclei is given in Table 12-1.

Example 12-1 *The mass of $^{56}_{26}$Fe is 55.9934934 u. Calculate the binding energy per nucleon for this nucleus, i.e. derive the entry for $^{56}_{26}$Fe in Table 12-1.*

The mass of 26 p and 30 n is

26 × 1.0078252 u + 30 × 1.0086652 u

= 56.4634112 u.

The binding energy

$= 56.4634112 \ u - 55.934934 \ u = 0.5284772 \ u$

$= 0.5284772 \ u \times 931.5 \ MeV/u$

$= 492.277 \ MeV$

The binding energy per nucleon $=$ $492.277/56 = 8.791 \ MeV$ ∎

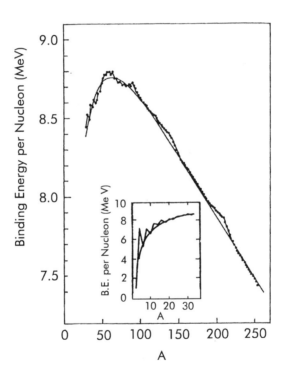

Figure 12-1:   Binding Energy per Nucleon.

Figure 12-1 and Table 12-1 show that there are, in principle, two ways to extract energy from the nucleus: If a very heavy nucleus like uranium were to split into two almost equal pieces, (with A ~ 50) the resulting two lighter nuclei would have more binding energy per nucleon and so energy would be released; this process is called "*nuclear fission*". In the second case if two very light nuclei were to join together to form one heavier nucleus, again the binding energy per nucleon would be increased and energy would be released; this process is called "*nuclear fusion*" and is discussed in Chapter 13.

Table 12-1: Binding Energy per Nucleon

Nucleus	B.E. (MeV)	Nucleus	B.E. (MeV)
$^{2}_{1}H$ (Deuterium)	1.112	$^{56}_{26}Fe$	8.791
$^{3}_{1}H$ (Tritium)	2.827	$^{95}_{38}Sr$	8.552
$^{3}_{2}He$	2.572	$^{100}_{40}Zr$	8.531
$^{4}_{2}He$	7.074	$^{139}_{54}Xe$	8.314
$^{6}_{3}Li$	5.332	$^{140}_{54}Xe$	8.295
$^{7}_{3}Li$	5.606	$^{235}_{92}U$	7.590
$^{7}_{4}Be$	5.371	$^{236}_{92}U$	7.586
$^{9}_{4}Be$	6.463	$^{238}_{92}U$	7.570
$^{10}_{5}B$	6.475	$^{239}_{92}U$	7.558
$^{12}_{6}C$	7.680	$^{239}_{94}Pu$	7.560
$^{14}_{7}N$	7.475	$^{240}_{94}Pu$	7.556

12.3 Spontaneous Nuclear Fission

Let us examine a possible case of nuclear fission of ^{235}U into a nucleus of ^{140}Xe and ^{95}Sr; the equation for the reaction is

$$^{235}_{92}U \rightarrow {}^{140}_{54}Xe + {}^{95}_{38}Sr \qquad (12\text{-}9)$$

Note that the rules given in Section 11.3 for balancing nuclear equations are obeyed. Using Table 12-1, the energy released in this fission is:

$$140 \times 8.295 + 95 \times 8.552 - 235 \times 7.590 = 190 \text{ MeV}$$

If a ^{235}U nucleus were to split in this way, 190 MeV of energy would be released which would appear mostly as kinetic energy of the two resulting *fission fragments*, that is, they would fly away from each other at high speed. When these fragments collide with other atoms the energy quickly becomes transformed into vibratory motions of the surrounding atoms — in other words as heat.

This type of *spontaneous fission* actually does occur but it is rather rare; it was only observed several years after the observation of *induced fission*.

12.4 Induced Nuclear Fission

It is the induced fission that is technologically important, and as will be seen, it is induced fission in the isotope ^{235}U (which only constitutes 0.72% of natural uranium) rather than the more plentiful ^{238}U that can provide useful energy. In induced fission a stray neutron encounters the nucleus and in the process of becoming bound to the nucleus causes it to fission.

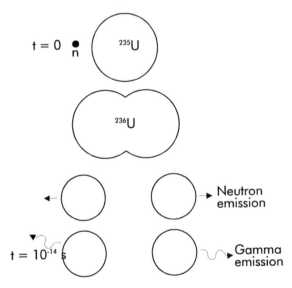

Figure 12-2: The "Liquid Drop" Nucleus Undergoing Induced Fission.

It is possible to understand the process of induced fission on the basis of what is known about binding energy and the Liquid Drop Model. It is important to realize that the stability of the nucleus is a result of the balance between forces. The short range nuclear forces are acting to attract their nearby neighbours, but the electrical (Coulomb) forces of the protons are acting to push the nucleus apart. If the heavy nucleus happens to capture a neutron, that neutron becomes bound with an energy of about 7 MeV. This amount of energy is therefore released to the nucleus and sets it into violent vibration. The vibrations may result in the nucleus splitting into two fragments as illustrated in Fig. 12-2.

Let us examine the neutron-induced fission of ^{235}U. Calculations show that an energy of 6.6 MeV is required to cause uranium to fission. If a neutron is absorbed by ^{235}U, a short-lived nucleus of ^{236}U will be created. From Table 12-1 the total binding energy of ^{236}U is 1791 MeV, whereas for ^{235}U it is 1784 MeV. The act of binding the neutron has released 7 MeV which is more than enough to cause the nucleus to fission. In this case the neutron need not even bring in any kinetic energy of its own; its binding energy is more than sufficient.

When neutron-induced fission takes place it is most effective if the neutrons are moving slowly, that is, they should have a small kinetic energy of the order of 1/40 eV; such neutrons are said to be "*thermalized*" because their kinetic energy is roughly the same as the kinetic energy of room temperature air molecules. A neutron has no charge so it is not affected by the charge on the protons. If it moves slowly, it will spend more time in the vicinity of the other nuclear particles increasing its chance of being captured by the short-range nuclear force. A rapidly moving neutron spends only a short time near the nucleons and its probability of capture is greatly reduced. This fact is very important in the operation of nuclear reactors.

12.5 Chain Reaction

For neutron-induced fission to take place there must be a source of neutrons; this is provided by the fission reaction itself. When the fission takes place the fission fragments, which are discussed in more detail below, are so neutron-rich that they quickly emit two or three neutrons. These neutrons have large velocities but if they can be slowed down, or thermalized, by some moderation process they can go on to induce further fissions.

The fact that more than one neutron is produced, on average, for each neutron absorbed means that the process can, under the right circumstances, grow in intensity; this is the so-called "*chain reaction*". For this to happen it is necessary to slow the neutrons down and to ensure that they are not lost by other processes. Losses can take place by the neutrons passing out of the reaction volume or by being absorbed by nuclei such as cadmium which have a large cross section for neutron absorption but do not fission. The absorption process is one way of controlling the chain reaction.

12.6 Fission Fragments

Returning to the neutron-induced fission of ^{235}U the reaction is

$$^{235}_{92}U + n \rightarrow \ ^{236}_{92}U \rightarrow X + Y \qquad (12\text{-}10)$$

where X and Y are the two nuclei produced by the fission; they are called the "fission fragments". Many different X,Y pairs are possible, grouped around the region where the mass is one half of the mass of the uranium. The distribution of abundance of fission fragments is shown in Fig. 12-3. As can be seen from the figure there is actually a minimum at half the mass; the nucleus is more likely to fission into unequal fragments of about 40% and 60% of the mass. Some of the important fission fragments which will be examined later have their high abundance as a result of this fact: ^{90}Sr, ^{137}Cs, ^{135}Xe, etc.

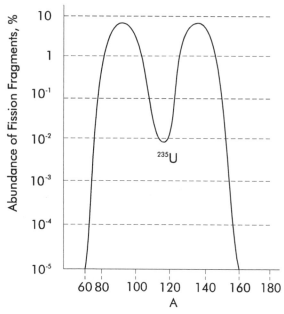

Figure 12-3: The Abundance of Fission Fragments for $^{235}_{92}$U.

After the initial fission there is still more energy available. The neutron-rich fragments emit several neutrons and gamma rays almost instantaneously ($\sim 10^{-12}$ s) as they rearrange their internal structure to reach some temporary degree of stability. They still have a neutron excess, however, which makes them highly radioactive and over a long period of time will undergo a series of β decays. These decays are sometimes accompanied by γ emission until eventually two stable nuclei are reached. The intermediate nuclei in this process constitute the radioactivity hazard which results if a nuclear reactor leaks.

12.7 Fission Energy

Table 12-2 summarizes the various forms of energy released during the entire process of fission and subsequent decay of the fragments of ^{235}U.

Table 12-2:   Typical Energy Released in ^{235}U Fission

Fragment Kinetic Energies	170 MeV
KE of 2 or 3 neutrons emitted	5
Prompt γ-rays (~5 gammas)	6
~7 β-particles	8
Fission Fragment Decay	
Antineutrinos	12
Gammas	6
Total	207 MeV

The total energy, 207 MeV, agrees very well with the 190 MeV predicted earlier. Of this energy, that of the antineutrinos that accompany the β decay (see Section 11.3) is lost, but all the rest will be captured and converted to heat. On the average, about 200 MeV is released in each fission of a uranium nucleus.

■

Example 12-2 *How many fission events of uranium nuclei must occur to produce one joule of energy?*

$$1 \ J = 1 \ J/(1.6 \times 10^{-13} \ J \cdot MeV^{-1}) = 6.25 \times 10^{12} \ MeV$$
$$= 6.25 \times 10^{12} \ MeV/(200 \ MeV \ per \ fission) = 3 \times 10^{10} \ fissions$$

■

The number of fissions in Example 12-2 might appear to be a very large number but remember that there are a very large number of nuclei in even a modest piece of uranium.

■

Example 12-3 *How many joules of energy are produced by the fissioning of all of the* ^{235}U *nuclei in 1/4 kg of natural uranium?*

$1/4 \ kg$ *contains* $(1/4)(1000 \ g \cdot kg^{-1}/238 \ g \cdot mole^{-1}) \times 6.02 \times 10^{23} atoms \cdot mole$
$= 6.3 \times 10^{23}$ *nuclei of which 0.72% or* $0.0072 \times 6.4 \times 10^{23}$
$= 4.5 \times 10^{21}$ *are* ^{235}U

The energy produced is

$$4.5 \times 10^{21} \times 200MeV/1.6 \times 10^{-13}J \cdot MeV^{-1} = 1.4 \times 10^{11} J$$

■

12.8 The Atomic Bomb

The atomic bomb, of the type exploded over Hiroshima and Nagasaki in Japan in 1945, utilizes the fission of either ^{235}U or ^{239}Pu. Provided there is enough fissionable material (i.e. a critical mass) to prevent the escape of too many neutrons, the fission rate grows rapidly, because of the multiplication of neutrons, and the temperature rises rapidly to very large values. A simple bomb consists of two sub-critical pieces of uranium with an explosive device to drive them together very quickly. A small amount of neutron-producing radioactive matter is present to supply the initial triggering neutrons.

Atomic bombs are rated in kilotons of TNT which would have the same explosive effect. The heat of combustion of TNT is 3600 Cal/kg and a kiloton is about 10^6 kg.

■

Example 12-4 *How much ^{235}U is required to produce the explosive effect of 1 kiloton of TNT?*

Assume that all the nuclei undergo fission (actually an unreasonable assumption)
Energy = 3600 Cal · kg^{-1} × 10^6 kg = 3.6 × 10^9 Cal = 1.5 × 10^{13} J
Number of U^{235} fissions   = 1.5 × 10^{13} J/(200 MeV × 1.6 × 10^{-13} J · MeV^{-1})
                      = 4.7 × 10^{23}
Mass of ^{235}U is 4.7 × 10^{23} × 0.235 kg · mole^{-1}/6.02 × 10^{23} atoms · mole^{-1} = 0.18 kg

■

12.9 Controlled Fission

Returning to more constructive uses of nuclear power, let us compare the energies released by 1 kg of uranium and of coal. The heat of combustion of coal is 5600 Cal/kg, so the energy produced by one kg of coal is 2.3×10^7 J.

One kg of ^{235}U produces

$$[(1 \text{ kg}/0.235 \text{ kg} \cdot \text{mole}^{-1}) \times 6.03 \times 10^{23} \text{atoms} \cdot \text{mole}^{-1}] \times$$
$$200 \text{ MeV} \times 1.6 \times 10^{-13} \text{ J} \cdot \text{MeV}^{-1} = 8 \times 10^{13} \text{ J}$$

Since only 0.72% of uranium is ^{235}U, then one kg of natural uranium produces $0.0072 \times 8 \times 10^{13} = 5.9 \times 10^{11}$ J. Weight for weight, uranium produces 25000 times more energy than coal. This number is not just a curiosity. Among other environmental factors, this number is an indication of how much less strain is placed on the transportation infrastructure by nuclear fuels compared with coal. In addition, since the mass of waste (solid and gaseous) produced by either coal-burning or nuclear fission is of the same order of magnitude as the mass of the fuel, this amount indicates that the quantity of waste from fission also is much less than that from coal.

It can be seen in Table 12-2 that the energy released in the radioactive decay of the fission fragments is responsible for about 9% of the total energy. This is important in the case where the cooling of a reactor may fail. The safety systems will shut down the fission reactions but the fragments keep on producing energy by means of their radioactivity. In a 300 megawatt electrical reactor operating at 33% efficiency there are 1000 Megawatts of heat being produced in the reactor core. If the reactor is shut down, there will still be about 90 Megawatts of fission-fragment heat. This is the most important reason for back-up cooling systems in nuclear reactors.

12.10 Uranium - 238

In the case of ^{238}U, the very abundant isotope, the situation is quite different. From Table 12-1, if ^{238}U captures a neutron to form ^{239} the total binding energy of ^{239}U is 1759 MeV whereas for ^{238}U it is 1754 MeV, a difference of only 5 MeV; this is not enough to cause fission. Nevertheless ^{238}U is not just a passive spectator in the ^{235}U fission process. It has a nuisance value in that it absorbs neutrons, removing them from availability to fission ^{235}U. This fact makes it very difficult (but not impossible) to have controlled fission in natural uranium. As a result, a large industry has grown up in some countries (particularly those with nuclear weapons programs) to enrich the concentration of ^{235}U in the uranium to be used in reactors.

There are several techniques used for this enrichment: The oldest technique, originally developed for weapons production, is to convert the uranium to its hexafluoride (UF_6) which is a gas. The gas is allowed to diffuse through a large number of small apertures. The $^{238}UF_6$ component diffuses a little bit more slowly than the lighter one, and after many successive treatments an enrichment can be achieved. The process is very expensive, requiring immense plants to process the tons of ore necessary for the operation of reactors. At present only a few countries have gaseous diffusion plants in operation.

High speed centrifugation of uranium compounds has also been used to separate the heavy and light fractions of uranium ore. Research is proceeding in other methods such as selective optical excitation in laser beams, but none are yet in the production stage.

When the ^{238}U absorbs neutrons, it is transformed into ^{239}U, which β-decays within minutes ($T_{1/2}$ = 23.5 min) to ^{239}Np which in turn β-decays ($T_{1/2}$ = 2.35 days) to ^{239}Pu. ^{239}Pu is a fissionable nucleus, and so this process transforms the useless ^{238}U into useful fuel. This process is called "*breeding*"; it can be used in a reactor to produce more fuel than is actually consumed. (Breeding will be discussed in more detail in Section 12.15.) Some of the ^{239}Pu produced in a reactor undergoes fission and contributes to the energy produced by the reactor.

Canada has followed a unique line of development in its nuclear reactors. By careful design and elimination of other sources of neutron loss, the Canadian reactors are able to achieve controlled fission with virtually no fuel enrichment. The CANDU reactor is discussed more fully in Section 12.13.

12.11 Nuclear Power Reactors

Nuclear reactors are devices for establishing and maintaining, in a controlled way, a chain reaction in a fissionable nuclear fuel, usually ^{235}U. Nuclear reactors are not all the same: Some are designed to maximize the production of some nuclear species like tritium or ^{239}Pu. Others, like the NRX and NRU reactors at Canada's Chalk River Laboratories, are designed to produce a maximum flux of neutrons for research purposes. Most reactors in the world have been designed to extract the maximum amount of heat from their cores for the production of electricity. While many features of reactors are common, in this Section we will concentrate on power reactors.

Did you know? The first attempt to construct a nuclear reactor was made in Canada by Dr. G. Laurence in 1940. It failed because of a lack of pure materials.

The essential features of a nuclear reactor are shown in Figure 12-4. The reactor core consists of a chamber containing a critical mass of the fuel elements surrounded by a *moderator* (discussed below). Through the core a cooling fluid (liquid or gas) is passed to remove the heat which is developed. Not shown in the drawing are the control devices necessary to operate the reactor and the safety and backup systems necessary for safe operation. Finally, of course, the reactor is enclosed in massive shielding to contain the ionizing radiation.

Moderator

As pointed out in Section 12.4, fast neutrons are ineffective in producing fission in ^{235}U; only slow, or thermalized, neutrons are effective. On the other hand the fission process, which produces the neutrons in the first place, gives fast neutrons. Some method must be found to slow the neutrons down so that they are neither absorbed nor lost by escaping.

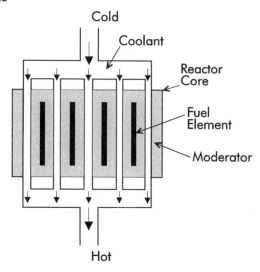

Figure 12-4:   Nuclear Reactor.

The neutrons are slowed down by allowing them to collide with other nuclei and give up some kinetic energy at each collision. The collisions of neutrons with other nuclei are "perfectly elastic": that is, momentum is conserved in the collision and the neutron loses energy only by transferring it to the colliding partner. (In other words, kinetic energy is also conserved by the colliding pair.) If we apply conservation of momentum and kinetic energy to the simple case of a head-on collision between a neutron (mass = 1, speed = v) and a nucleus (mass = A, speed = 0) it is easy to show that the *moderating ratio* of the initial and final kinetic energies of the neutron is given by

$$KE_{final}/KE_{initial} = [(A-1)/(A+1)]^2 \qquad (12\text{-}11)$$

(See Problem 12-12.) From Eqn. 12-11 it is clear that effective moderators (that is, those which make this ratio small) have small values of A. The most effective moderator is hydrogen with A = 1 giving a value of zero for the ratio, that is, the neutron loses all its energy in only one collision. The value of the moderating ratio is given for several nuclei in Table 12-3 along with the most common practical form of the moderator.

Table 12-3:   Moderating Ratio

Nucleus	Common Form	Ratio	No. of Collisions To Thermalize
Hydrogen	Normal Water	0	19
Deuterium	Heavy Water	0.11	32
Carbon	Graphite	0.72	110

The last column of Table 12-3 gives the average number of collisions necessary to thermalize the neutron. Why is the number for hydrogen not 1? This is because very few of the collisions are head-on; the value (19) shown in Table 12-3 includes the average over all the angles of impact, a much more complicated analysis.

Why do we even need to consider more than one moderator? Why do not all reactors use the most efficient and cheap one — that is, light (or normal) water? In fact, each moderator has certain advantages and disadvantages.

Light Water

Water is, of course, very cheap and can be highly purified. Moreover it can also serve as the coolant for the reactor, doing double duty. Unfortunately, the hydrogen nucleus, which is a proton, has a rather large cross-section for reaction with the neutron to produce a deuteron according to the equation, (see Section 11.9 and Problem 11-9)

$$ {}_1^1H + {}_0^1n \rightarrow {}_1^2H \tag{12-12} $$

Thus, although water is an efficient moderator, it also eliminates so many neutrons from the reactor core as to make it impossible to achieve a chain reaction with natural uranium, which has only 0.7% of fissionable ${}^{235}U$. If water is to be used, the amount of ${}^{235}U$ in the fuel must be increased, i.e. enriched.

Heavy Water

Deuterated or heavy water (D_2O) is the ideal moderator for a reactor using *natural* uranium as its fuel, because the cross section for the reaction

$$^2_1H + ^1_0n \rightarrow ^3_1H \qquad (12\text{-}13)$$

is very small and so neutron loss is small, (see Section 11.9 and Problem 11-9). The disadvantage is that heavy water is very expensive because of the large capital investment required in equipment for its separation from normal water and the large running costs since it is an energy-intensive process. Another problem lies in the end product of Eqn. 12-13. A contamination of ^3_1H, or tritium, slowly accumulates in the form of tritiated water DTO. Since tritium is radioactive and water is ubiquitous in living organisms, the tritium must be carefully controlled or removed.

Graphite

Graphite is not so effective a moderator as the above but it has the advantages that it is inexpensive, is structurally useful in the reactor since it is a solid, and does not strongly absorb neutrons (particularly if it is very pure). Enrico Fermi's very first reactor constructed at the University of Chicago in 1942 used a graphite moderator. Its disadvantages are that it will burn in air, and cannot also serve as the reactor coolant.

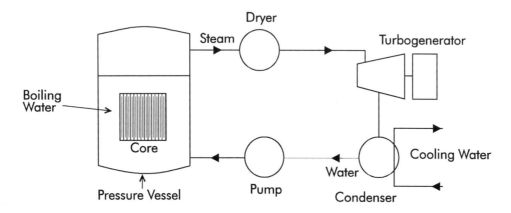

Figure 12-5: Boiling Water Reactor.

12.12 Light Water Reactors

Most of the power reactors in the world are of this type. These include many of the reactors of the U.S.A., France, United Kingdom, Germany, Russia and Japan. These reactors are mostly based on original American designs. Any country choosing this option is forced to develop isotope enrichment technology or to buy fuel from some country which has. This technology is closely related to isotope separation technology essential for nuclear weapons, so it is not surprising to find that all the nuclear-weapons powers have light water reactors.

Light water reactors are of two basic types: the earlier and simpler Boiling Water Reactor (BWR) and the more sophisticated Pressurized Water Reactor (PWR); these will be discussed separately.

1. Boiling Water Reactors

A schematic drawing of a BWR is shown in Figure 12-5. Fuel elements in the form of long rods containing enriched uranium are held in an open array in a stainless steel tank filled with water. The water, which acts as both moderator and coolant, is heated by the nuclear reactions and boils. The steam at a pressure of about 7 MPa (70 atmospheres) is passed to the turbine to drive the generators. It is then condensed, usually with cooling water from a river or lake, and pumped back into the reactor vessel.

The temperature of steam at 7 MPa is 285°C, which is low for turbine systems, so these plants are not very efficient. Fossil fuel plants which generate steam at about 550°C can have efficiencies as high as 40%; in contrast, BWR plants are usually about 30% efficient.

2. Pressurized Water Reactors

In order to increase the efficiency of a power plant, the temperature of the steam must be raised. The only way to do this is to raise the pressure. The schematic of a PWR is shown in Figure 12-6. The reactor core is contained in a vessel which operates at a pressure of 14 MPa (140 atmospheres). At this pressure the primary coolant water is prevented from boiling and remains liquid at 320°C. This hot pressurized water is pumped through a heat exchanger where it gives up its heat to the secondary coolant loop making steam for the turbines. In the PWR the primary coolant water does not leave the reactor vessel and so keeps radioactive contaminants inside.

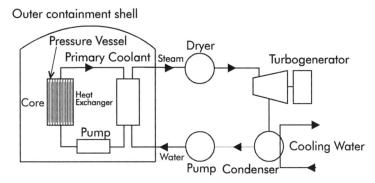

Figure 12-6:   Pressurized Water Reactor.

In each of these reactors the vessel can be opened only from above; this is done about once per year to renew the fuel rods. Both types of pressure vessel are housed within a steel containment vessel, which in turn is inside a reinforced concrete shield. Early reactors were in the 400-800 MWe range but newer ones have been built in excess of 1000 MWe.

For these reasons, PWR systems generally have a lower efficiency, resulting in increased thermal pollution of the water body used to condense the steam. For a 1000 MWe plant operating at 30 % efficiency about 50 m^3 per second of water is drawn, and raised through 10°C, before being discharged. On the other hand there are no emissions of sulphur or nitrogen oxide, or particulates, and only a few truckloads of uranium per year as opposed to trainloads of coal per day are needed to supply fuel.

Light water reactors continuously release radioactivity to the environment. Most of the fission products remain within the fuel elements and only small quantities of radioactive material diffuse out into the water; these include various inert gas isotopes and tritium. A further contribution arises from traces of corrosion products and other impurities in the water; these include iron, cobalt, and manganese. Solids are continuously removed by filtration, dissolved materials by passage through ion exchange resins, and dissolved gases by gas strippers. In a BWR several short-lived rare gas isotopes are carried by the steam to the turbines; these are held back for a period of time to allow the shortest-lived ones to decay before being released up a stack.

12.13 Heavy Water Reactors (CANDU)

The Canadian heavy water reactor is very different in concept from the enriched-uranium, light-water reactors. Instead of one vertical pressure vessel containing all the moderator-coolant, it has about 480 small horizontal pressure vessels in a honeycomb structure called a "calandria" which, in turn, is contained in a concrete vault as shown in Figures 12-7 and 12-8. The heavy water coolant and moderator are separate systems; the pressure tubes are surrounded by moderator, but coolant heavy water flows at a pressure of 10 MPa through the pressure tubes which contain the natural uranium fuel elements. The coolant leaves the pressure tubes at a temperature of $310\,^{\circ}C$ and its heat is transferred in a heat exchanger to normal water to make steam. A small annular space on the outside of the pressure tubes carries a flow of cooled nitrogen gas to prevent heat transfer from the cooling system to the moderating system.

Figure 12-7:   CANDU Calandria and Honeycomb Fuel Array.

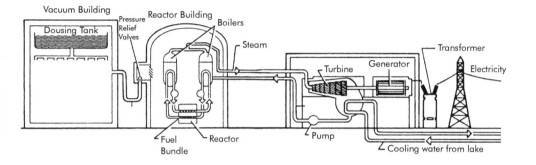

Figure 12-8:   CANDU Power Reactor.

Earlier versions of the CANDU could stop the reaction in 30 seconds in an emergency by dumping the moderator into a tank below the reactor. Unlike the light water reactors, depriving a CANDU of its moderator does not deprive it of its cooling through the pressure tubes. However, the moderating power of the cooling water is inadequate to sustain the chain reaction. Newer versions such as those at Darlington retain the moderator fluid for additional cooling in an emergency but inject a solution of gadolinium nitrate into the moderator to "poison" it. Since gadolinium is a strong absorber of neutrons (see Table 11-3) this method effectively, and permanently, stops the nuclear reactions.

Another unique feature of the CANDU stations is the vacuum building, also shown in Fig 12-8. This is a reinforced concrete building connected to all reactor buildings by a pressure-relief duct. The pressure in the vacuum building is kept below one atmosphere. If any reactor accident should cause a pressure buildup, relief valves open up into this building. It has the capacity to accommodate all the heavy water transport fluid which would be required should a reactor inlet pipe be completely severed. This guards against release of radioactive material to the environment in case of an accident.

Each of the fuel elements in the pressure tubes can be replaced independently without shutting down the reactor. The fuel elements are even moved from tube to tube in a planned sequence to maximize the utilization of the fuel. As a result the CANDU has the lowest fuelling cost of any reactor system.

Other countries also operate heavy water reactors: India, Korea, Pakistan, and Romania. Almost all are based on the Canadian technology.

12.14 Graphite Reactors

Power reactors using ultra pure graphite (carbon) as the moderator have been used in several countries, particularly Britain and Russia. Of course some fluid must also be used as the coolant and the British reactors, which are the most successful, use carbon dioxide (CO_2) gas at high pressure. Hot CO_2 is corrosive to many materials and so the fuel elements must be clad in a non-corrosive material capable of withstanding high temperatures. The first generation of British reactors used a beryllium/magnesium alloy called "Magnox" which lent its name to the reactor. Since the first one at Calder Hall in 1956 the efficiency steadily increased from 19% to 31% at the Wylfa installation in 1970. The CO_2 coolant enters the reactor at 247°C and leaves at 402°C at a pressure of 2.8 MPa.

A new generation of British reactors called "AGR" or "Advanced Gas Cooled Reactor" was next developed which have an upper temperature of 675°C at a pressure of 3 MPa. This is achieved with the use of slightly enriched uranium fuel elements. The efficiency of these reactors is as high as 42%, but the U.K. ended construction of AGRs in favour of PWRs.

In the United States the "HTGR" or "High Temperature Gas Cooled Reactor" has been under development but none has been built for power production. This design uses elaborately clad fuel elements in an ultra pure graphite moderator. The cooling is by means of helium gas which is raised to a temperature of 770°C. The efficiency approaches 40%. The safety of the HTGR is enhanced by its materials and construction. There is no metal in the core; all structural elements are graphite which increase in strength as the temperature rises. Nothing in the core will melt or vaporize at temperatures up to 3600°C.

Russian reactors of the type used at Chernobyl in the Ukraine use graphite as the moderator but pass the water directly over the fuel elements to turn it to steam for use in the turbines. This eliminates the use of heat exchangers to extract the heat from the reactor, but the apparent simplification is the source of the safety problem with these reactors, as will be discussed in Chapter 15.

12.15 Breeder Reactors

In a conventional reactor some of the ^{238}U is converted to ^{239}Pu which is fissionable. After a year or so of operation of a CANDU reactor more of the heat produced actually comes from the ^{239}Pu than the ^{235}U, since the latter is depleted whereas the neutron flux continues to build up the former. This suggests the idea of building reactors which produce more fuel than they consume. There are two so-called "fertile" materials that can be used: ^{238}U and ^{232}Th. The reactions are

$$^{238}_{92}U + n \rightarrow \; ^{239}_{92}U \rightarrow \; ^{239}_{93}Np + \beta^- \rightarrow \; ^{239}_{94}Pu + \beta^- \qquad (12\text{-}14)$$
$$T_{1/2} = 24 \text{ min} \qquad T_{1/2} = 23 \text{ d}$$

$$^{232}_{90}Th + n \rightarrow \; ^{233}_{90}Th \rightarrow \; ^{233}_{91}Pa + \beta^- \rightarrow \; ^{233}_{92}U + \beta^- \qquad (12\text{-}15)$$
$$T_{1/2} = 22 \text{ min} \qquad T_{1/2} = 27 \text{ d}$$

Breeding can be achieved if 1 fission event releases, on average, 1 neutron to maintain the chain reaction, about 0.2 neutrons to account for escape and absorption losses, and 1 more for breeding, for a total of 2.2. Figure 12-9 shows a graph of the number (n) of neutrons per fission as a function of the neutron energy.

In the low energy region (thermal neutrons) the curves are mostly at, or below, the critical value of 2.2. This means that breeding with thermal neutrons is inefficient. However, at high energies the curves, for all reasonably fertile materials, rise rapidly above 2.2 with increasing energy and breeding becomes efficient. The obvious choice as a fertile material is ^{238}U, the large portion of natural uranium that does not contribute substantially to the operation of thermal reactors. Unfortunately breeding cannot be done in a moderated reactor since the moderator quickly thermalizes the neutrons. A special "fast" reactor which operates on fast neutrons must be constructed to do this.

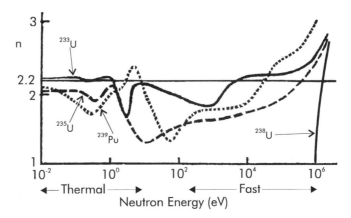

Figure 12-9:   Number of Neutrons per Fission.

Such a reactor is the French Phénix fast breeder reactor which uses a ^{239}Pu core operating at a very high power density to produce fast neutrons which breed more ^{239}Pu via Eqn. 12-14. Of course water cannot be used as a coolant in these reactors since water is effective in thermalizing neutrons and that would defeat its breeding with fast neutrons. The coolant in the Phénix is liquid sodium metal which has given rise to great controversy as to the safety of these reactors. Sodium burns spontaneously on contact with water and although the technology has been used for decades on a smaller scale in the reactors aboard ships and submarines, land-based reactors of the scale of Phénix and Super-Phénix are quite another thing. The operation of these reactors is even much more controversial than thermal power reactors because of their reliance on plutonium, a highly toxic substance and a nuclear weapons material. Breeder programs have a low priority in most countries except France where the program is proceeding. One of the important uses of breeder reactors could be to "burn" up the plutonium in nuclear weapons in the present era of nuclear disarmament; this is one of the only sure ways to accomplish this.

There are a few other breeder reactors in operation around the world as well: a low power unit in Kazakhstan was once part of the U.S.S.R. nuclear effort. One larger unit is also in operation in Russia with three others under construction. Great Britain also operates one low-power unit.

Breeding of thorium, represented by Eqn. 12-15, has only been investigated in a limited way, theoretically and experimentally, mostly in Canada; the program is in abeyance.

12.16 Reactor Control and Stability

As discussed in Chapter 11, each fission event produces, on average, more than one neutron. This increase of neutrons causes the power of the reactor to rise very rapidly as each generation of neutrons produces even more in the next generation. This rise will continue until the rate of production of neutrons is equal to the rate of loss by absorption in non-fission producing ways or by leakage of the neutrons from the core of the reactor. A reactor is controlled by adjusting the absorption of neutrons through control-rods of neutron-absorbing materials which are inserted into, or withdrawn from the core.

It may not be obvious that this will work, however, since most of the neutrons are produced within a time of 10^{-14} seconds (the so-called "prompt neutrons"). If changes in the neutron population can take place this quickly then how is it possible — using slow-moving mechanical devices — to control them? Fortunately only 99.3% of the neutrons are prompt; 0.7% are delayed, on average, for 14 seconds, which gives an average neutron lifetime of 0.1 s $[(99.3 \times 10^{-14} + 0.7 \times 14)/100 = 0.1]$. It is via these "delayed" neutrons that control is exercised. The *reactor constant* is defined as

$$k = \frac{\text{Number of prompt neutrons in the (n + 1)st generation}}{\text{Number of prompt neutrons in the nth generation}} \quad \text{(12-16)}$$

For a reactor that is just critical on prompt neutrons alone, k = 1. It is not difficult to show that the *response time* of a reactor — that is, the time to increase its rate of reactions by a factor e — is given by

$$T = \tau/(k - 1) \quad \text{(12-17)}$$

where τ is the neutron lifetime (see Problem 12-13). Clearly if $\tau = 10^{-14}$ s then the reactor will be uncontrollable; a small random increase will run away so fast that no control system could cope with it. However if the reactor is adjusted to be sub-critical with the prompt neutrons, then control can be exercised through the delayed neutrons. If the fraction of delayed neutrons is β then the condition for safe operation is

$$(1 - \beta)k = 1 \qquad\qquad (12\text{-}18)$$

For ^{235}U, $\beta = .007$ and, from Eqn. 12-18, k = 1.007. Thus if control rods are inserted into the reactor such that k = 1.007 the reaction time will be about 0.1 s; if the value of k is adjusted below 1.007 the response time can be made even longer so that mechanical control can be exercised very precisely. The value of k is usually set so that the response time is about 100s.

A very important aspect of the different nuclear reactors is their stability in power output when fluctuations occur in certain parameters. For example, what happens in a reactor if there should be an upward fluctuation in temperature? (Will the reactor naturally damp the fluctuation out, or will it require outside intervention to control it?) If it is the latter, then it is important that the fluctuations be slow enough that the mechanical devices (control-rods etc.) can operate in time. Compare for a moment the light water reactors and the water-cooled graphite-moderated reactors, as these will be important for an understanding of the Three Mile Island and Chernobyl accidents discussed in Chapter 15.

PWR and BWR reactors are inherently stable against temperature fluctuations. If the temperature should rise suddenly the water decreases in density or even boils. This reduces the moderating effect and with less thermalized neutrons the fission rate decreases. The reactor is said to have a *negative void coefficient*. This stability is so absolute that research reactors have been built in which the control rod is driven out very rapidly by compressed air so that the reaction rate will rise rapidly, however the reactor just as quickly shuts itself down. The result is to produce, in a perfectly safe manner, short but intense pulses of neutrons for research purposes.

A Chernobyl-type reactor is moderated by graphite but cooled by water flowing directly over the fuel rods. There is not enough water in the reactor at any one time to contribute significantly to the moderation but it does act as a significant *neutron poison*, removing neutrons by the reaction of Eqn. 12-12. If the temperature should rise and the water boil this poison is reduced and the neutron flux will increase, thus increasing the fission rate. Such a reactor is inherently unstable to temperature fluctuations and has a *positive void coefficient*. So long as the time constant of the reactor is maintained sufficiently long by keeping it critical on the delayed neutrons, then the reactor can be operated in perfect safety. If however anything occurs to make the reactor *prompt critical* then a fluctuation will grow and outrun any mechanical control.

12.17 Safety Systems

The safety of power reactors will be discussed specifically in Chapter 16 where the Three Mile Island and Chernobyl incidents will be discussed in detail. Safety systems in reactors are not there to prevent nuclear explosions; that type of accident is an impossibility. The various types of failure that may occur can have only one final result of public importance and that is the release of radioactive material. Safety systems are designed to eliminate or minimize this.

Reactors are designed with safety systems to cope with every reasonable type of failure. The most serious accident which can happen to any power reactor is to have a *loss of coolant accident* (LOCA). In the event of a loss of coolant all reactors have *emergency core cooling systems* (ECCS) for backup. For light water reactors, with their negative void coefficient, there is no danger of the reactor running out of control; the necessity is to provide cooling of the core to remove the heat of radioactivity of the fission fragments, otherwise the build-up of heat in the core can melt or warp some of the fuel elements and structural members. In a reactor with a positive void coefficient the reaction rate must be controlled while the ECCS is brought into action. This presents no problem for a reactor whose time constant is on the order of hundreds of seconds; this situation was intentionally circumvented at Chernobyl.

12.18 Uranium Resources

Table 12-4: World Resources of Uranium (tonnes)*

Country	1990 Production		Reserve[1]	Requirements
Australia	3520	8.5%	480,000	0
Canada	8706	20.9%	132,000	1402
France	2816	6.8%	42,000	8179
Germany	2972	7.1%	1,000	3333
Japan	0	0	0	5761
South Africa	2481	6.0%	317,000	209
UK	0	0	0	2030
USA	3387	8.1%	107,000	15778
Former USSR	5000	12%	465,000	5417
Rest	13021	30.6%	697,000	9453

* Adapted from data supplied by the Canadian Nuclear Association.
[1] Low Cost Reserves-Production Cost < US$80 per kg.

Uranium is actually a rather common element in the Earth's crust, being 30% more plentiful than tin. It is present in all rocks and soils as well as rivers and oceans. As a result, there are even traces in animal tissue. Granite, which accounts for about 60% of the Earth's crust, contains about 4 parts per million (ppm) and some rocks have as much as 400 ppm; this is particularly true of some phosphates which are used for fertilizers. Coal contains about 3 ppm and this is a major concern considering the enormous volume of coal which we mine and burn, putting many contaminants into the atmosphere (including uranium) in the process.

Natural uranium exists as the oxide in ores containing other elements such as nickel, cobalt and vanadium. Any ore with a concentration greater than 1000 ppm is considered economic to mine for its uranium. Uranium is mined by both shaft and open-pit techniques. It has all the difficulties, hazards and environmental costs of all hard-rock mining. One added hazard is a higher concentration of radon gas, particularly in underground shaft mining. As a result, uranium mines are copiously ventilated and workers use respirators in some cases. The total amount of radon ventilated from uranium mines is negligible compared to the natural outgassing of the continental rocks into the atmosphere. A uranium mine produces two kinds of waste: rock and water. The rock is deposited in controlled areas and contains little uranium and is no more radioactive than other types of rock. It does exhibit the hazards common to all mine tailings, for example, the susceptibility to chemical leaching which must be controlled. The water used to wash the ore does have traces of uranium and radium. The radium is valuable and so the water is treated with barium chloride to remove it and the treated water is mixed with the discarded part of the ore called "*mill tailings*".

The known and estimated reserves of uranium in the major source countries are given in Table 12-4.

By far the largest producer of uranium is Canada, but the largest requirement is that of the USA. The United Kingdom and Japan, while generating substantial amounts of electricity by nuclear technology, must import all of their fuel. France and the USA also import a substantial fraction of their requirements.

12.19 Future Nuclear Power Production

At the start of 1992 there were 414 reactors in operation or under construction around the world, providing some 400 GWe of power. Over 50% of these reactors are PWRs and some 20% are BWRs; the rate of new projects is very small. In part this reflects the concern caused by the Chernobyl accident, a concern that ignores the very great differences in design and safety features between the majority of reactor types built by European and North American suppliers and the small number of graphite-moderated types still used in Russia and some parts of eastern Europe.

The Brundtland Report recognized the potential of nuclear power as an energy source that would not contribute to the greenhouse effect, but it stressed the need for strict international standards of design and operation, together with enforceable control mechanisms, to be applied with no exception.

The best route to maximum safety lies in development of standardized reactors of modest size, employing simple design and passive safety features. (A proliferation of different designs and, as a result, different operating protocols, may be in part the reason for the disappointing "down-time" (50-60%) of U.S. reactors compared to Swiss units (85%) or Finnish units (90%).) Simplified designs enhance reliability, for example, by replacing massive bundles of copper cables by compact fibre-optic systems, and by physically separating safety systems from all other components. This minimizes maintenance, since there are fewer parts to service. Passive safety features rely on nature, e.g. gravity and natural convection, rather than on technology, such as pumps which demand power and are subject to malfunction.

Various designs are emerging for a new generation of such reactors. They have even more massive containment than present designs, to ensure retention of radioactivity in the event of an accident. A much increased ratio of reactor to core volume ensures a greater rate of cooling. Water reservoirs sited above the core provide emergency cooling without the need for power. Core heat can be removed by using it to evaporate water stored for this purpose in an elevated pool. The intent of these designs is that following an accident, several days can pass without human intervention, the passive safety systems ensuring both cooling and containment.

One problem remains: A global capacity of 400 GWe will use up the world's low-cost uranium reserves in 25 years. (This will hardly make a dent in the generation of global warming by fossil fuels.) Only with breeder reactors is there any potential for an impact on this problem. Breeders are not restricted to the 0.7% of ^{235}U, but can utilize a large fraction of the 99.3% of ^{238}U; they can therefore exploit the much greater reserves of low-grade uranium ore. However, breeders are much more complex than conventional reactors, and research on them has been scaled back sharply in recent years.

Problems

12-1 Using the nuclear masses given below, calculate the binding energy per nucleon for the nuclei of $^4_2$He, $^{12}_6$C, and $^{235}_{92}$U as given in Table 12-1.

Nucleus	Mass (u)
$^4_2$He	4.002603
$^{12}_6$C	12
$^{235}_{92}$U	235.04394

12-2 Consider the fission of ^{239}Pu according to the equation

$$^{239}_{94}\text{Pu} \rightarrow {}^{100}_{40}\text{Zr} + {}^{139}_{54}\text{Xe}$$

Calculate the fission energy released in this reaction. Use Table 12-1.
  [Ans. 2×10^2 MeV]

12-3 If the nucleus $^{239}_{94}$Pu absorbs a neutron what is the excess binding energy? Use Table 12-1. [Ans. 6.6 MeV]

12-4 A common neutron detector is made from a tube filled with boron trifluoride (BF$_3$). The isotope $^{10}_5$B, which constitutes 20% of natural boron, reacts with a neutron according to

$$n + {}^{10}_5\text{B} \rightarrow {}^7_3\text{Li} + {}^4_2\text{He}$$

What is the energy released in this reaction? [Ans. 2.8 MeV]

12-5 When a ^{235}U nucleus fissions, about 200 MeV of energy is released. What fraction of the mass of the nucleus is converted into energy? [Ans. 0.09%]

12-6 The heat of combustion of carbon is 94 Cal/mole. What mass is converted to energy when one mole of carbon is burned? Express the mass loss as a fraction of the initial mass. [Ans. 4×10^{-8}%]

12-7 The mass of the proton is 1.6726×10^{-27} kg and that of the neutron is 1.6750×10^{-27} kg. The mass of the $^{206}_{82}$Pb nucleus is 341.92×10^{-27} kg.

What is the binding energy per nucleon of the $^{206}_{82}Pb$ nucleus? [Ans. 7.99 MeV]

12-8 For a reactor that is critical with delayed neutrons find the value of k that gives a reactor response time of 100 s. [Ans. 1.001]

12-9 What mass of <u>natural</u> uranium is required to operate a nuclear reactor for one day based on the fission of ^{235}U? The reactor produces 1.00×10^3 MW of electrical power at 32% efficiency. Assume that the energy available from one fission event is 2.0×10^2 MeV, and that all of the ^{235}U undergoes fission. [Ans. 4.6×10^2 kg]

12-10 In the reactor of problem 12-9, how many radioactive fission product nuclei are created in 1 day? [Ans. 1.7×10^{25}]

12-11 A reactor produces 3500 MW of heat.
a) How many fissions are occurring each second? [Ans. 1.09×10^{20}]
b) Actually 16 of every 100 neutrons captured by ^{235}U does not produce a fission event but nevertheless uses up a fissionable nucleus. How many atoms of ^{235}U are consumed each second. [Ans. 1.3×10^{20}]
c) For every 10 atoms of ^{235}U consumed 3 atoms of ^{239}Pu are bred. How much plutonium is produced in one day? [Ans. 1.33 kg.]

12-12 A neutron (mass=1) collides head-on, at speed v, with a nucleus of mass A at rest. The neutron rebounds with speed v'. If the collision is elastic (energy is conserved) show that $v/v' = (A-1)/(A+1)$, thus verifying Equation 12-11.

12-13 If the reactor constant k is slightly greater than 1 then the neutron population in the reactor grows in n generations from N_0 to N according to

$$N = N_0 k^n$$

If T is the elapsed time and τ is the neutron lifetime (i.e., $n = T/\tau$) show that the value of T to increase the number of neutrons by a factor e is approximately given by Equation 12-17. (The series expansion for the natural logarithm will be needed: $\ln x = (x-1)-\frac{1}{2}(x-1)^2+\frac{1}{3}(x-1)^3...$)

⑬ | *ENERGY FROM THE NUCLEUS - FUSION*

In Section 12.2 it was mentioned that there was a second method by which energy can be extracted from some nuclei. If very light nuclei are combined the binding energy of the resulting nucleus is increased and energy is released; this is called "*nuclear fusion*". The process has not yet been achieved in a controlled way (as at this writing), but much research effort and money has been expended on it in the hope of developing controlled fusion reactors. These may provide a clean, almost boundless source of energy for the future, but that future is surely still quite distant.

In this chapter we will explore the physics and technology of nuclear fusion.

13.1 The Physics of Fusion

In Fig. 12-1 it can be seen that if two very light nuclei, such as two protons, are combined into a heavier nucleus, the binding energy per nucleon is increased and energy is released. For reasons which will be evident later, the reaction which has been most extensively studied is that of deuterium ($^2_1$H) with tritium ($^3_1$H).

$$^2_1\text{H} + ^3_1\text{H} \rightarrow ^4_2\text{He} + ^1_0\text{n} \tag{13-1}$$

Using Table 12-1 the energy released is
$$4 \times 7.074 - (3 \times 2.827 + 2 \times 1.112) = 17.6 \text{ MeV}.$$

How do we know that this is even possible? This and many other fusion reactions have been observed in nuclear laboratories using particle accelerators and the reaction of Eqn. 13-1 has been made to work on Earth albeit in a destructive way, that is, in the hydrogen bomb. In addition there is very strong evidence that the energy produced in the Sun and all stars results from the fusion of hydrogen into helium.

The technical difficulty arises from the fact that the attractive nuclear forces which bind the two hydrogen nuclei together into the helium nucleus — releasing energy in the process — have a very short range and so the particles must be brought very close together. Opposing this is the electrical repulsion caused by the positive charge on each of the hydrogen nuclei. Very large forces must be exerted on the two nuclei in order to push them close enough together so that the very much stronger nuclear force can take over and bind them together.

One of the methods chosen to accomplish this is to raise the temperature of a 50-50 mixture of tritium and deuterium to very high levels. As the temperature (T) is raised the mean speed (v) of the particles rises according to the equation

$$v = (3kT/m)^{1/2} \qquad \qquad (13\text{-}2)$$

where m is the mass of the particle and k is Boltzmann's constant. At sufficiently high temperatures enough of the particles will make sufficiently strong head-on collisions to overcome the electric repulsion and fuse. The temperatures required to accomplish this are very high; it requires tens and even hundreds of millions of degrees to produce a significant number of fusing collisions.

This requirement of high temperatures causes other severe difficulties. A relatively high density of material is required in order to get a significant number of fusions but gases at high temperature exert high pressures if the densities are high. This can be seen from the Ideal Gas Law

$$pV = NkT \qquad \qquad (13\text{-}3)$$

Very high pressures would require a very strong container for the plasma, and would contribute so substantially to the heat loss that it could not be maintained. In fact, it is not possible to use a material container at all!

Since it is not possible to work at very high densities, then some arrangement must be made to keep the particles confined in a restricted volume for a long time so that there is ample opportunity for the particles to make many collisions and thus increase the chance of having one of the rare head-on fusing ones. But the high temperature has made this difficult as well: any increase in the particles' speed will also increase their tendency to escape the reaction area. The problem reduces itself to one of finding a way to have a sufficiently high temperature and a sufficiently high product of the density (n in particles per m^3) and confinement-time (τ in s). This condition is known as the "*Lawson Criterion*"; it states

$$n\tau > 10^{20} \text{ s} \cdot \text{m}^{-3} \tag{13-4}$$

The technical solution will be to achieve high temperatures and long confinement times at as high a density as possible. So far this has been beyond the capabilities of modern technology.

13.2 Fusion Technology

At the temperatures where fusion will take place all materials are vaporized and completely ionized, that is, all of the orbiting atomic electrons have been stripped away from the nuclei. In ordinary matter the material is electrically neutral at the atomic level. Now, although the material is still electrically neutral in bulk, the individual particles are not. Matter in this form is called a "*plasma*".

The technology that has been most pursued is that of *magnetic confinement*. It takes advantage of the fact that charged particles in the presence of a magnetic field are trapped on the magnetic field lines and so can be contained without actually using a physical container. The most advanced device of this type is called a "*tokamak*"[1]. The elements of a tokamak are shown in Figure 13-1.The plasma is free to move in a highly evacuated "doughnut"-shaped toroidal chamber. The toroidal and vertical field coils carry electric currents which produce a magnetic field, keeping the plasma confined along the axis of the toroid and prevents it from touching the walls and losing its heat. The plasma, being conducting, is itself a single loop of electrical conductor and inserted through the centre of the toroid is another magnet acting as a transformer (See Fig. 13-1); as current rises in the transformer primary coil it induces a rising magnetic field in the magnet which in turn induces a current in the plasma, heating it up by "ohmic heating".

---

[1] "tokamak" is a Russian acronym for "a toroidal chamber in a magnetic field". Such devices were first built in Russia.

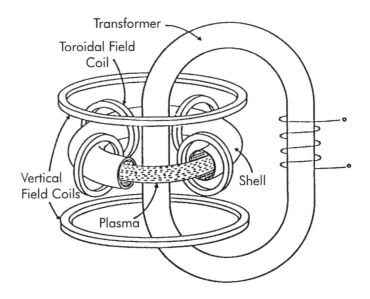

Figure 13-1: The Elements of a tokamak.

Many tokamaks have been built in ever-increasing size and complexity. None of these devices has been built to actually produce useful power. They have only had the objective of *"breakeven"*, that is, to produce as much heat energy from fusion as was actually put into the device to run it. In spite of much expenditure and effort this has not yet been achieved. The most recent devices are very large as can be seen from Figure 13-2 of the "Joint European Torus, (JET)", at Harwell, England. The best performance of some of these devices is shown in Figure 13-3 where the plasma temperature is plotted against the Lawson number. The region of "breakeven" is indicated; in 1991 JET was very near breakeven. Breakeven alone does not make a fusion power reactor — for that it is necessary to achieve *"ignition"* where the reaction is self-sustaining. It is clear that we are a long way from a working fusion reactor.

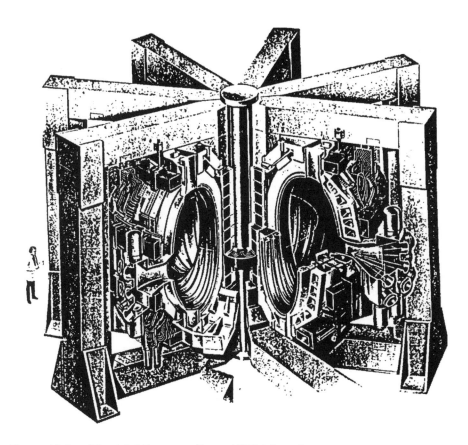

Figure 13-2:   The Joint European Torus (JET) tokamak.

Another completely different method of achieving fusion has been proposed and received considerable funding and attention in the 1970s and '80s: *inertial confinement*. The idea here is to irradiate a small spherical pellet of solid D-T mixture from all sides with a very powerful pulse of light from a laser. The explosion of material from the surface of the pellet creates a shock wave that compresses and heats the remaining core of the pellet. The compression must take place in less than 10^{-9} s and so the Lawson criterion says that the density must be increased by a factor of about 10^4 to 10^5. Enormous lasers have been constructed to investigate this possibility but such compressions are still a long way off. Research in this area has not been well funded in the 1990s and there is, at present, an air of disenchantment, within the research community, with inertial confinement.

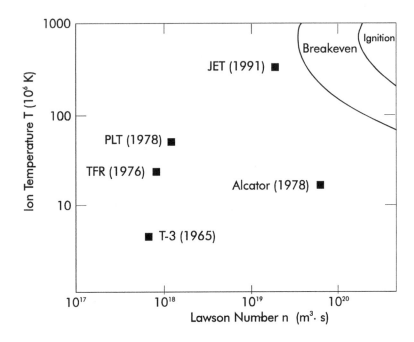

Figure 13-3: Performance of Various tokamaks.

Even if all of the foregoing were possible by whatever method of confinement there is still a formidable problem in extracting the energy from the device. When the reaction given in Eqn. 13-1 takes place, most of the energy is taken by the neutron. When the D and T nuclei collide head on they do so with almost zero total momentum. Since momentum is conserved, the fusion products ^{3}He and $_0^1$n also have zero total momentum. This means they recoil from each other but with unequal speeds because of their unequal masses. A simple calculation (see problem 13-8) shows that the velocity of the neutron is four times that of the helium nucleus and it carries 80% of the reaction energy. It is hoped that the 20% of the energy remaining with the ^{3}He will sustain the heat of the plasma and the neutron's energy can be extracted. Since the neutron has no charge it is not confined by the magnetic field and passes out of the plasma. As we have seen earlier, neutrons are difficult to capture but if this is effected then the neutron's energy will be transferred to the capturing nucleus and appear as heat in the bulk material. How this is to be effected bears on the subject of "fusion fuel" and is discussed in the next section.

13.3 Fusion Fuel

It is clear that if the present research on fusion ever bears fruit the fuel will be a mixture of deuterium (D) and tritium (T). One of the popular misconceptions about fusion on Earth is that when its proponents talk about "hydrogen fusion" they mean the fusion of ordinary hydrogen as takes place on the Sun. This is definitely not the case, and some of the proponents have been disingenuous in not making this clear in their publicity. Of course ordinary hydrogen is almost limitless in supply in ordinary water but the reaction that takes place on the Sun is, in fact, quite different in that it depends on a different aspect of the nuclear force. The solar reaction will probably never be within the grasp of technology for very fundamental reasons: only the Sun (and other stars) can do what the Sun does.

The proposed fusion fuels D and T are isotopes of hydrogen but are much rarer; in fact only one of them, D, occurs naturally in any appreciable quantity.

Deuterium is the stable isotope of hydrogen and occurs in about one part in 7000 in normal hydrogen and so about one in 3500 water molecules on Earth is HDO (see Problem 13-2). Only one in 50 million water molecules is actually D_2O. Deuterium is extracted from ordinary water by a series of hydrogen exchange reactions to build up the concentration of D_2O to about 6%, and then to further enrich it to 99% or higher by fractional distillation. The latter is possible, since the boiling points of ordinary and heavy water are slightly different. All of these processes are very energy intensive and often use the waste heat from a power plant to carry them out. The price of heavy water is rather high because of the large amount of energy required and the complexity of the separation plants. Among the nations of the world Canada is the most advanced in the technology of producing heavy water because D_2O is required as the moderator and coolant in the CANDU power reactors.

Tritium does not occur in nature except for a very small amount produced by the action of cosmic rays on water in the atmosphere. It is, in any case, radioactive with a half life of only 12.3 yr. so it must be manufactured continuously for use as a fusion fuel. For experimental and start-up purposes Canada could again be a major source as tritium is produced in the CANDU nuclear reactors (see Section 12.13) and a tritium separation facility has been built in conjunction with the Darlington Nuclear Power Station in Ontario.

In a fusion power reactor it is proposed to solve the problem of energy capture and tritium fuel production in one step. It is proposed to blanket the reactor with thick layers of some lithium compound. About 7% of lithium is ^6Li and this nucleus will absorb a neutron according to the equation

$$^6_3Li + {}^1_0n \rightarrow {}^3_1H + {}^4_2He \tag{13-5}$$

This reaction also produces an extra 4.8 MeV of energy. The net result is to heat up the lithium and to produce one of the necessary fuels (^3_1H) for further operation. Both the heat and the ^3_1H must be extracted from the lithium blanket by continuous processing. The 93% of ^7Li can also assist via the reaction

$$^7_3Li + {}^1_0n \rightarrow {}^3_1H + {}^4_2He + {}^1_0n \tag{13-6}$$

In this case, however, the reaction extracts energy from the neutron but with 80% of the reaction energy the neutron has more than enough. Needless to say, none of this technology has been carried out on a large scale.

It is often stated that fusion energy production will be the clean safe way to produce limitless supplies of energy. "Clean" is interpreted to mean 'without the radioactive waste of fission reactors'. While it is true that the only waste of the fusion reaction is the harmless inert gas helium, the process is not without its radioactive hazard. The tritium budget of a fusion reactor will be very large and an accidental release could be quite serious. Most fission reactor wastes are not gases but tritium is, and so it is easily dispersed. In the form of radioactive water it would be very damaging in the biosphere. In addition the intense neutron fluxes of fusion reactors will render many structural members of the reactor radioactive, and it is a challenge to technologists to find ways to minimize or control this.

13.4 The Future for Fusion

At the end of the twentieth century the future utilization of fusion looks further away than ever. The proponents of the technology have had a long history of presenting optimistic predictions of the imminence of a working system - this has always met with disappointment. Every step in research or development has been progressively more costly. Many governments have ended financial support, owing to the discouraging results so far. Those nations still interested have come to realize that only international co-operation can supply the money needed to continue and that success is probably further away than has been realized. The scale of research has been reduced everywhere particularly as the energy crises of the 1970s and '80s have receded. Useful, controlled fusion energy may yet become available,but in the twenty-first century at the earliest.

13.5 Cold Fusion

The sheer weight, size and complexity of fusion technology is very daunting; it is not now clear whether it will ever be economically feasible. Imagine the surprise, shock and even incredulity on the part of physicists and engineers when, on Mar. 23 1989, two chemists announced that they had carried out nuclear fusion in a test tube at room temperature with a large release of energy! The form of the announcement was a press conference, a form that angered and alienated almost all of the scientific community right at the beginning. The accepted way of announcing scientific results is in the scientific literature; results are submitted for scrutiny by scientific peers before public announcements are made, especially truly important ones.

The claim was made that by passing an electric current through heavy water (which contains deuterium) between electrodes made of palladium, more heat was evolved than could be accounted for from the electric energy input. The claim was also made that only the fusion of deuterium nuclei within the palladium lattice could account for this energy. There were additional claims that some of the signatures of fusion were also present, such as the emission of nuclear particles and radiation. Many scientists were sceptical but at the same time there was a rush to the laboratory on the part of electrochemists and nuclear physicists everywhere; the promise of cheap boundless energy from apparatus which cost almost nothing was too good to pass up.

In the first few weeks there were many experiments that failed to confirm the claim but a few that provided partial confirmation. As time went on however the sceptics were proven correct as one after another of the positive results was explained away. There may indeed be something peculiar about electrochemistry with palladium and heavy water, but there seems no substantial evidence that nuclear fusion is involved. The exact nature of the initial experiments has never been fully disclosed and the motives of the proponents of cold fusion in choosing the method of press releases over peer review has never been explained. Though a few die-hard believers still claim that the effect is real (although much weaker than first proposed), the vast majority of scientists have dismissed the claims and relegated the phenomenon to one of the more bizarre episodes in science.

Exercises

13-1. What is the energy released in each of the following reactions? (Use Table 12-1)

a) $^2_1H + {}^2_1H \rightarrow {}^3_1H + {}^1_1H$ [Ans. 4.03 MeV]

b) $^2_1H + {}^2_1H \rightarrow {}^3_2He + {}^1_0n$ [Ans. 3.27 MeV]

c) $^2_1H + {}^3_2He \rightarrow {}^4_2He + {}^1_1H$ [Ans. 18.4 MeV]

Problems

13-1 Of the two reactions involved in making tritium by neutron capture on lithium (section 13.4) one produces energy and the other takes energy from the neutron. Calculate the energy for each case (use Table 12-1). [Ans. 4.79 MeV; -2.5 MeV]

13-2 Deuterium (D) occurs as 1 part in 7000 in normal hydrogen (H). Show that for every molecule of D_2O in normal water there are 14×10^3 molecules of HDO and 49×10^6 molecules of H_2O.

13-3 How many molecules of D_2O and HDO are in 1 m^3 of ordinary water? [Ans. 6.8×10^{20}; 9.6×10^{24}]

13-4 In one cubic metre of water, what is the mass of a) D_2O; b) HDO; and c) D ? [Ans. 2.3×10^{-5} kg; 0.30 kg; 0.032 kg]

13-5 If all of the D in 1 m^3 of normal water was fused with tritium how much energy would be released? [Ans. 2.7×10^{13} J]

13-6 If all the D in 1 m^3 of normal water was fused with tritium how long would it run a 1000 MWe power plant operating at 33% efficiency? [Ans. 2.5 hr.]

13-7 Considering the power plant in problem 13-6, what would be the activity of the tritium required to run it for one year? The half life of tritium is 12.3 yr. [Ans. 6.1×10^{19} Bq]

13-8 A tritium and a deuterium nucleus, with a total momentum of zero, fuse to produce a helium nucleus (mass = 4u) and a neutron (mass = 1u). They recoil from each other with speeds v_{He} and v_n. Show that

$$KE_n/KE_{Total} = 0.8$$

13-9 a) What mass of deuterium would be required daily to power a reactor based on D-D fusion (See Exercise 13-1(b))? The reactor produces 1.00×10^3 MW of electrical power at 50% efficiency. Assume that the energy released in one fusion event is 4.0 MeV. [Ans. 1.8 kg]

b) In this reactor, what mass of tritium is created daily? What is the activity of this freshly- produced tritium? The half life of tritium is 12.3 years. [Ans. 1.3 kg; 4.8×10^{17} Bq]

13-10 The oceans contain about 1.3×10^{24} cm^3 of water. Deuterium constitutes 0.028% by mass of natural hydrogen. What is the total energy (in joules) available from this deuterium via D-D fusion? Assume 4.0 MeV per fusion event. For how many years could fusion reactors of 50% efficiency supply a power requirement of 2.0 million MW? (This requirement is expected to be reached early in the next century.) [Ans. 3.9×10^{30} J; 3×10^{10} yr!!!]

THE INTERACTION OF RADIATION WITH MATTER AND ITS BIOLOGICAL EFFECTS

14.1 Absorption of Radiation

When alpha, beta or gamma radiation enters matter it interacts with the atoms and molecules of the matter via numerous processes. In these interactions the energy of the radiation is transferred to the atoms and molecules which may become altered in important ways, such as the ionization of atoms or the production of free radicals from molecules.

The mechanisms of energy loss are important to understand from the point of view of shielding from radiation and the effect on the absorbing matter has important consequences for radiation damage. These subjects are the focus of the present chapter.

Radioactive materials are always stored behind sufficient shielding material (or at least they should be) to reduce the level of radiation outside the container to a tolerable level. Shielding is important in the manufacture and use of radioactivity and, of course, proper shielding of accumulated fission product wastes from nuclear reactors is crucial.

In the previous chapter we saw that α and β particles can be stopped by a cm or so of plastic. This makes shielding an almost trivial matter in radioisotope or research laboratories, or factories where source activities of 10^4 Bq (μCi) up to 10^{18} Bq (100 MCi) are normally encountered. When radioactivities reach the order of 10^{19} Bq (1000 MCi) or greater, two problems arise. First the material is weakened by the constant damage to its structure, and second the kinetic energy deposited by the particles as they slow down in the shielding ends up as heat and the shielding may warp or even melt. As shown later, some fission wastes have to be stored in metal tanks which are constantly cooled.

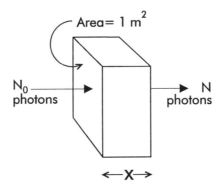

Figure 14-1: Absorption of Gamma Rays.

γ rays present a totally different situation. Being E-M radiation, these are much more penetrating than α or β particles. By definition, a photon travels with the speed of light or it ceases to exist. Consider a beam of high energy photons striking a steel wall as shown in Fig. 14-1. There are several processes — which are beyond our scope to examine — which remove photons from the beam, so that as the beam proceeds through the material the number of photons in it steadily decreases. Actually the number decreases exponentially with distance so that if N_0 photons strike a unit area of the wall per second, then the number surviving a distance x downstream is

$$N = N_0 \, e^{-\mu x} \tag{14-1}$$

where μ is a constant called the *"linear attenuation coefficient"* and is a characteristic of the material and the photon energy.

This is analogous to the radioactive decay law. Here a number of photons are attenuated by thickness; there unstable nuclei were attenuated by time. Define a thickness $L_{1/2}$ that removes one half of the photon beam;

this *half thickness* is given by

$$L_{1/2} = 0.693/\mu \qquad\qquad (14\text{-}2)$$

Material of high atomic number Z has a larger μ and smaller $L_{1/2}$ than material of low Z. It is for this reason that lead is an effective and convenient shield against γ rays.

In the case of stored radioactive liquids, lead is too soft to be structurally sound and stainless steel tanks are preferred. Such tanks may be surrounded by lead to reduce the intensity of γ rays that escape the steel.

It is important to realize that heavy lead shielding in radiation structures and equipment is there not for the α and β particles but for the γ rays. If the gammas are adequately shielded the alphas and betas will automatically be taken care of.

Example 14-1 *What thickness of lead is needed to reduce a flux of 1.5 MeV γ rays by a factor of 100? (For 1.5 MeV γ rays, $\mu = 55$ m^{-1} in lead).*

Since $N/N_0 = 0.01$, we have

$0.01 = e^{-\mu x}$, or

$-\mu x = ln(0.01) = -4.6$

therefore, $x = 4.6/\mu$

For lead, $x = 8.4 \times 10^{-2}$ m = 8.4 cm.

In the energy range of 1 to 2 MeV the attenuation of γ rays is almost all due to interaction with the orbital electrons of the atom, and the linear attenuation coefficient is roughly proportional to the material's density ρ. If the linear attenuation coefficient μ is replaced by a product $\mu_m \rho$ then the quantity μ_m, called the "*mass attenuation coefficient*", will be nearly a constant for all materials. Equation 14-1 then becomes

$$N = N_0 e^{-\mu_m \rho x} \qquad\qquad (14\text{-}3)$$

In Table 14-1 the mass attenuation coefficient at 1.5 MeV for a number of materials is given. It is evident that it is almost constant and

$$\mu_m \approx 0.0050 \text{ m}^2 \cdot \text{kg}^{-1} \tag{14-4}$$

Table 14-1: Mass Attenuation Coefficients

Material	ρ (kg$\cdot$ m^3)	μ_m (m$^2 \cdot$ kg^{-1})
Air	1.3	0.0058
Al	2700	0.0053
Cu	8900	0.0047
NaI	3200	0.0047
Pb	11000	0.0052

Example 14-2 *Repeat Example 14-1 for steel which has a density of 7800 kg$\cdot$ m^{-3}.*

Using Eqn. 14-3,

$$N = N_0 e^{-\mu_m \rho x}$$

or 0.01 = $e^{-0.0050 \cdot 7800x}$

ln0.01 = -0.0050 $\cdot$ 7800x

x = 0.12m = 12 cm

Neutrons require a rather different analysis. Neutrons have no electric charge so they don't ionize atoms like alphas and betas. They are not E-M radiation, so our discussion of photons is not relevant. In fact, fast-moving neutrons are very hard to stop; slow-moving neutrons, on the other hand, (E ~ 1 eV) are copiously absorbed by the nuclei of certain elements (e.g., boron and cadmium) as a result of peculiarities of the nuclear structure in these cases. The task then reduces itself to slowing the neutrons down and then absorbing them with boron or cadmium. The slowing-down process is simply to let them bounce around in large layers of a light substance like water or graphite. At each bounce the neutron loses energy and slows down. This process is called "*moderation*" and since it is important in the operation of nuclear reactors it was examined quantitatively in Chapter 12.

14.2 Radiation Absorbed Dose

When fast-moving α or β particles travel through material they undergo a very large number of collisions with the orbiting electrons of the atoms but they rarely collide with the nuclei. As a result, nuclear reactions are rare in irradiated matter; the flux of the radiation must be very large, such as is found in the core of a nuclear reactor. The reason for this is, of course, purely geometric; the electrons are spread out over a volume of the dimension of an atomic diameter ($\sim 10^{-9}$ cm), whereas the nucleus is tiny by comparison, having a diameter of $\sim 10^{-13}$ cm.

The particle is slowed down by these collisions and is eventually brought to rest. The α particle is rather like a snooker ball ploughing through ping-pong balls; it moves in a straight line smashing everything out of its way. The lighter β particle tends to scatter around somewhat. In every collision with an electron the electron receives some of the kinetic energy of the incoming particle and may even receive enough to be torn from its parent atom; in other words the atom is ionized. This type of damage can be classified in two ways in living matter.

1. The ionization process may break a valence bond in a macromolecule such as DNA. The resulting rearrangement of bonds in the molecule, or subsequent chemical reactions with the disturbed site on the molecule, may then disrupt the proper functioning of the molecule and the cell.

2. Much of the material in a cell is water. Incoming particles may disrupt the water molecule leaving molecular fragments called "*free radicals*" such as H and OH; these are chemically very reactive, and attack biological molecules doing great damage. In fact, most radiation damage to living material is of this type.

The amount of biological damage is determined by the amount of energy deposited by the particle. An α is stopped in a few microns (10^{-6} m) of tissue, a β particle in a few mm. Both have fairly well defined ranges which increase with the initial particle energy. Both deposit all their energy within that range. *Dose* (D) is defined as the amount of energy deposited per unit mass. Absorbed doses are now usually expressed in *grays* (Gy) and absorbed dose rates in *grays per second*, where

$$1 \text{ gray} = 1 \text{ J/kg} \qquad \textbf{(14-5)}$$

An older unit of dose is the *rad* which is defined as

$$1 \text{ rad} = 0.01 \text{ J/kg} \qquad \textbf{(14-5a)}$$

Clearly from Eqn. 14-5 and 14-5a, 1 rad = 10^{-2} Gy.

Example 14-3 *An individual ingests 3.7×10^4 Bq (1 μCi) of a 2.0 MeV β emitter and it distributes uniformly throughout his body. The effective half-life is 28 yr. Calculate the initial absorbed dose rate in Gy/s and rad/hr and the total absorbed dose received in the first 28 years. Take the person's mass as 70 kg.*

$$
\begin{aligned}
\textit{The energy per second} \; &= \; 3.7 \times 10^4 \; \textit{decay/s} \times 2.0 \; \textit{MeV/decay} \\
&= \; 7.4 \times 10^4 \; \textit{MeV/s} \times 1.6 \times 10^{-13} \; \textit{J/MeV} \\
&= \; 1.2 \times 10^{-8} \; \textit{J/s} \\
\textit{The absorbed dose rate} \; &= \; 1.2 \times 10^{-8} \; \textit{J/s/70 kg} \\
&= \; 1.7 \times 10^{-10} \; \textit{J} \cdot \textit{kg}^{-1} \cdot \textit{s}^{-1} \\
&= \; 1.7 \times 10^{-10} \; \textit{Gy/s} \\
&= \; 1.7 \times 10^{-10} \; \textit{Gy/s} \times 3600 \; \textit{s/hr/0.01 Gy/rad} \\
&= \; 6 \times 10^{-5} \; \textit{rad/hr}
\end{aligned}
$$

At the start the number of atoms is $N = A/\lambda = 4.7 \times 10^{13}$

In 28 years one half of these emit 2-MeV betas for a total absorbed dose of

$(1/2) \times 4.7 \times 10^{13}$ decays × 2 MeV/decay × 1.6×10^{-13} J/MeV/70 kg

$$
\begin{aligned}
&= \; 0.11 \; \textit{J/kg} = 0.11 \; \textit{Gy} \\
&= \; 11 \; \textit{rads}
\end{aligned}
$$

The situation is more complex with γ rays. Since these are radiations whose speed is always that of light, they are not slowed down by successive collisions as charged particles are. Also, since they are so very penetrating, many can pass through the body without interaction. When they do interact, they lose all of their energy, or a large fraction of it, in a single interaction. A full treatment of this is complex and beyond the scope of this study; the result is that the absorbed dose rate due to a beam of n γ rays per second per m^2 of tissue is 3.3×10^{-3} nE J$\cdot$kg$^{-1}\cdot$s^{-1} (Gy/s), where E is the energy in joules of one γ ray photon.

Example 14-4:   *An individual stands 1.0 m from a 3.7×10^7 Bq (1 mCi)source of 0.50 MeV γ rays. What absorbed dose does the person receive by remaining there for one hour?*

First work out the intensity (number, n, per m^2) of the γ rays at a distance of 1 m. In a sphere of 1 m radius, clearly all the γ rays must go through the surface of that sphere and the number per unit area per sec. is given by
  $3.7 \times 10^7/4\pi r^2$ *where r = 1.0 m.*

n $= 3.7 \times 10^7/4\pi = 2.94 \times 10^6$ *per s*

E $= 0.50 \ MeV = 0.50 \times 1.6 \times 10^{-13} = 0.80 \times 10^{-13} \ J$

Dose $= 3.3 \times 10^{-3}nE = 3.3 \times 10^{-3} \times 2.94 \times 10^6 \times 0.80 \times 10^{-13}$

     $= 7.8 \times 10^{-10} \ Gy/s = 7.8 \times 10^{-8} \ rad/s$

In 1 hr the absorbed dose D is $7.8 \times 10^{-10} \times 3600 = 2.8 \times 10^{-6}$ Gy.

14.3 Effective Dose

When radiation enters living tissue the damage it produces — whether directly or as a result of making chemically active species — has various consequences for the organism. Some damage is repairable; it would be remarkable if this were not so since living organisms have had to evolve in an environment that includes natural sources of ionizing radiation. Some damage is not repairable, however, and this is the topic of the following sections.

A simple calculation of the absorbed dose in grays (or rads), as was done in the last section, does not tell the whole story insofar as biological effects are concerned. The biological effects also depend on the spatial distribution of the energy deposited. For every electron knocked out of its parent atom the incident α or β particle uses up an amount w of its kinetic energy, where w is about 3 eV. If the initial kinetic energy is E, then the number of ionizing events caused by a single α or β is

$$n = E/w \tag{14-6}$$

Thus one particle with an energy of 1 MeV produces about 30,000 ionizing events. This is true for both the α and the β particle. The difference is that the easily stopped α does all this damage in a region a few microns deep, whereas the damage caused by the β is more sparsely distributed over a depth of a few millimetres. In the α case, a cell can be damaged at several places so that it is unable to recover, but repair of the less concentrated damage caused by the β may occur.

The quantity that accounts for this difference in the relative effect of radiation has had several names in the past; among these are the "Quality Factor" (QF) and "Relative Biological Effectiveness" (RBE). The most recent name is the "*Radiation Weighting Factor*" (w_R). Although these quantities have slightly different definitions, for our purposes they are the same and simply give, as a dimensionless multiplication factor, the relative effectiveness to produce biological damage. The quantity w_R has been determined by comparing the dose needed to produce a specific effect in comparison to an arbitrary standard. This standard is taken to be the dose of X-rays and γ-rays which are among the least damaging of radiations; the w_R of this standard is taken to be 1. We expect then, that values of w_R will be numbers equal to or greater than 1. With this standard it is found that alphas require only one-twentieth the dose to produce the same damage as X-rays, therefore for alphas $w_R = 20$. The values of w_R are summarized in Table 14-2 for all the radiations of biological interest and concern.

The "*Equivalent Dose*" (H_T) is defined using w_R, and is a more accurate measure of biological damage than is the absorbed dose. The equivalent dose must be summed over all the radiations incident on the tissue ("T") of interest. Thus

$$H_T = \sum_R w_R \cdot D_R \tag{14-7}$$

Table 14-2: Values of w_R*

Radiation	w_R
Photons, all energies	1
Electrons and muons, all energies	1
Neutrons, energy < 10 keV	5
10 keV to 100 keV	10
100 keV to 2 MeV	20
2 MeV to 20 MeV	10
> 20 MeV	5
Protons, energy > 2 MeV	5
α particles, fission fragments	20

* Adapted from 1990 Recommendations of the International Commission on Radiological Protection (ICRP).

The unit of Equivalent Dose, for a dose in grays, is the "sievert" (Sv) (if the dose is in rads the Equivalent Dose is in rems).

A further complication arises because different radiations affect the different tissues of the body in different ways. For example, high-energy γ-rays pass completely through every tissue in the body and so expose them all. On the other hand α particles, when incident on the body externally, have a small effect since many of them are absorbed in the dead layers of the skin, whereas the same particles from radon gas inhaled into the lungs are much more damaging. It is not realistic to simply add up all the radiation to which the body is exposed and use that number as a measure of risk. The irradiation of the body's most sensitive tissues must be taken into account; this is done by multiplying each equivalent dose by a *tissue weighting factor*, w_T, and summing over all the tissues of the body to produce the *Effective Dose*, still measured in sieverts (or rem). The Effective Dose E is given by

$$E = \sum_T w_T \cdot H_T \tag{14-7}$$

where the summation is over all the tissues that make up the body. In Table 14-3 the values of w_T are given.

Example 14-5 *A small point-like radioactive source of 3.7×10^4 Bq is spilled on a researcher's hand. The source is ^{210}Po which is an α emitter with an energy of 5.4 MeV. The range of 5.4 MeV α particles in skin (essentially water) is 0.002 cm. What is the absorbed dose, equivalent dose, and effective dose received in 1 hour?*

For a source on the surface of the skin only one half of the radiation enters the skin; the volume affected is that of a hemisphere of volume

$$V = (1/2)(4/3)\pi r^3 = (2\pi/3)(2 \times 10^{-5})^3 = 1.7 \times 10^{-14} m^3$$

The mass affected $= 1.7 \times 10^{-14} m^3 \cdot 1000 kg \cdot m^{-3} = 1.7 \times 10^{-11} kg$

The activity $= 3.7 \times 10^4$ *Bq*

The energy per second $= 3.7 \times 10^4 s^{-1} \times 5.4$ *MeV* $\times 1.6 \times 10^{-13} J \cdot MeV^{-1}$

$$= 3.2 \times 10^{-8} J \cdot s^{-1}$$

Dose rate $= 3.2 \times 10^{-8} J \cdot s^{-1} / 1.7 \times 10^{-11} kg = 1900$ *Gy* $\cdot s^{-1}$

In 1 hr the Absorbed Dose, D $= 1900$ *Gy* $\cdot s^{-1} \times 3600$ *s* $= 6.9 \times 10^6$ *Gy*

Using Table 14-2, the Equivalent Dose, H $= w_r \times D = 20 \times 6.9 \times 10^6$
$= 1.4 \times 10^8$ *Sv*
Using Table 14-3, the Effective Dose, E $= w_t \times H = 0.01 \times 1.4 \times 10^8$ *Sv*
$= 1.4 \times 10^6$ *Sv*

14.4 Sources of Environmental Radiation

Living things on the planet are subject to a continual irradiation by high energy particles and γ rays. There are a number of natural sources, such as cosmic rays, and in the modern technological era there are a number of ubiquitous artificial sources. Under all normal circumstances, the dose delivered by natural sources is greater than from artificial sources. (It is worth emphasizing that "natural" radiation is just as damaging as "artificial" radiation, but more plentiful. This should be obvious, but the point is often missed by opponents of nuclear power.)

Table 14-3:   Tissue Weighting Factors w_T*

Tissue or Organ	w_T
Gonads	0.20
Bone Marrow, Colon, Lung, Stomach (each)	0.12
Bladder, Breast, Liver Oesophagus, Thyroid (each)	0.05
Skin, Bone Surface (each)	0.01
Remainder	0.05

* Adapted from 1990 Recommendations of the International Commission on Radiological Protection (ICRP).

Table 14-4 lists a number of sources of radiation, both natural and artificial, with their annual average dose for the average resident of the United States.

There are several points of interest in Table 14-4: the annual exposure to radiation from the natural internal sources (mostly ^{40}K) is larger than any artificial exposure except for medical X-rays. The exposure from the nuclear fuel cycle is almost unmeasurable and all artificial sources account for less than one fifth of the total. One natural source of radiation — recognized only in recent years — is radon which contributes over half of our exposure.

A source of radiation not included in the table is that encountered by passengers and aircrew in high flying aircraft. While flying at an altitude of 10,000 m a passenger in northern latitudes can receive a dose of 0.1 mSv in an 8 hr trip. This is about equal to a diagnostic chest X-ray. Air crew can receive between 6 and 9 mSv annually which is above that recommended for radiation workers. Epidemiological studies are now underway for aircrew.

Table 14-4:   Sources of Radiation and Doses*

Source of Radiation	Annual Effective Dose	
	mSv	%
Natural		
Radon	2.0	55
Cosmic Rays	0.27	8
Terrestrial	0.28	8
Internal	0.39	11
Total Natural	3.0	82
Artificial		
Medical		
X-ray diagnosis	0.39	11
Nuclear Medicine	0.14	4
Consumer Products	0.10	3
Other		
Occupational	<0.01	<0.03
Nuclear Fuel Cycle	<0.01	<0.03
Fallout	<0.01	<0.03
Miscellaneous	<0.01	<0.03
Total Artificial	0.63	18
Total	3.6	100

* Adapted from "Health Effects of Exposure to Low Levels of Ionizing Radiation", BEIR V 1990.

Well-defined rules have been set up by national and international bodies regulating the maximum radiation dose that an individual should receive. Related tables have also been compiled defining the maximum amounts of various radionuclides which can be ingested "safely"; of course no one is supposed to ingest *any*. The objective of the ICRP system of dose limitation is to promote the use of safety precautions whenever radioactive materials are handled and to ensure that external and internal doses are controlled within the ICRP recommended limits for occupational exposure.

The limits recommended by the ICRP for radiation workers and the general public are given in Table 14-5.

Comparison of Tables 14-4 and 14-5 shows that the recommended maximum level of 1 mSv/yr is lower than the environmental sources now encountered on average in, say, the U.S.A. This average value is dominated by the large figure for radon; the figure is very variable from location to location and indeed even from house to house. Modern well-insulated and well-sealed houses are likely to have high, and even unacceptably high, concentrations of radon. This is a problem that has only recently been recognized. It is clear that if this figure were much reduced, for a person with no medical X-rays the recommended international standard could be met.

Table 14-5: Recommended Annual Dose Limits*

	Dose Limit	
	Radiation Worker	Public
Effective Dose	20 mSv[1]	1 mSv
Equivalent Dose in:		
Lens of the Eye	150 mSv	15 mSv
Skin	500 mSv	50 mSv[2]
Hands and Feet	500 mSv	-

1. Averaged over 5 years; not to exceed 50 mSv in any one year. Additional restrictions apply to pregnant women.
2. Averaged over any 1 cm^2 of skin regardless of the area exposed.
* Adapted from 1990 Recommendations of the International Commission on Radiological Protection (ICRP).

14.5 Somatic Effects: Large Doses

Somatic effects due to *large* one-time ionizing radiation doses are many and varied. Among the effects that are visible shortly after irradiation are blood changes, vomiting, and skin damage. Doses up to 0.25 Sv (250 mSv) produce no detectable effect. Between 0.25 and 1 Sv (from a single dose) there is measurable blood and bone-marrow damage, but the person feels little immediate effect. Above 1 Sv, treatment with antibiotics may be necessary to cope with malaise, vomiting and blood changes. Above 3 Sv internal haemorrhaging and diarrhea may occur, and blood transfusions may be necessary. At the level of 5 Sv, there is a 50% chance of death.

Delayed somatic effects are defects which occur months or years after the dose. They include eye cataracts, leukemia, and other cancers; in the case of leukemia it is delayed about 6 years, as determined from studies of Japanese atom-bomb survivors. From the studies of these effects, an almost linear relationship has been established between absorbed dose and its effects, for equivalent doses under 3 Sv. This relationship is shown in Fig. 14-2, which shows the relation of radiation-induced leukemia to dose as determined in the Japanese studies. It is very important to note that the linear relation is only established for sudden, high doses. Unfortunately we do not know if the line can be extrapolated linearly back to predict the effect of small doses which would result if thousands of nuclear reactors were in operation around the globe.

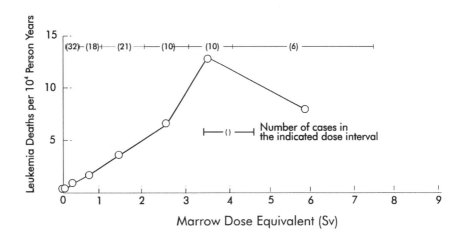

Figure 14-2: Radiation-induced Leukemia in Japanese A-bomb Survivors.

14.6 Somatic Effects: Small Doses

This uncertainty in the low-dose behaviour of the graph in Fig. 14-2 is a source of great dispute by those on opposite sides of the nuclear power debate. It is easy to see why we have very little information on low-dose effects. At 3 Sv, as can be seen in Fig. 14-2, the fraction of the population who will be stricken with leukemia in a given year is about 7×10^{-4}. If the linear hypothesis holds then a dose of 0.01 Sv would produce 2 new cases in one million people. However, the natural incidence of leukemia is about 60 cases per million, and fluctuates statistically. It is out of the question to experimentally determine any significant effect from this small dose. (The "experiment" necessary to detect these effects would require the irradiation of millions of people, followed by long-term medical studies. Clearly, such an experiment is morally and practically indefensible.)

Some radiation biologists believe, on scientific evidence, that there is a threshold dose below which some damage is repaired by the body, leaving no effect. This is probably true for cases where whole cells are killed in an organ which continuously eliminates dead cells and generates new ones. An example of this type would be the depression of the blood count after irradiation and subsequent recovery from the radiation-induced anemia as new cells are produced.

Other types of radiation injury, however, might affect only certain sensitive molecules or structures in individual cells. Such effects might include chromosome aberrations or induction of malignancy. These processes are less likely to have a threshold and the natural occurrence of some mutations and cancers may depend wholly or partly on the natural background radiation. Two cell structures which may be particularly important are:

Genes. The most critical molecule in the cell is undoubtedly DNA since damage of a single gene may profoundly alter or even kill a whole cell.

Chromosomes. Radiation can also have effects on the chromosomes *as a whole* by breaking or rearranging them. One important effect is to interfere with the normal division of cells during mitosis. In tissues where cell replacement is continuous, this sensitivity of dividing cells to radiation can be particularly damaging.

14.7 Genetic Effects

Experiments on individual cells strongly suggest that there is a repair mechanism operating for, at least, single lesions to DNA molecules. This seems a reasonable point of view in light of the fact that all organisms live and have evolved in an environment of natural background radiation. (The children of A-bomb survivors have shown no increase in inherited abnormalities). Animal experiments indicate that a dose of more than 1 Sv would be required to double the natural frequency of inheritable mutations in humans. This means that about 2% of diseases which have a genetic origin may be caused by the natural background radiation.

14.8 Incidence of Human Cancer

One of the most frightening words in the medical lexicon is "cancer" and it is certainly known that radiation, while being one of the most important tools in the treatment of malignant tumors, can also induce them in healthy organisms. Unfortunately, misguided and even malicious publicity has created the impression among many that artificial radiation is a major cause of the disease.

Normal cancer incidence varies greatly, depending on cancer type, age and sex. For any one type of cancer there is a great variation among countries and even districts within countries; stomach cancer varies between 60 and 700 per million over different countries. Since natural radiation background varies by less than ten times this spread, it can only be responsible for a small fraction of these cancers. Many factors contribute, probably in a synergistic manner, to the incidence of types such as bronchial and lung cancer; and these are presently increasing due to environmental factors such as chemicals, smoking, air pollution, etc.

It is very difficult to predict the increase in cancers to be expected from radiation in excess of the background. The only data available are for subjects who received very high doses in one, or just a few, shots. The most important are the 120,000 Japanese A-bomb survivors, 30,000 Canadian women treated with X-rays for tuberculosis between 1930 and 1952, and 14,000 British spondylitis patients treated by spinal irradiation between 1935 and 1954. The A-bomb survivors showed an increased death rate, due to leukemia, of 0.25 deaths per million person-years per mSv over a period of 5 - 25 years after exposure. This figure corresponds to the slope of the low exposure portion of the graph in Fig. 14-2.

It is not known how this data extrapolates to much lower exposures delivered over a protracted time rather than in one large burst. Furthermore it is difficult, if not impossible, to obtain the relevant data. The best that can be done is to use the linear extrapolation of Fig. 14-2, although many think that this is a gross overestimation of the case. Other more elaborate extrapolations are used to form the basis for national and international recommendations regarding exposure.

Rather sophisticated analyses have been made on many occasions by both government and private groups and reported extensively. These reports list data on other irradiated groups; they examine the linearity hypothesis, and estimate the "absolute risk", defined as the number of extra deaths per million of population per Sv of exposure. Making detailed calculations of the risk of a certain age group is rather complex and requires the input of much medical radiation data. If calculations are done for all age groups one can conclude that a 1 mSv/yr continuous lifetime exposure will result in 5800 extra cancer mortalities per million people which is about 3% of the total cancer mortality rate. If all of this dose were received at once instead of over a lifetime the figures are not much altered. A 100 mSv single dose raises the excess cancer mortality to about 5%.

What are the current doses from nuclear power? The current effective dose due to power and processing plants is less than 0.01 mSv per year; in fact, around most plants the dose is unmeasurable and has to be calculated on the basis of known releases of radioactivity. An estimate of the total cancer rate in a population of one million is given in Table 14-6.

There are persistent reports in the media that increased incidence of certain cancers has been detected in populations living around nuclear installations. These reports usually refer to one particular study of the incidence of leukemia in children living in the neighbourhood of the Sellafield nuclear installation in England where six cancer deaths in children aged 0 to 24 years occurred between 1968 and 1974 when, on average, 1.4 is the expected value. The hypothesis advanced to account for this is that the effects of radiation to the parents is passed on to the children via the germ cells and increases their chance of contracting cancer

Table 14-6: Lifetime Cancer and Leukemia Deaths per One Million Population

Total normal incidence	190,000
Radiation Background, 1 mSv/yr.	5,800
Medical X-rays, 0..39 mSv/yr.	2,300
Nuclear effluent, <0.01 mSv/yr.	<60

This hypothesis is contradicted by the fact that there is no corroborating evidence for such an influence; for example, the children of the A-bomb survivors show no such effect. In addition, a study of childhood leukemia around Canadian nuclear facilities found no increase in the rate of the disease comparable to that found near Sellafield.

There are several difficulties with studies of this kind. One of the major ones for the epidemiologist and the statistician is that the population to be studied is defined after the effect is discovered and not before. What should be the boundaries of the population under study? Should it be expanded or contracted? If this is done will the effect vanish? In proper epidemiological studies the population is defined before it is studied. Another difficulty is that the site — in this case a nuclear station — has been singled out for attention in a mood of public hysteria. Other installations, let's say automobile manufacturing plants, are not under the same scrutiny. If they were, would we find unusual statistical fluctuations in certain diseases around them as well? Further, it is not clear that all the relevant facts have been identified. Is the fact that it is a nuclear site the important parameter, or is it because it is a newly collected group of people of a certain narrow age and socio-economic status? The latter is just as likely since similar excesses in childhood leukemia have been found around *prospective* sites for nuclear plants, as well as other places where there is a large population influx. It should be noted that Sellafield is a nuclear reprocessing plant, something far different from a nuclear plant; is it clear that the nuclear aspect is the cause? (Reprocessing plants use large amounts of benzene, a known carcinogen.) Finally it should be noted that there are far more nuclear installations which show no such effects, and these observations must not be excluded. Epidemiological problems like this are very difficult to resolve and require very careful and lengthy study and analysis. This problem is similar to the one discussed in Section 10.3.

PROBLEMS

14-1 A quantity of a fall-out radioactive beta emitter is taken up by a growing plant and distributed uniformly throughout the tissue. A 200 gram plant is found to contain 3700 Bq (0.1 μ Ci). What dose does the plant receive in one day if the average beta energy is 0.10 MeV?
[Ans. 2.6 $\times$ 10^{-5} Gy]

14-2 What dose is received in a one hour exposure at 1.0 m from a 3.7 $\times$ 10^7 Bq (1 mCi) radium source? Each disintegration yields two gamma rays whose average energy each is 1.0 MeV.
[Ans. 2.7 $\times$ 10^{-4} Gy]

14-3 Repeat Example 14-5 assuming the source is ^{210}Bi, a β source of mean energy 0.40 MeV and range 0.25 cm in tissue.
[Ans. D = 0.26 Gy, H = 0.26 Sv, E = 0.01 Sv]

14-4 A small drop of solution of ^{131}I is spilled on the surface of a researcher's hand. The area of the spot is 1.0 cm^2 and it contains 0.51 μCi (1.9$\times$10^4 Bq). The half life of ^{131}I is 8.0 days and the average energy of its β particles is 0.606 MeV with a range of 0.20 cm in water. The spot is not removed for 16 days.
a) What is the number of ^{131}I nuclei present?
   [Ans. 1.9$\times$10^{10}]
b) How many βs enter the flesh of the researcher?
   [Ans. 1.3$\times$10^{10}]
c) What is the dose, equivalent dose and effective dose?
   [Ans. 8.0 Gy, 8.0 Sv, 0.4 Sv]

14-5 The U.S. regulations for the maximum permissible concentrations of radionuclides in water give the values:

Species	Max. Conc.	$T_{1/2}$
^{131}I	0.011 Bq/cm^3	8 days
^{239}Pu	0.19 Bq/cm^3	2.4×10^4 year
^{90}Sr	0.011 Bq/cm^3	28 year

Calculate the corresponding concentrations of these elements by mass in parts per million (ppm).
[Ans. 3×10^{-12} ppm, 8×10^{-5} ppm, 2×10^{-9} ppm]

14-6 By retaining the effluents for 60 days, a 1000 MWe pressurized water nuclear reactor emits 37×10^5 Bq (10 μCi) of ^{85}Kr per second into the atmosphere. Assume a population of 500 such reactors. The halflife of ^{85}Kr is 10.8 yr.

a) If the atmosphere is 10 km high, what is the activity of ^{85}Kr that accumulates after 1 year of operation? Neglect the decay. The Earth's radius is 6400 km.

b) What dose rate in Gy/s is given to a typical person in this one year accumulation? [Hint: Since β particles travel 1 metre in air, only the ^{85}Kr within 1 m has any effect. The mean β energy is 0.3 MeV.] Assume a man of height 2 m and weight 90 kg; even so you will have to make some approximations.
[Ans. a) 1.1×10^{-2} Bq/m^3., b) 10^{-16} - 10^{-17} Gy/sec.]

ENVIRONMENTAL EFFECTS OF NUCLEAR ENERGY

This chapter examines the environmental effects of nuclear energy, from the initial mining of uranium, through a reactor's operation, to the final decommissioning of a reactor.

15.1 URANIUM MINING AND PROCESSING

As discussed in Section 12.18, uranium mines present all the hazards of any hard-rock mining, with the additional danger of radioactive radon gas. However, since the energy content of uranium ore is greater than that of coal by about a factor of 30, the total hazard per unit of energy produced is considerably lower in uranium-mining than in coal-mining[1]. The radon hazard gives uranium miners a higher than normal chance of contracting lung cancer.

The mill tailings (the ore discarded after removal of the uranium) are slightly radioactive because of long-lived daughter isotopes from uranium which have already decayed in the ore. This material is usually in the form of liquid sludge, and after drying is susceptible to dispersal to the environment through erosion (by wind and water) and by leaching. Therefore, the preferred place of disposal is in empty mines or in lined pits dug for the purpose. Another problem associated with tailings is the emission of radon gas to the atmosphere (Problem 15-1), which can be inhibited by covering the tailings with compacted earth, clay, concrete, or asphalt.

The concentrated ore from the mills is sent to refineries (then to ^{235}U-enrichment facilities if the uranium is to be used in LWRs), next to fuel-fabrication plants, and finally to reactors. At all stages in this transportation, there are strict requirements on the design of shipping containers to prevent escape of radioactivity to the environment in the case

---

[1] W.F. Vogelsang and H.H. Barschall, "Nuclear Power," in *The Energy Sourcebook*, Ed. by R. Howes and A. Fainberg, American Institute of Physics, New York, 1991, p. 142

of an accident en route. In the various plants, there are rigorous accounting procedures to ensure that no uranium is lost, stolen, or spilled into the general environment. These procedures are in place for a number of reasons: the uranium is expensive, it could be used as a source of material for weapons if enriched sufficiently in ^{235}U, and it is radioactive, although not strongly. (Because of uranium's long half-life, its activity is relatively low. The activity of 1 kg of natural uranium is 3×10^{-4} Ci, whereas the activity of 1 kg of ^{60}Co, an isotope of half-life 5.3 y that is used in cancer therapy, is 1×10^6 Ci.)

15.2 POLLUTION AND RADIOACTIVE WASTE FROM REACTORS

Although a nuclear reactor does generate radioactive waste, it produces no greenhouse gases, no SO_2, no NO_X, no ozone, and no particulate matter. Hence, a nuclear plant is much "cleaner" than a fossil-fuel electrical plant. However, it creates thermal pollution of the cooling waters (more than does a fossil-fuel plant), as discussed in Section 6.2.

All energy-generating technologies have waste products to cope with, and the waste from nuclear reactors has attracted much attention from the public and media. What is not often appreciated, however, is the physical scale of the problem for the various technologies. Let us first, then, compare the volume of waste which must be handled in a coal-burning and a nuclear plant of 1000 MWe capacity. We will assume that the nuclear fuel is in the form of UO_2 which has a specific gravity[2] (SG) of 11, whereas coal has SG = 1.5. Since only nuclear energy is extracted from the UO_2, the mass and volume of the waste is almost the same as the fuel. For coal, the ash comprises about 10% of the volume but of course the other 90% is exhausted up the stack in the form of CO_2; thus coal-fired emissions include large amounts of that invisible and very important greenhouse gas, not to mention the nitrogen and sulphur oxides. Table 15-1 summarizes the situation for the two cases, for solid wastes alone.

---

[2] Specific Gravity (SG) is the density of a substance relative to that of water. Thus the SG of water is, by definition, 1. A substance with a SG of 2 has a density twice that of water.

Table 15-1: Fuel and Solid Waste from a 1000 MWe Power Plant

Type of Plant	Annual Fuel (Tonnes)	(m³)	Annual Waste (m³)
Coal	3 000 000	2 000 000	200 000
Nuclear	200	20	20

Every year the coal-fired plant must dispose of 200 000 m² of toxic ash — this is equivalent to the volume of a large office building ($\sim$ 60 m on a side). The nuclear plant must deal with 20 m³ of very toxic used fuel occupying a volume about the size of a very small office ($\sim$ 2.5 m on a side). The nuclear waste problem can be dealt with in two stages: short-term and long-term storage.

Short-term Storage

After the fuel elements are first removed from the reactor, it is necessary to store them in such a manner that they are secure and adequately cooled. Heat is still being generated in the fuel by radioactive fission fragments. The elements are usually stored in ponds under recirculated water which provides both shielding and cooling. There are a large number of very short-lived isotopes in the spent fuel and the activity decreases by about 60% in the first hour. After that, the activity is dominated by the longer-lived isotopes. After three months, one tonne of spent fuel has an activity of 23×10^{16} Bq and generates heat at a rate of 27 kW. Of this activity, 98% is due to fission products and 2% to actinides (Pu generated by breeding, for example). After ten years, the activity has decreased to 9×10^{16} Bq and the rate of heat production is 1 kW. At this stage the necessity for close-monitored cooling is over and long-term storage can be considered.

Long-term Storage

No nation has yet adopted an accepted method of long-term storage of spent reactor fuel. There are many reasons for this as there are several conflicting interests; some people persist with an idea formulated early in the development of nuclear power, which envisaged setting up reprocessing plants to strip out of the spent fuel all the remaining fissionable fuel (^{235}U, ^{239}Pu, etc.) in order to maximize the efficiency of the power generation. (This is an elaboration of the fuel enrichment technology that has grown out of the nuclear weapons programs.) With all the fears of nuclear terrorism, plutonium toxicity, and weapons proliferation, the plan has met with almost universal opposition. Although a few small reprocessing plants have been built in England and France, it is doubtful that this industry will prosper in the foreseeable future.

It is much more likely that each nuclear nation will opt for some form of long-term storage of the fuel with little, or no, preparation and processing. One option is to do nothing other than provide supervised above-ground storage. Supporters of this option argue that "out-of-sight is out-of-mind" and it is better to keep the material in view and look after it properly. Detractors of this view argue that it is unlikely that institutions can be put in place to tend it for 1 000 or 10 000 years.

The most probable option will be to develop some sort of permanent storage in below-ground vaults, either supervised or more likely sealed, deep in continental rock. Large ancient rock shields like the Canadian Shield seem particularly suitable for this purpose. There seems to be no technological reason that such vaults and containers cannot be designed and constructed to contain the wastes for a sufficiently long time to reduce the radiation to levels consistent with that of the rock itself. Several countries are engaged in this type of research and it seems only a matter of time until the first scheme is adopted, quite possibly at a site in Nevada.

Canada has been involved since 1978 in an active research program to study long-term storage of nuclear waste, and has created an Underground Research Laboratory (URL) located 420 m below the earth's surface in undisturbed granitic rock in Manitoba. Experiments are being performed to determine how rock and groundwater behave at depth, and how they would be affected by heat given off by used fuel elements. Other experiments study the mobility of weakly-radioactive tracers released underground.

Low-Level Radioactive Discharges from Reactors

Reactors continuously deliver radioactivity to the environment. The radioactive isotopes created in reactors are of three types:

- fission fragments,
- isotopes produced from neutron bombardment of impurities and corrosion products in the coolant-moderator water as it passes through the reactor core, and
- isotopes produced in the entire reactor structure by neutron bombardment.

These last isotopes, although radioactive, are not released to the external environment.

Most of the fission products remain in place within the fuel bundles, but some can diffuse or leak through the fuel cladding into the coolant-moderator. The most common elements that diffuse are isotopes of cesium, iodine, xenon, krypton, rubidium, and bromine, as well as tritium[3]. The radionuclides in the coolant-moderator from fission products and from impurities and corrosion can exist in gaseous form, as dissolved solids, or as suspended solids. The suspended solids can be removed by filtration, and the dissolved solids by ion-exchange beds; these materials can then be disposed of, typically by burial. Gaseous waste can be stripped from the coolant-moderator, and stored for up to three months to allow short-lived isotopes to decay, after which it is less radioactive and is vented to the atmosphere. The coolant-moderator water, after having the solids and gases removed, can be re-used in the reactor or discharged (with only slight radioactivity) to the environment.

The gaseous fission products in BWRs are dealt with in a different way. These gases, mainly krypton and xenon, boil off with the steam and pass through the turbine into the condenser. There they are removed, along with air that has leaked into the condenser and hydrogen and oxygen formed by the radiolysis of water in the reactor. In older BWRs, these gases were vented to the atmosphere with a holdup time of only a few minutes, during which time very little radioactive decay could occur. Nowadays, the gases are held up considerably longer to allow for virtually complete decay of short-lived isotopes (Problem 15-2).

The tritium that is generated in reactors usually becomes bound in water molecules, replacing one of the hydrogens to produce HTO, which

---

[3] M. Eisenbud, *Environmental Radioactivity from Natural, Industrial, and Military Sources (Third Ed.)*, Academic Press, Orlando, 1987, p. 210.

would be relatively difficult to remove. However, the concentrations of tritium in normal emissions of tritiated water from a reactor present a negligible health hazard. In part, this is because tritium emits only low-energy β radiation, and does not bio-accummulate in biological systems. (However, a large leak of heavily tritiated water would be considered a serious problem, because the tritiated water could become incorporated into plants and animals in the area.)

Occasional leaks (particularly of radioactive gases) from valves, pipes, etc. also result in discharges of weakly radioactive materials to the environment. The radiation dose to the general public as a result of all these emissions is extremely small —— recall from Table 14-4 that all aspects of the nuclear power industry contribute only 0.03% of a person's average annual effective dose.

Radiation from Coal

Although not directly related to nuclear power, the radioactive emissions from coal-burning are an interesting environmental pollutant, and are discussed here. Coal contains naturally radioactive uranium and thorium, plus radioactive daughters of these elements, and ^{40}K. When coal is burned, the radioactive nuclides tend to concentrate in the small solid particles known as "fly ash." In former times, the fly ash was allowed to escape up the smokestack into the atmosphere, and as a result the radioactive emissions from a coal-burning electrical plant were actually greater than from a nuclear plant[4]. However, fly ash is now collected by electrostatic precipitators, bag houses, and scrubbers, and the radioactive emissions from a typical coal-burning plant are now only about 1/60 of those from a plant with no emission controls, and are less than the emissions from a nuclear plant.

15.3 NUCLEAR SAFETY AND ACCIDENTS

Probably the most serious concerns about nuclear reactors are related to the possibility of a nuclear accident. The total inventory of radioactive

---

[4] Ref. 3, pp. 152-155.

materials in a reactor is so large that it would represent a serious threat to life if released into the environment, and could render a large area of land uninhabitable. Recall that a reactor cannot undergo a nuclear explosion ——— the concentration of ^{235}U is not large enough. However, the heat produced by a reactor could, under certain circumstances, produce melting of the fuel, thus allowing fission products into the coolant (and also making the reactor inoperable). In an extreme situation, a steam explosion could result.

Safety Mechanisms

A reactor is a producer of heat, and if the heat produced cannot be removed from the reactor core, then overheating can lead to a serious accident. Reactors have various safety devices in place to prevent this occurrence. In case of a problem with a reactor, there are two issues that need to be addressed: shutting down the fission process (which produces most of the heat), and removing the residual heat produced by the fission products. The various types of reactors have safety mechanisms that differ slightly in their details, but all reactors in Western countries share some common features in their approaches to safety.

In order to stop the nuclear fissions, neutron-absorbing rods are inserted into the reactor core. These rods can be special shutoff rods, or just the usual control rods that are normally used to adjust the reactor's output. By absorbing neutrons, these rods cause the number of fissions to decay, shutting down the reactor. In the event that these rods fail to operate properly, a neutron-absorbing material (a "neutron poison") can be injected into the coolant-moderator.

Once a reactor is shut down, normal circulation of the coolant removes heat produced by decay of fission products. If this circulation is somehow interrupted in a loss-of-coolant accident, an emergency core-cooling system is turned on which provides a completely separate circulation of cooling water. A final safeguard is provided by the structure of the containment building surrounding the reactor; this building is typically constructed of thick concrete with a steel-plate lining, and is designed to withstand a steam explosion. (The Chernobyl reactor had a much flimsier containment building, and this feature contributed to the severity of the accident, discussed in Section 15.5.)

15.4 THREE MILE ISLAND

The first accident at a nuclear power reactor occurred in March 1979 at a PWR plant on Three Mile Island (TMI) in the Susquehanna River in Pennsylvania. (A small number of accidents had occurred previously[5] at experimental reactors, or at reactors producing plutonium for weapons.) The TMI accident occurred as a result of a combination of mechanical problems, design features which gave inadequate monitoring and display of reactor conditions to the operators, and human errors.

The first step leading to the accident occurred when the feedwater pumps supplying the steam generators shut down, leading to automatic shutdown of the reactor and the steam-generating system. Heat from fission products in the fuel caused the temperature and pressure in the core area to rise, and a pressure relief valve opened. This valve should have closed within a few seconds, but it stuck open, allowing coolant water to drain from it. A loss-of-coolant accident was in progress, and the automatic safety systems came into action. Three auxiliary feedwater pumps began to operate, but the water could not reach the reactor because two valves had incorrectly been left closed after a maintenance check. After a couple of minutes, emergency high-pressure water pumps began operating automatically, but a reactor operator stopped one pump and reduced the output of the second drastically. The operator was apparently misled by an indicator that showed a high water level in the reactor, when in fact the level was too low. Bubbles of steam began to form in the coolant system, but this problem was misdiagnosed, and operators began to drain coolant water even more. As a result of all these actions, part of the core became uncovered, the fuel cladding failed and a portion of the fuel melted. Radionuclides from the fuel passed into the coolant water, which was draining from the pressure relief valve and filling a drain tank. A seal on this tank ruptured, and water began to collect in the reactor basement. Finally, operators recognized what was happening, and about six hours after the accident began, the core was again covered with water. The core temperature had risen so high that zirconium in the fuel cladding reacted with steam to produce hydrogen. Roughly 24 hours after the accident had begun, a small hydrogen explosion occurred, but produced

---

[5] For a full discussion of these accidents, see Ch. 14 of Ref. 3

little damage.

The accident liberated some radionuclides into the environment. When the contaminated cooling water flowed into the reactor basement, pumps started up and sent this fluid into an auxiliary building, from which some radioactivity was released into the atmosphere. The total population dose received by people within an 80-km radius is estimated to be about 30 person·Sv (the product of the number of people and the average per capita dose).[6] This dose is expected to result eventually in about two additional cancer deaths in the population.

There are views of the TMI accident. Some people argue that in spite of mechanical problems and human mistakes that led to a loss-of-coolant accident, the radioactivity was largely contained and the effects on the environment and on the surrounding population were minimal. Others express the conviction that the accident serves to illustrate the sorts of problems that the nuclear industry can have, and should serve as a warning that nuclear energy is inherently dangerous. Certainly the biggest effect of the TMI accident was psychological. Many people in the U.S.A. lost their trust in nuclear technology, and the U.S. nuclear industry has never recovered.

15.5 CHERNOBYL

The disaster in April 1986 at the Chernobyl reactor in the Ukraine region of the former U.S.S.R. was much more serious than the TMI accident. An explosion and fire released a great deal of radiation to the environment, and a number of people were killed in the initial blast.

The reactor that exploded was one of four at the plant, each producing about 3200 MW of thermal power, generating about 1000 MWe. Each reactor, a BWR with graphite moderator, had two 500-MW turbine-generators for the production of electricity, and was housed in a separate building having only a standard industrial factory roof instead of concrete and steel containment. (There was better containment on the sides and bottom of the reactor.) The shutdown rods were extremely slow-moving,

---

[6] Ref. 3, p. 372, and *Physics Today*, June, 1979, p. 78

requiring about 18 s to come into place.

The control of the reactor was much more manual than in a Western plant, and the operators had no specific training for dealing with accident situations. Operators had no training on simulators, unlike the common practice in Western countries. There were no independent safety audits, and operators were allowed to work 36-hour shifts. The reactor that exploded had been very reliable and had a high capacity factor (Section 4.5), and the operators had perhaps become somewhat overconfident and complacent.

The accident occurred as operators were performing a test on one of the turbine generators; they wanted to determine its ability to generate electricity while "spinning down" after the steam supply was shut off. A previous test showed that the voltage fell off long before the mechanical energy of the rotor had been dissipated. Engineers wanted to see if a special magnetic-field regulator for the generator would help to maintain the voltage of the turbine while it was coasting down. (The reason for doing this was that some of the pumps in the emergency core-cooling system were normally powered by the turbine, and in the event of a reactor shutdown it was desirable that this system could still be powered by the turbine for a short time (30 s or so) until backup diesel generators could come into operation.) The reactor staff was very eager to complete the test successfully on this particular occasion —— the reactor was to be shut down for scheduled maintenance in a few days, and future tests would have to wait for a year.

Figure 15-1 shows the sequence of events leading to the explosion. At 1 a.m. on April 25, the power began the decline from its normal level of 3200 MW toward the level of 700-1000 MW planned for the test. By 1 p.m. the reactor was at half-power, and one of the two turbogenerators was disconnected. Instead of continuing the power descent immediately, the operators had to maintain half-power until 11 p.m. because of unexpected electrical demand. At 2 p.m. they disconnected the emergency core-cooling system, in violation of operating procedures, because they did not want this system coming on during the test and thereby draining energy from the turbine.

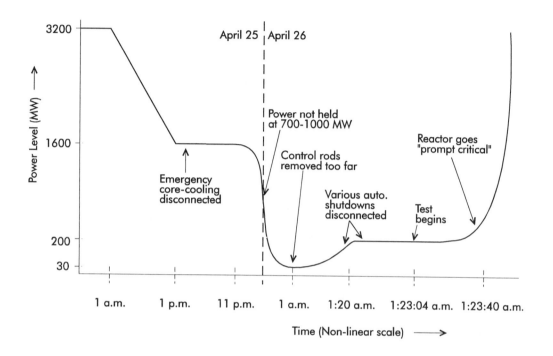

Figure 15-1:   Events leading to the Chernobyl accident.

At 11 p.m. the power descent resumed, but at 12:28 a.m. on April 26, the operators made an error in entering an instruction into the computer and instead of the power levelling off between 700 MW and 1000 MW, it continued to drop to 30 MW, which was much too low a power level for the test. At such low power, there is a buildup of the fission product ^{135}Xe in the fuel; this isotope is a strong absorber of neutrons and tends to decrease the power level even more. In addition, at the 30-MW level, the water in the core area was not boiling as much as usual, and absorption of neutrons by the liquid water was also tending to decrease the power. In order to increase the power, at 1:00 a.m. the operators withdrew most of the control rods from the reactor core beyond the allowable limit and managed to get the power to 200 MW by about 1:20. This control-rod withdrawal was a huge violation of operating procedure; the control rods are also used as emergency shutdown rods, and if they are withdrawn beyond a certain limit, it takes too long for them to return to the core in case of an emergency. (As we discuss later, there was an additional unexpected problem that arose as a result of withdrawing the control rods too far.)

The operators were having other problems, since the reactor was not designed to run at such low power. They had to switch to manual control of the water flow returning from the turbine, and had to make many adjustments of steam pressure and water flow. These adjustments were difficult, and the reactor was on the verge of shutting down automatically because of fluctuating flow and pressure. Because such a shutdown would abort the test, the operators disconnected the automatic shutdowns related to steam pressure and water flow (another violation of operating procedure). By 1:22 a.m. the operators felt that the reactor was as stable as it was going to be, and decided to begin the test. But first they turned off the safety device that shuts down the reactor when the turbine is disconnected, again violating proper procedures; they wanted the reactor not to shut down in case they needed to perform the test a second time.

Let us review the situation. The reactor was in a highly unstable state, requiring much more manual control than normal; it should have been shut down when the power reached 30 MW. Most of the control/shutoff rods had been withdrawn too far, and many of the automatic shutoff signals had been disabled. Recall also from Section 12.16 that a graphite-moderated water-cooled reactor has a positive void coefficient; as more water boils, the bubbles (voids) of vapour absorb fewer neutrons, and the fission rate accelerates. As long as the reactor is critical on the delayed neutrons, this situation is acceptable, but if it becomes "prompt critical" (able to continue a chain reaction using only prompt neutrons), then the fission rate increases uncontrollably.

At 1:23:04 the turbine was disconnected to begin the test. As it slowed down, the regular water-coolant pumps that it powered also slowed down, decreasing the flow of cooling water over the core. This increased boiling in the water, and the power began to rise slowly, then more quickly. (Remember the positive void coefficient.) At 1:23:40 an operator noticed this increase and pushed the "scram" button to drive the emergency shutoff rods into the core. As will be explained in the next paragraph, because the shutoff rods had been withdrawn so far from the core, their movement back led not to a decrease in power, but to an increase. The reactor had become "prompt critical," and within four seconds, the power output rose to about 100 times normal.

The control/shutoff rods travel in vertical tubes, and in normal operation, part of the rod projects above the core, and part below (Fig. 15-2 (a)). The rods for this particular type of reactor are rather unusual in design: the top portion is standard, constructed of a neutron-absorbing material, but the bottom section is graphite, a neutron moderator. Raising the rod removes neutron-absorbing material from the top of the core, and introduces graphite at the bottom, thus increasing the rate of fission. When the rod is lowered, graphite moves out of the core area at the bottom, neutron-absorber comes in at the top, and the reaction rate decreases. Just prior to the Chernobyl explosion, the rods had been raised so far that their bottoms were above the bottom of the core and the lower parts of the tubes contained water (Fig 15-2 (b)). Water is both a neutron absorber and a neutron moderator, but compared to graphite, it is a more effective absorber than moderator. When the "scram" button was pushed and the rods were lowered (very slowly, remember), the effect in the bottom section was to displace water in the tubes with graphite, thus increasing the fission rate (temporarily, but significantly).

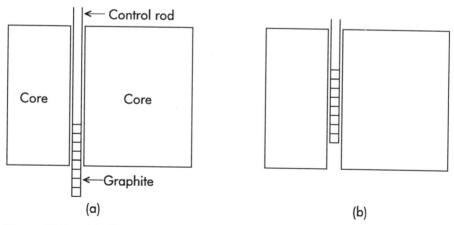

Figure 15-2:　(a)　Normal control-rod position in Chernobyl-type reactor.
　　　　　　　　(b)　Control-rod position prior to explosion.

The sudden burst of heat in the fuel broke it into small pieces and caused the cooling water to boil extremely rapidly; the resulting high steam pressure blew the roof off the reactor building, sending flaming fuel and graphite onto the roofs of adjacent buildings. Local firefighters extinguished the exterior fires by 5 a.m., but many of them suffered lethal doses of radiation. However, graphite in the core was still burning and the updraft was carrying radioactive particles high into the atmosphere. Starting on April 28, military helicopters dropped about 5000 tonnes of boron, dolomite, sand, clay, and lead onto the exposed core. This decreased emissions for a few days, but also thermally insulated the core, raising its temperature and increasing emissions again. On May 4 and 5, nitrogen was pumped under pressure into the space beneath the reactor, cooling it effectively. The reactor is now entombed in concrete; it is cooled by air, which is filtered afterward.

Effects on People

The accident produced 203 cases of acute radiation doses in reactor workers and firefighters. As of August 1986, 31 people had died —— one from the explosion, one from steam burns, and 29 from radiation. Approximately 100 MCi of radiation was released to the environment, half of this during the initial explosion, and half later as the core continued to burn. Some of the radiation was carried great distances by wind; indeed, the first indications outside the immediate Chernobyl area that there was a problem came from nuclear power plants in Sweden and Finland, where radiation monitors detected fission products on the clothes of workers *entering* the plants and also on air filters.

Within a 30-km radius of the plant, 135000 people were evacuated. The average effective dose received by each of these people was about 120 mSv, equivalent to about 60 years' worth of natural radiation at 2 mSv/yr; a few were as high as 400 mSv[7,8]. The dose decreased with distance from the plant; in the population of 75 million in the Ukraine, the typical

---

[7] B.G. Levi, *Physics Today*, December 1986, p. 19.

[8] V.G. Snell and J.Q. Howieson, *Chernobyl —— A Canadian Perspective*, AECL, 1991, p. 14.

effective dose[9] ranged from about 8 mSv to 32 mSv, and the average dose to people in Western Europe was less than 2 mSv. In the evacuated population of 135 000, it is estimated that there will be about 1300 additional cancer deaths as a result of the accident[10]; this is an increase of roughly 5% in the expected number of 24 000 cancer deaths. In the regional population of 75 million, the future additional cancer deaths are much more difficult to estimate because of the low doses involved. The cancer death rate could have an increase[10] in the range of 0.4% to 1%, that is, an increase of 50 000 to 130 000 deaths above the expected 13 million. The effects in other countries are highly uncertain because of the extremely low doses; there will certainly be some extra cancer deaths, especially in "hot spots" in Scandinavia, Germany, and Poland where rain deposited a lot of radioactive material.

The Chernobyl accident showed what can happen when a reactor and its containment are not designed with enough attention to safety, when operators are given insufficient training for accidents, and when operating procedures can be violated easily. Not surprisingly, steps have been taken to improve safety in Chernobyl-type reactors. Better training is being given to operators, more effective displays of information are being installed in reactor control rooms, and much more stringent controls are being placed on the disabling of safety systems. The structure and operation of shutoff rods have also been improved, and the composition of the fuel has been changed (enriched further in ^{235}U) to decrease the magnitude of the positive void coefficient.

15.6 DECOMMISSIONING OF NUCLEAR PLANTS

Nuclear reactors have useful lifetimes of about thirty years, and then need to be shut down permanently. The remaining fuel, containing

---

[9] Ref. 8, pp. 14-15.

[10] The number of additional cancer deaths was determined using single-exposure risk factors from *Health Effects of Exposure to Low Levels of Ionizing Radiation (BEIR V)*, National Academy Press, Washington, 1990, p. 172. The number of expected additional deaths is higher by about a factor of 6 than estimates based on the earlier BEIR III Report (1980).

radioactive fission products, will be removed and disposed of, but the entire building structure is also somewhat radioactive from neutron bombardment, and pipes, valves, etc. are contaminated with corrosion and wear products that have been neutron-activated, and also with fission products that have leaked from the fuel assemblies. What can be done with all this radioactive material? There are essentially four options, ranging from immediate dismantling of the plant to using the normal reactor containment structure as a long-term storage facility.

Immediate dismantling would require precautions to ensure that the personnel engaging in the work do not receive large doses of radiation, since the radioactivity will have had little time to decay. Some of the work would probably have to be done by robots. The material would have to be stored elsewhere in a secure site, likely underground. Proponents of quick dismantling argue that this approach minimizes radiation exposure to the general public.

A second option is to wait perhaps 30 to 50 years before dismantling, to allow the radioactivity to decrease. This would reduce both the danger to workers and the amount of radioactive material that would have to be stored permanently. An obvious disadvantage is that the site would have to be made secure during the waiting period.

A third alternative is to entomb the entire reactor building in a strong encasement, and simply leave it for a very long time until the radioactivity has decayed to a level where the site poses no danger to anyone. Again, site security and maintenance of the structure would be required. A fourth choice, which is a variant of this entombment, would simply be to seal the existing reactor containment structure and use it as the storage site. Supporters of this option point out that it would be the least expensive. Yet both of the entombment approaches result in long-lasting eyesores on the landscape.

In the U.S.A., the Nuclear Regulatory Commission has endorsed the first two options as being preferred, and has required that plant operators set aside sufficient funds for dismantling. The amount required varies from about $100 million for PWRs to $140 million for BWRs (in 1986 U.S. dollars).

15.7 NUCLEAR POWER AND NUCLEAR WEAPONS

Although a reactor cannot undergo a nuclear explosion, there is at least an indirect connection between nuclear power and nuclear weapons. As stated in Section 12.8, any fissionable material such as ^{235}U or ^{239}Pu can be used to create a nuclear bomb; however, the concentration of the fissionable material must be enriched to 90% or more. In the case of ^{235}U, this enrichment is neither easy nor inexpensive. However, the ^{239}Pu created by neutron bombardment of ^{238}U in a nuclear reactor (Section 12.10) is easily separable by chemical means from the uranium fuel and fission fragments, and could be used to construct a nuclear weapon. Thus, any country with nuclear power plants could produce nuclear weapons. (And, in addition to plutonium's danger in nuclear weapons, it is an extremely toxic chemical.)

In 1957-58 a Nuclear Non-Proliferation Treaty was signed by more than one hundred countries. One of the main objectives of the treaty was to require the processing of used fuel elements to come under the surveillance of the International Atomic Energy Agency in order to help prevent the spread of nuclear weapons. However, a number of countries failed, or refused, to sign the treaty; examples include Brazil, Argentina, and Spain, all of which already have nuclear reactors. Some other non-signing countries possess the technical ability to construct reactors, but have not yet done so.

There is also the danger of construction of a crude nuclear weapon from plutonium in used fuel assemblies stolen by a terrorist group, for example. If breeder reactors come into use, plutonium would be generated in larger amounts, and the possibility of someone stealing plutonium for non-peaceful means would increase.

EXERCISES

15-1 Table 15-2 shows the principal fission-product radionuclides that are contained in a LWR core after the reactor has been shut down for one day

after two years of steady operation[11]. If these radionuclides are allowed to decay for two years, which ones will then be making the most significant contributions to the overall activity? [To answer this question, do not perform any calculations; just look at the original activity for each isotope in Table 15-2, and think about whether this activity will decrease a little or a lot in two years, given the half-life.]
[Ans. ^{144}Ce, ^{137}Cs, ^{90}Sr]

Table 15-2: Principal Fission-Product Radionuclides 1 Day After 2 Years of Operation of a LWR (3000 MW thermal) (for Exercise 15-1)

Isotope	Activity (MCi)	Half-life
^{3}H	0.013	12.3 y
^{85}Kr	0.75	10.7 y
^{89}Sr	72	51 d
^{90}Sr	5.4	28.9 y
^{90}Y	5.4	64 h
^{91}Y	96	58.8 d
^{99}Mo	120	66.6 h
^{131}I	84	8.06 d
^{133}Xe	162	5.3 d
^{134}Cs	1.8	2.06 y
^{132}Te	102	78 h
^{133}I	66	20.8 h
^{136}Cs	2.2	13 d
^{137}Cs	7.2	30.2 y
^{140}Ba	138	13 d
^{140}La	147	40.2 h
^{144}Ce	105	284 d

---

[11] Ref. 3, p. 207

PROBLEMS

15-1 The U.S. Environmental Protection Agency has set a limit[12] on the emission of radon-222 from mill tailings: 20 pCi· m^{-2}· s^{-1}.   For an uncovered pile of tailings of dimensions 35 m × 25 m × 5.0 m, sitting on the ground and emitting radon-222 at the maximum rate allowed, what mass of radon-222 is emitted to the atmosphere per day? The half-life of radon-222 is 3.8 d. [Ans. 1.6 × 10^{-11} kg]

15-2 BWRs continually release several Kr and Xe radionuclides to the atmosphere. Table 15-3 shows the Kr isotopes that are normally released. (An "m" if front of the Kr indicates an excited nucleus that is metastable, i.e., one that has a sufficiently long half-life to be studied independently from the ground-state nucleus.) The isotopic half-lives are listed, along with the typical activity[13] that would be released from a 3400-MW (thermal) BWR in one year if the gases are held up for only 30 min. Charcoal beds are often used to hold the gases for an additional 45 h. Calculate the activity of each isotope that would be released each year after this additional holdup, and comment on the benefit (or lack of it) of this holdup to radioactive contamination of the atmosphere.
[Ans. 0.0033 Ci/yr for ^{83m}Kr, etc.]

Table 15-3:   Typical Annual Releases of Krypton From a BWR After 30-min Holdup (for Problem 15-2).

Radionuclide	Half-life	Ci/yr emitted
^{83m}Kr	1.9 h	44000
^{85m}Kr	4.4 h	84000
^{85}Kr	10.8 y	290
^{87}Kr	76 min	240000
^{88}Kr	2.8 h	280000
^{89}Kr	3.2 min	2800

---

[12] Ref. 3, p. 179

[13] Ref. 3, p. 214

16

ALTERNATIVE ENERGY SOURCES

A general recognition that the poisoning of the atmosphere with carbon dioxide cannot be allowed to continue until the last kilogram of fossil fuel is burned, combined with a reluctance to embrace the nuclear alternative, has focused the public attention on other sources of energy that produce less pollution. In this chapter various other energy sources are explored — the sun, wind, geothermal sources, tides, waves, and biomass — emphasizing the advantages and disadvantages of each.

16.1 ENERGY FROM THE SUN

The sun pours out prodigious amounts of energy, more than we could ever use. Why not capture it and use it directly? Probably no area of alternative energy generation has had so much research and media attention with so little return. The first four sections of this chapter focus on some of the possibilities and the problems with the large-scale and small-scale utilization of solar energy. Much of the material follows on from the first three sections of Chapter 7.

As mentioned in Section 7.5, the sun supplies energy at the position of the earth at an average rate of 1372 W/m^2 at the top of the atmosphere; about 65% of this, or 900 W/m^2, is delivered to the earth's surface. For any location on the surface, this figure must be adjusted for the inclination of the surface to the sun's direction, then averaged over the day and night hours, and finally averaged over the year.

For example, in southern Ontario (latitude 44° N) the annual averaged solar radiation energy (or "insolation," R) has a value of R = 3.5 kW· h/m^2 per day. If this was delivered continuously over the 24-hour day, the intensity would be 146 W/m^2; of course, in reality it is a higher value at noon and zero at night. There is a great seasonal variation; in Toronto the daily insolation varies between 1 and 6.4 kW· h/m^2 as shown in Fig. 16-1. About 90% of this energy is from the direct sunlight, and 10% from diffuse light. On normal cloudy days the light is all diffuse, and the insolation is about 25% of the value on clear days.

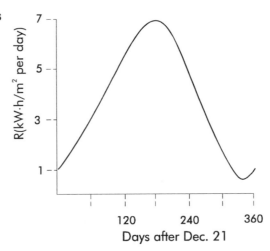

Figure 16-1:   Mean daily solar insolation on a horizontal surface near Toronto (Lat. 44°N).

As expected the insolation is lowest at the winter solstice (Dec. 21) and highest at the summer solstice (June 21). Figure 16-2 is a recording of the insolation at the University of Guelph's Elora Research Station on June 19, and December 19, 1990, very close to the summer and winter solstice. The details of the shape of the curves have to do with the daily variations of local weather. For example, on Dec. 19 there was precipitation in the morning which reduced the insolation but cleared the atmosphere of particulate matter. After the clouds moved away there was an unusually clear sky with higher-than-average intensity for that time of year.

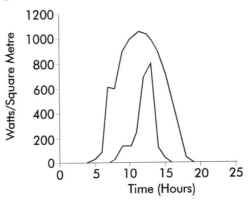

Figure 16-2:   Solar intensity at Guelph, Ont., (Lat. 44°N) at summer and winter solstice. Courtesy-R.H.Stinson

Going south from the Canada-U.S.A. border, R increases from 3.5 to 7 $kW \cdot h/m^2$ per day at the U.S.A.-Mexico border. The U.S.A. average is 4.8 $kW \cdot h/m^2$ per day. The American southwest, where $R = 7 \ kW \cdot h/m^2$ per day, has a distinct advantage for the utilization of solar energy.

16.2 SMALL-SCALE DIRECT UTILIZATION OF SOLAR HEAT

Schemes to use solar heat directly generally are of three types: passive heating, active collector panels, and solar ponds.

Passive Heating

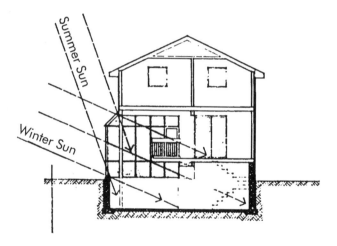

Figure 16-3:   A passive solar-heated house.

Many styles of house have been designed to optimize the direct absorption of solar heat. This, combined with careful insulation, can greatly reduce the amount of additional energy required for comfortable temperatures, at least in temperate climates. The challenge to the architect is to design a house that collects solar radiation maximally in the winter but excludes it in the summer; otherwise undue amounts of air-conditioning energy would be required. Figure 16-3 shows a typical house of this type; it has been designed for a latitude of $46°$. At the summer solstice the sun rises as high as $67°$ ($90° - 46° + 23°$, where $23°$ is the tilt of the earth's axis). Shades on the sloping part of the solar windows can exclude most of this light but still permit illumination of the rooms. At the winter solstice the sun rises $90° - 46° - 23° = 21°$ above the horizon, and the solar windows are designed to maximize entry of this low-angle radiation. Furthermore, in summer the radiation falls on reflective floors and is largely returned to the outside. In the winter the radiation falls on massive concrete structures which store it to be given up at night. Houses of this type are more expensive to build but can eliminate 20% to 50% of annual heating costs. In the early 1980s when energy costs were greater there was great interest in this type of house. As energy costs dropped drastically, the payback times became unattractively long and interest has waned. It will surely return as the cost of energy inevitably rises.

Active Collector Panels

Because of high cost, there is very little economic incentive to build and install solar collectors that track the sun, or even those that focus the light. For this reason most solar collectors are of the fixed flat-plate type mounted at the best angle and facing south. In Ontario, mounting the plate at $60°$ to the horizontal reduces the seasonal variation of Fig. 16-1 (which was for a horizontal plate) to the range 1.5 to 4.6 kW· h/m^2.

Flat-plate collectors consist of blackened absorbing surfaces which transfer heat to pipes containing water; an outer glass cover admits visible light but prevents re-radiation of the emitted infrared radiation (cf. the greenhouse effect —— Chapter 7). Efficiency varies with design; for a given collector, the hotter the surface the greater the heat losses and the less the efficiency. Obviously, improving the insulation and even evacuating the space under the glass improves the efficiency, but both are costly. The hallmark of small domestic solar collectors is usually cheapness and simplicity at the expense of small gains in efficiency; otherwise the monetary cost is not worth it.

In many countries with long periods of clear sunny skies (Israel, Australia, Japan, Southern U.S.A.) domestic water and space-heating have been accomplished for years with solar collectors. Let us consider the possibilities for a location like southern Ontario. A typical single family home has a roof space sufficient for about 30 m^2 of south-facing collector. At 33% efficiency, this would provide about 15 kW·h (0.33 × 1.5 kW·/m^2 × 30 m^2) per day in the winter and about 45 kW· h per day in the summer. At an average of 30 kW· h per day this is 11,000 kW· h per year. According to the Ontario Ministry of Energy, a two-storey 1800-ft^2 house in Toronto uses 30 000 kW· h of energy for heating per year; existing homes cannot be heated solely by solar energy. The situation is even worse because in the winter when heat is needed, there is not as much solar energy available, as shown in part (c) of Example 16-1 below.

Example 16-1 *As a case study, one of the authors' homes in Guelph Ontario has a floor area of 1800 ft^2 and uses 2615 m^3 of natural gas per year for heating and hot water. The consumption is distributed over the year as shown in Fig. 16-4, as reported directly on the Union Gas Company's bill.*

(a) *Estimate how much gas is used each year for space-heating.*
(b) *What is the energy equivalent of this gas in kilowatt hours?*
(c) *Compare the gas consumption in December with the energy that could be obtained from a solar collector of area 30 m^2.*

(a) *From the summer readings in Fig. 16-4 it can be estimated that hot water uses about 2 m^3 per day for a yearly consumption of 2 m^3 × 365 = 730 m^3. This leaves 1885 m^3 for space-heating.*

(b) *The energy content of natural gas is given in Appendix II as 38 × 10^6 J/m^3. Therefore, the energy equivalent of 1885 m^3 of natural gas is*

$$1885 \; m^3 \times 38 \times 10^6 \; J/m^3 = 7.2 \times 10^{10} \; J$$

To convert this to kilowatt · hours, we use the conversion factor from Section 1.5:

$$1 \; kW \cdot h = 3.6 \times 10^6 \; J$$

Thus, $7.2 \times 10^{10} \; J \times \dfrac{1 \; kW \cdot h}{3.6 \times 10^6 \; J}$

$$= 2.0 \times 10^4 \; kW \cdot h$$

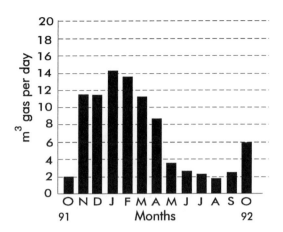

Figure 16-4: Yearly gas consumption for a representative two-storey house.

(c) *Looking at December we see that the gas consumption was 11 m³ × 30 = 330 m³, which is equivalent to about 3500 kW·h. A 30-m² solar collector could supply at most 1.5 kW·h/m² per day. (This value is provided in the previous text material.) Multiplying this by 30 m² and by 30 days gives about 1400 kW·h, which is 40% of the energy required in the house.* ∎

To some extent the inefficiency of solar heating can be improved if a method of storage can be implemented to save the heat from times of plenty to be used when needed. For single-family dwellings this is usually not practical or economical for periods exceeding 24 hours, but daytime heat can be stored to use at night. Season-long storage is only practical in large buildings as discussed in Chapter 17. When houses do have heat storage it is usually in rock or water or heavy concrete walls designed for the purpose.

In most of the populated areas of North America the effectiveness and economic prospects of domestic space-heating with solar energy are not great (See problems 16-1 and 16-4 for example). Some of the apparently successful "solar houses" are really "super-insulated" houses, and it is the insulation and passive solar aspects that are worth enshrining in building codes, not active-solar aspects. The retro-fitting of existing houses is difficult, expensive and usually inefficient unless there is massive reconstruction to incorporate super-insulation. After a flurry of activity in the 1970s when solar heating was oversold by its advocates, enthusiasm has considerably cooled on the part of the public and governments.

Simple domestic water-heating systems look rather more promising, since the need for hot water is year-round. Even here though there is little financial incentive unless the system is installed on a do-it-yourself basis.

Solar Ponds[1]

Large-area bodies of water absorb solar energy very efficiently but, because of convection, do not build up large temperature differences with depth. The energy is absorbed in the first metre or so below the surface and because of thermal expansion the water becomes less dense and starts to rise. This convective process results in the water being well mixed and of uniform temperature. However, in 1902 A. Kalecsinsky first reported an unusual phenomenon in the Medve Lake in Transylvania where the temperature of the water 1.3 m below the surface was 70°C higher than at the surface. The same phenomenon has been reported in small lakes in the United States and even Lake Vanda in Antarctica which, while frozen on the surface, had a temperature of 25°C at a depth of 65 m.

Of course something must be suppressing the natural convection of the water. This comes about because there is a varying concentration of salt dissolved in the water, with more salt at the bottom than at the top. The presence of the salt increases the density of the water so that it cannot become light enough to rise by thermal expansion alone. Such *solar ponds,* if made artificially, can support temperature gradients so steep that the bottom water is near the boiling point.

Many countries have shown an interest in solar ponds, including the United States and Australia, but nowhere have they been more developed than in Israel. Artificial ponds one to two metres deep with a blackened (absorbing) bottom and areas of thousands of square metres have been constructed which produce temperature differences of 85 to 90°C. For power production the temperatures are still not very high so the thermodynamic efficiency is very low; the theoretical maximum is about 20% and half of this is a more likely performance. Nonetheless small thermal pond power plants have been built. A 7000-m^2 pond at Ein Bokek on the shores of the Dead Sea provides a peak electrical power of 150 kW.

---

[1] H.Tabor, "Solar Ponds", *Solar Energy* **27**, pp. 181-194 (1981)

Solar ponds have also been used as a source of industrial heat to desalinate water and to produce salt.

16.3 LARGE-SCALE DIRECT UTILIZATION OF SOLAR HEAT

The basic problem with solar energy is that it is so dilute; it has to be collected from a large area if it is to be useful as a substitute for our other large-scale energy systems, such as electric utilities. This generally precludes inhabited and agricultural areas, since large tracts of land are unavailable or expensive. The earth has many desert and semi-desert areas, however, where the concentration of solar energy is at a maximum. Clearly solar plants, of whatever type, must be used to manufacture a high-value energy currency to ship to wherever the energy is needed; this means electricity or hydrogen.

In a solar "farm" a vast area is covered with steerable mirrors which collect the sunlight and direct it to a central receiver. This receiver is usually a steam boiler on a high tower in the centre of the mirror array, constructed so as to absorb the energy and transfer it to flowing water in steam tubes. In principle this concentration of light can produce steam at temperatures around 500 °C, which compares to the best coal-fired plants; the high temperatures result in high efficiencies. Although the receiver may be complicated and expensive, the mirrors must be cheap since so many of them are required, and they must be steerable. After much experimentation in the late 1970s and early 1980s a few small power plants were built.

The first was the Eurelios plant at Adrano in Sicily which develops one MWe under clear skies; it was opened in December 1980. The energy is collected by 182 heliostats (steerable mirrors) with a total area of 6400 m². The motor-drives track the sun to an accuracy of 0.2°; this high accuracy is necessary because the angular size of the sun is only 0.5°. The receiver is a conical shell lined with water tubes on a tower 55 m high. The steam at a pressure of 100 atmospheres is directed either to the turbines or to storage. When it opened, Eurelios produced power at twice the current cost and the cost structure has not improved since that time. Several other plants in the 1 to 10 MWe range have been built since in the U.S.A., Spain and Japan.[2]

Distributed Collector Plants

The most promising type of solar electric plant so far is the distributed collector plant pioneered in California by the LUZ Corporation[3]. These plants are built in modules of 30-MWe output, each covering 6×10^5 m^2 of land. A single unit consists of a parabolic collector mirror 52 m long with an aperture of 2.5 m as shown in Fig. 16-5; 1600 of these units constitute one 30-MWe module. The mirrors concentrate the light on steel tubes encased in an evacuated glass tube with a concentration ratio of 61:1. Oil is pumped through the steel tubes where it is warmed from 245°C to 310°C and then produces steam at a temperature of 250°C and a pressure of 38 atmospheres. This is a rather low temperature for operation of turbines, and the resulting thermodynamic efficiency would be low. The steam is further heated to 415 °C in a gas-fired superheater before being sent to the turbine. This system is then a hybrid system; the solar collector serves as a preheater. By the beginning of 1990 the total installed capacity was 285 MWe.

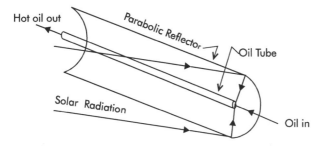

Figure 16-5:   Parabolic solar collector.

---

[2] C.J. Weinberg and R.H. Williams, Sci. Am. **263**, No. 3 (Sept. 1990), pp. 146-155.

[3] Ingersol, "Energy Storage Systems", in *The Energy Sourcebook*, Amer. Inst. of Physics, New York, 1991, p. 220.

The installation cost is less than $3000 per kWe which is less than the cost of nuclear installations and the estimated cost of the electricity is $0.08 per kW·h, which is competitive. Of course, it again must be stressed that this is only feasible where sunlight is plentiful and land is cheap.

16.4 DIRECT CONVERSION OF SOLAR ENERGY TO ELECTRICITY

Solar cells are based on semiconductors, of which silicon is the most widely used. Silicon is a material of valence 4 which forms a cubic crystal with each silicon atom covalently bonded to four other silicon atoms. This is a very stable structure since the only electrons that are free are the few that are momentarily loosened by the vibration of the parent atoms on their sites; the material is a very good insulator at room temperature.

Silicon is made into a semiconductor by doping (adding a small impurity, say one part in 10^8, of an element of valence 3 or 5). For example, phosphorus with a valence of 5, has 5 outer electrons available for covalent bonding; it uses 4 of these to fit into the crystal lattice and its fifth electron is surplus. This extra electron is free to provide conduction in the lattice. Phosphorus is a "donor" atom and the semiconductor is called "*n-type*" because it has negative charge carriers. The substance is *not* negatively charged; it is just that the electrons in the solid are not all needed for the covalent bonding, and so a few are free to move in the presence of an applied electric field.

A different type of semiconductor can be made by doping with a valence-3 atom like boron. Boron is called an "acceptor" atom. This "*p-type*" semiconductor ("p" for positive charge carriers) lacks one electron for covalent bonding at each boron atom; this gap is called a "hole." As the atoms vibrate on the lattice the few free electrons can hop from one site to another; if a hole is filled with an electron, a new hole is created at the site where the electron came from. It is as if the holes moved in the opposite direction, behaving as though they were positive charges. The holes will also "move" in the presence of an electric field.

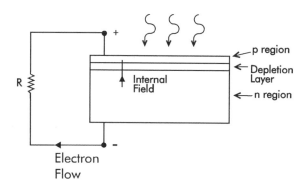

Figure 16-6:   A silicon solar cell.

Now suppose a slab of p-type silicon is placed in contact with a slab of n-type as in Fig. 16-6. At the junction the holes in the p-type will be filled with electrons from the n-type and this region will become rapidly depleted of both free electrons and free holes. The assembly as a whole is still electrically neutral. The depletion layer is thin because the depletion of holes in the p-type builds up an overall negative charge there, while a positive charge is left behind in the n-type due to a loss of electrons. An electric field is created, directed from the n- to the p-type. A state arises, essentially instantaneously, where a hole trying to enter the n-region is unable to do so because of the force due to the electric field; a similar argument holds for electrons going the other way. The depletion region is very narrow ($\sim 10^{-4}$ cm), and the voltage that appears across it is about 0.4 V.

A Si-photodiode, designed for use as a solar cell, is just such a p-n junction. One surface is doped with boron to about 0.002 mm depth; the remaining 1 mm is doped with arsenic to make it n-type. When the device is exposed to sunlight, a photon may transfer its energy to a bound electron, ejecting it from a valence bond and leaving it free to move as a conduction electron and also creating a hole. If this happens in the depleted region the electric field will separate the electron-hole pair and the electron will flow around the external circuit to refill a hole on the p side. (See Fig. 16-6.) The n- and p-layers act essentially as the terminals of a battery and the photoelectric interaction in the depleted layer is the

357

analogue of the chemical processes in chemical batteries which release electric charge. A silicon solar cell will produce electron-hole pairs by absorption in the wavelength range 350 nm to 1100 nm — a range which covers the visible spectrum and some of the near infrared. This corresponds to photon energies of 3.5 eV to 1.1 eV.

Example 16-2 *Given that the maximum wavelength absorbed in a np-Si solar cell is 1100 nm, verify that the corresponding photon energy is 1.1 eV (as stated in the previous paragraph).*

The maximum absorbed wavelength is 1100 nm or 1.1×10^{-6} m.

The energy of a photon is given by Planck's equation:

$$E = \frac{hc}{\lambda}$$

where h is Planck's constant, c is the speed of light, and λ is the wavelength of the electromagnetic radiation.

Substituting numerical values, and then converting joules to electron-volts:

$$E = \frac{(6.63 \times 10^{-34} \ J \cdot s)(3.0 \times 10^{8} \ m/s)}{1.1 \times 10^{-6} \ m} = 1.81 \times 10^{-19} \ J$$

$$1.81 \times 10^{-19} \ J \times \frac{1 \ eV}{1.60 \times 10^{-19} \ J} = 1.1 \ eV$$

Not all of the solar spectrum has photons of sufficient energy to release valence electrons, and so a Si solar cell has a theoretical maximum efficiency of 29%. In practice the efficiency is much less than this for various reasons: Some of the light is reflected from the surfaces and some photons are absorbed in such a manner that they do not produce electron-hole pairs but only excite vibrations in the crystal lattice, simply producing heat. The cell also has electrical resistance, so there is ohmic heating when a current is drawn from the cell. The practical efficiency is of the order of

12%. Much of modern research in this area is directed to improving the efficiency of solar cells. Efficiencies approaching the theoretical limit have been achieved in the laboratory but not yet in the field.

The difficulty of producing monocrystalline silicon has led to the development of thin layers of amorphous silicon which is very efficient at absorbing light. A layer of amorphous silicon 1 μm thick absorbs as much light as a 50 μm layer of crystalline silicon. Amorphous layers have found considerable use in small power applications such as calculators but the silicon suffers damage with continued exposure to bright light and has not yet proven itself for large power applications.

In 1980 solar cells in large quantity could be manufactured for about $10 per peak watt; today this is slightly better at about $5. If we take the best location in the American southwest with 300 W/m^2 the installation cost is $16 per watt or $16,000 per kWe. These costs would have to fall by at least a factor of six to compete with conventional sources. In Canada where the solar flux is one-half, the cost is worse by a factor of two. The cost of the generated energy is likewise uneconomically high; Figure 16-9 in the next section summarizes the cost of electricity from various sources, and it can be seen that in 1990 photovoltaic power was the highest-priced form of power, exceeding conventional power by a factor of four; in addition, its price is predicted to be the slowest declining.

In spite of this gloomy economic forecast several large photovoltaic power plants have been built in the U.S.A., Germany, Switzerland, Finland, Austria and Britain with capacities ranging from 15 kWe to 6.5 peak MWe at Carissa Plains in California. Fortunately, the energy cost of making Si-crystals is not too large; a 1-kW solar device requires 7000 kW·h of energy for manufacture. This must be supplied by conventional power, but takes only about one year to "pay back" using energy produced by the solar cells themselves.

Photovoltaic power does, of course, have its uses. In regions where it is difficult or expensive to deliver power in other ways, photovoltaic power does have a place; for instance, telephone repeater amplifiers have been installed on poles in remote sunny areas to provide almost trouble-free power. Another example is illustrated in Figure 16-7, which shows a common sight along highways in California —— a solar-powered emergency call box.

Figure 16-7: This emergency call box in California is powered by the solar-cell array at the top.

Research is also being done on other photovoltaic compounds such as the copper-indium-chalcogenides (e.g., $CuInS_2$) which have the promise of higher efficiencies, but they are some years away from practical and economic application. Another approach is to construct systems which mimic photosynthesis, the process in living systems that converts sunlight into usable energy. In these systems a metal-organic dye molecule, chosen to maximize the efficiency of sunlight absorption, is grafted onto a titanium dioxide semiconductor. These devices show promise but, again, are still in the early laboratory phase of development. All of these opportunities merit intensive research.

16.5 WIND ENERGY

The energy of the wind has been harnessed for centuries by sailing-ships, windmills, etc. Now wind energy is being explored as a source of electrical energy. One important consideration when contemplating an alternative energy source is whether the power it can provide is large enough (on a global or national basis) to replace a significant amount of fossil fuel. Without question, the source that can provide the greatest power is the sun; it has been estimated[4] that the ground-level solar power incident on the U.S.A. is about 500 times the U.S. total power consumption. The next "strongest" source available is the wind —— the estimated power in surface winds in the U.S.A. is about 30 times U.S. power requirements. All other sources, such as photosynthesis (to give biomass), geothermal energy, tides, and waves could each make only a very small contribution, much less than the sun and wind.

Wind Power

Figure 16-8:   Typical wind turbines. A generator is housed in the box connected to the blades at the top of each tower.

The power P available from a horizontal-axis wind turbine such as those shown in Fig. 16-8 depends strongly on the wind speed v and on the rotor-blade radius R:

$$P = \frac{1}{2}\eta\rho\pi R^2 v^3 \tag{16-1}$$

where η is the efficiency of conversion of the wind's kinetic energy to electrical energy, ρ is the density of air, and π is the usual 3.14... The derivation of Eqn. 16-1 is left as a problem (16-5), but we can readily understand the origin of some of its terms. The power of a wind turbine depends on the volume of air that it can intercept, which in turn is proportional to the area swept out by the blades, that is, πR^2. The energy contained in a mass of air passing through this area is the familiar kinetic

---

[4] J.M. Fowler, *Energy and the Environment (2nd Ed.)*, McGraw-Hill, New York, 1984, p. 364.

energy, $\frac{1}{2}mv^2$, which depends on v^2, and the mass intercepted in a given time interval is proportional to the speed of the wind, v. Hence, the power (energy per unit time) depends on v^3.

Because of this v^3-dependence, it is apparent that wind turbines should be sited in areas of high wind speed —— a doubling of v increases the power by a factor of eight. As well, the rotor-blades should have as large a radius as possible; doubling R quadruples the power. To have a large rotor radius, the turbines must be tall; but as a turbine is increased in height, its base must be built more securely (and expensively) in order to prevent the entire unit from being blown over.

The efficiency (η) of extraction of wind energy could never be 100% because the wind would then completely stop after passing the turbine blades. Indeed, if the efficiency is reasonably high, the turbine is extracting a lot of energy from the wind and acts as a large impediment to the air flow. Hence, some of the wind will divert itself around the turbine instead of flowing through the area of the blades, and some of the wind's energy will thus not be recoverable. A complete theoretical analysis shows that the upper limit for the efficiency of windmills is 59%. The most efficient designs at present have an efficiency of about 45%, i.e., about 3/4 of the theoretical maximum.

Advantages and Disadvantages of Wind

Wind energy has some obvious advantages: it generates no air pollution (NO_X, SO_2, CO_2) and no radioactive waste, and is renewable. Wind turbines can be prefabricated quickly in a variety of sizes, and easily installed at appropriate sites. The small unit size (in contrast with coal-fired and nuclear plants) results in rapid advances in construction technology and installation procedures. Perhaps most importantly, wind energy is now becoming economically competitive[5] with coal-fired electrical plants, as shown in Fig. 16-9. (Notice that the cost of solar energy is not expected to compete until the first decade of the 21st century.) Wind energy is already an economical source in remote areas where electricity would otherwise be produced by expensive diesel-fuelled generators.

---

[5] C.J. Weinberg and R.H. Williams, Sci. Am. **263** No. 3 (Sept. 1990), p. 154.

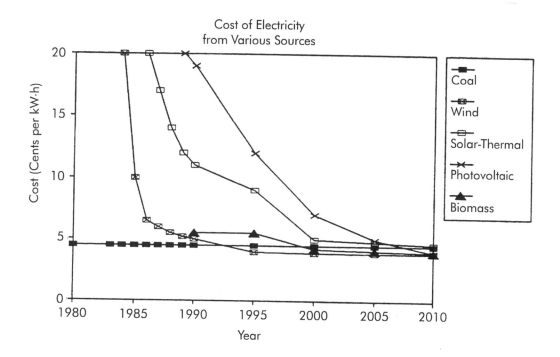

Figure 16-9: Wind energy is rapidly becoming economically competitive with coal-fired generation of electricity. (Costs after the year 1990 are projections made in 1990.)

As is the case with all energy sources and technologies, wind has disadvantages as well. Since the wind does not blow all the time, it does not provide energy continuously. Hence, it requires either energy storage or a backup energy source. In addition, sites with high winds usually are not located close to major population centres, thus necessitating very-long-distance transmission of electricity; in Canada, the windiest locations are remote coastal regions of Newfoundland, Quebec, and the Northwest Territories. Many turbine units are required to generate the power output of even one fossil-fuel or nuclear plant. Typical wind turbines presently in use have power ratings of 50 kW to 300 kW, although there are a few huge

2.5-MW units, standing 60 m high and having rotor-blades of diameter 100 m. If we assume an average power output for existing turbines of 100 kW, it requires 10000 of these units to replace a single 1000-MWe fossil-fuel plant. Hence, a great deal of land is required for large-scale "wind farms," though 90-95% of the land can still be used for agriculture[6].

In the early days of wind turbines noise was a concern, but modern turbines make little noise above the rush of the wind. Steel rotor-blades cause television interference, but most blades are now made of light non-metallic materials, such as fibreglass, which alleviate this problem. Bird-kills are a possible worry, but the most serious difficulty might simply be an aesthetic one —— many people view the turbines as an eyesore.

Example 16-3 *A typical home in southern Ontario uses about 40 kW·h of electrical energy daily during the winter. (This includes water heating, but not space heating.) The annual average windspeed is about 5.5 m/s (a moderate breeze) at a height of 10 m above the ground. In order to provide all the electrical power for a home, using a wind-electric generator of efficiency 35%, what diameter is required for the rotor-blades?*

We first convert 40 kW·h to joules, and then determine the electrical power (P) required by the home:

$$Energy\ E = 40\ kW \cdot h \times \frac{10^3\ W}{1\ kW} \times \frac{1\ J/s}{1\ W} \times \frac{3600\ s}{1\ h} = 1.44 \times 10^8\ J$$

$$P = \frac{E}{t} = \frac{1.44 \times 10^8\ J}{1\ day} \times \frac{1\ day}{24 \times 60 \times 60\ s} = 1.67 \times 10^3\ W$$

Now use Eqn. 16-1:
$$P = \frac{1}{2}\eta\rho\pi R^2 v^3$$

The power P has been calculated, η is given as 0.35, ρ is the density of air (which is 1.3 kg/m³), and v is given as 5.5 m/s. Calculation of R gives:

$$R = 3.74\ m$$

---

6 J. Chapman, "Wind as a Utility Generation Option," in *The Energy Sourcebook*, Ed. by R. Howes and A. Fainberg, American Institute of Physics, New York, 1991, p. 316.

Hence, the required diameter is 2 × 3.74 m = 7.5 m, which is rather large (and expensive) for a typical backyard. We have not included any "wind-shadowing" effects due to buildings, trees, or other wind turbines.

■

Vertical-axis Turbines

Most of the wind turbines in use today have the horizontal-axis design shown in Fig. 16-8. However, a few have a vertical axis; the most common version is the eggbeater-style Darrieus rotor (named after the French inventor G.J.M. Darrieus who patented it in 1920), shown in Fig. 16-10. A vertical-axis rotor has the advantage that the turbine and generator are located at the base, thus providing easy access for maintenance. It also is not sensitive to wind direction, unlike the horizontal-axis design which must have some means to keep it aligned with the wind. A disadvantage of the vertical-axis system is that it is less efficient, because each blade must move against the wind during part of the rotation. With properly shaped blades that present a smaller profile during upwind movement, efficiencies can reach 35%.

Figure 16-10: A Darrieus rotor.

Present and Future Wind Installations

As of 1992, the world wind-electric generating capacity[7] was only about 2350 MW; approximately 68% of this was in California, mostly at Altamont Pass (Fig. 16-11) where there are approximately 7500 wind turbines, producing about 1.5% of the state's electricity. (In the 1970s and 1980s, tax incentives produced a boom in development of wind energy in California.) Most of the remaining 750-MW capacity is in Europe, where about 3250 MW of new capacity is scheduled to be installed by the year 2000, compared to only an additional 750 MW being planned in the U.S.A.[7] As the cost of wind-generated electricity decreases, we might see a rapid increase in generating capacity in areas of high wind and close access to population centres. Several European countries are exploring the possibility of locating "wind farms" offshore.

Figure 16-11: A few of the thousands of wind turbines at Altamont Pass near San Francisco, California.

New designs will help to reduce the cost of wind-produced electricity through the use of advanced airfoil shapes for rotor-blades, computerized electronic controls that allow the turbine to turn at optimum speed for

---

[7] J.G. McGowan, "Tilting Toward Windmills," in *Technology Review*, July 1993, p. 42.

high efficiency under a wide variety of wind conditions, and lighter, stronger materials. Newer turbines will probably have power outputs in the range of 300-500 kW.

16.6 GEOTHERMAL ENERGY

We are all aware of dramatic displays of geothermal energy (for example, volcanoes and geysers —— Fig. 16-12), as well as its more passive manifestations, such as hot springs. Geothermal energy is derived from natural nuclear energy; the radioactive decay of elements such as uranium, thorium, and radium in the earth produce heat, which gradually makes its way to the surface of the earth (and is eventually radiated away into space as infrared radiation). In most areas of the world, this geothermal energy is very diffuse; the rate of heat transfer to the earth's surface is only about 0.06 W/m^2. A related aspect of geothermal energy is the increase of temperature with depth below the earth's surface, with the average thermal gradient being about $30\,^\circ$C/km.

Figure 16-12: A geyser in the Napa Valley in California.

Geothermal energy can be readily exploited in regions of the earth where the rate of heat transfer to the surface is much higher than average, usually in seismic zones at continental-plate boundaries where plates are colliding or drifting apart. For example, the heat flux at the Wairakei thermal field in New Zealand[8] is approximately 30 W/m^2, and there are regions of Iceland[9] where the thermal gradient is greater than 100°C/km.

Using Sources of Geothermal Energy

There are two general ways that geothermal energy is utilized: for generating electricity, and space-heating. The first geothermal electric plant began operation at the Larderello thermal field in Italy in 1904, but by 1985 the world's geothermal electrical generating capacity was only about 5000 MWe, with 188 separate power units in 17 countries[10]. The space-heating power provided worldwide[8] in 1985 was about 7500 MW, with 11 countries having substantial developments[10]; the major users are in Japan, Iceland, and Hungary.

Geothermal sources are categorized into various types:

- *Natural steam reservoirs* —— These are the most desirable sources, since the dry steam (containing no liquid water) can be used directly in steam-electric turbines. Such fields are being used in the U.S.A., Italy, and Japan. The technology is basically the same as for fossil-fuel electric plants, except that the temperature and pressure of the geothermal steam are much lower. For example, at the Geysers geothermal plant in California, the temperature is about 200°C and the pressure about 700 kPa (7 atm.). Hence, from the second law of thermodynamics (Sections 4.4-4.6), the efficiency of conversion to electricity is less than at a fossil-fuel plant; at the Geysers, it is only 15% to 20%. The generating units are also smaller, ranging in size from 55 MW to 110 MW at the Geysers.

---

[8] R. Howes, "Geothermal Energy," in *The Energy Sourcebook,* American Institute of Physics, New York, 1991, p. 239.

[9] R. Harrison, "Applications of Geothermal Energy," *Endeavour*, New Series, Vol. 16, No. 1, 1992, p. 31.

[10] Ref. 9, p. 33.

- *Wet steam reservoirs* —— These fields are much more common than the simple dry type. Here the field is full of very hot water, under such high pressure that it cannot boil. When a lower pressure escape route is provided by drilling, some of the water suddenly evaporates ("flashes") to steam, and it is a steam-water mixture that reaches the surface; the steam can be used to drive a turbine. The first flash is conducted under some pressure, and the pressurized hot water from the ground, along with the hot residual water from the turbine, can be flashed again to lower pressure, providing steam for a second turbine. Flashed power production is used in many countries, including the U.S.A., the Philippines, Mexico, Italy, Japan, and New Zealand. If the temperature of the original hot water is too low for effective flashing, the water can be pumped to the surface under pressure to prevent evaporation (which would decrease the temperature), and its heat transferred to an organic fluid which evaporates and acts as the working fluid in a turbine.

- *Hot water reservoirs* —— These reservoirs contain hot water at a temperature too low for electricity generation. However, the water can be used for heating of buildings such as homes, greenhouses, fish hatcheries, etc. This heating can be either direct or through the use of heat pumps. The most extensive use of geothermal heat is in Iceland, which lies in a region of continental-plate activity. Virtually all the homes and buildings in Iceland are heated geothermally, and there is also a small geothermal electric plant. Other important examples of geothermal heating are in France, one near Paris and one in the southwest. During oil-exploration drilling in the 1950s, hot water was discovered in the Paris region, but exploitation did not begin until the 1970s as a result of rapidly increasing oil prices. In France, the equivalent of 200 000 homes are being provided with space-heating and water-heating from geothermal sources. One interesting feature of the French geothermal sources is that they do not occur in regions of elevated thermal gradient.

- *Hot dry rock* —— In many regions of the world, hot rocks lie near the earth's surface but there is little surrounding water. Attempts have been made to fracture such rocks and then pump water into them to extract the thermal energy, but the technical difficulties in fracturing the rock have proven to be much more troublesome than anticipated, and there has been a problem with water losses. Consequently, progress in extracting energy from hot dry rocks is expected to be slow.

- *Normal geothermal gradient* —— In principle, the normal geothermal gradient produces a useful temperature difference anywhere on the globe. If a hole is drilled to a depth of 6 km (which is feasible), a temperature difference of about $180\,^\circ$C is available, but no technology has been developed to take advantage of this resource. At this depth, water is unlikely, and the problems of extracting the energy are similar to the difficulties encountered with hot dry rock near the surface.

- *Magma* —— The potential thermal resource represented by volcanoes and magma pools presents an immense technological challenge. The high temperatures produce obvious problems with melting and deformation of equipment.

Advantages and Disadvantages of Geothermal Energy

The main advantage of geothermal energy is that it can be exploited easily and inexpensively at several locations in seismic zones around the world. ("If you've got it, use it.") However, it is unlikely that geothermal sources will become economical in areas of normal thermal gradient in the foreseeable future.

One of the primary disadvantages of geothermal energy relates to dissolved salts and gases in the water and steam. Hot water dissolves salts from surrounding rocks, and this salt produces severe corrosion and scale deposits in equipment. In order to prevent contamination of surface water, the geothermal brine must be either returned to its source or discarded carefully in another area. A gas that is often found in association with geothermal water and steam is hydrogen sulphide (H_2S), which has an unpleasant odour (like rotten eggs), and is toxic at high concentrations. At the Geysers plant, attempts have been made to capture the H_2S chemically, but this job has proven to be surprisingly difficult; a working H_2S-extractor is now in place, but is too expensive to provide a long-term solution. Geothermal plants are also a source of the greenhouse gas CO_2 which is dissolved in the brines.

Most people tend to think of geothermal energy as being renewable, but in fact one of the major problems in choosing a geothermal energy site is estimating how long the energy can usefully be extracted. If heat is withdrawn from a geothermal source too rapidly for natural replenishment, then the temperature and pressure can drop so low that the source becomes unproductive. Since it is expensive to drill geothermal wells and construct power plants, a source should produce energy for at least 30 years in order to be an economically sound venture, and it is not an easy task to estimate the working lifetime beforehand.

Another potential problem, particularly if geothermal water is not returned to its source, is land subsidence. For example, there has been significant subsidence at the Wanaker field in New Zealand. Finally, an annoying difficulty with geothermal heat has been the noise produced by escaping steam and water.

16.7 TIDAL ENERGY

You probably know that in the past there have been grain-grinding mills powered by wind or by falling water in rivers. However, there have also been mills powered by the tides, dating from as early as the eleventh century in England, France, and Spain. One such mill[11] at Woodbridge, England, was mentioned in the local parish records in 1170, and operated until 1957. In North America, tide mills were used as early as 1617.

Nowadays, attention is being given to using the tides to generate electricity. Whether the tides power a mill or produce electricity, the basic principles are the same. A dam is built across a natural tidal basin, with sluice gates to allow the passage of tidal water into the basin. The incoming tide is allowed to enter, and the gates are closed at high tide. After the tide has fallen somewhat, the collected water is allowed to leave the basin, turning a wheel or turbine.

Tides are caused by the gravitational forces of the moon and sun on the waters of the earth. As the ocean waters move relative to the earth in twice-daily tides, energy is dissipated through friction with the coast and

---

[11] G. Duff, *Tidal Energy: or, Time and Tide Wait for No Man*, American Association of Physics Teachers, 1986, p. v.

also through internal viscous friction. The rate of energy dissipation by the tides has been estimated[12] to be 4×10^{12} W, or 4 million MW, and since this energy is drawn largely from the earth's rotational kinetic energy, the earth's rotation is gradually slowing down. The length of the day is increasing by about 0.002 s per century[13].

Only a small fraction of this tidal energy could possibly be tapped. A suitably large basin and a drop in water level of several metres are needed to make a tidal project worthwhile. It has been estimated that only about 60 000 MW of generating capacity might eventually be developed[14]. This is very small compared to, say, the estimated worldwide hydroelectric resources[15] of 2 million MW.

Present Use of Tidal Energy

As of 1992, there are only three tidal projects in operation in the world. A 240-MWe plant has operated since 1966 in the La Rance estuary in the northwest of France; on the Barents Sea north of Murmansk in the former U.S.S.R., there is a 0.4-MWe pilot plant, built in 1969; and at Annapolis Royal in Nova Scotia (Fig. 16-13), a 20-MWe pilot came into operation in 1984. With the current low price of oil, there is little activity in developing new tidal plants, which have high capital costs.

Advantages and Disadvantages of Tidal Energy

Tidal energy is renewable, and produces no air pollution and no radioactive waste. Although the operating costs are low and tidal plants have long lifetimes, they are expensive to build. An evident problem is the periodicity of the power output, since high tide occurs only twice per day.

---

[12] Ref. 11, p. 12. (Original source: D. Cartwright, "Oceanic Tides," Rep. Prog. Phys. **40**, pp. 665-708 (1977)).

[13] G. Abell, *Exploration of the Universe (4th Ed.)*, Saunders Publ., Philadelphia, 1982, p. 142.

[14] M. Sanders, "Energy from the Oceans," in *The Energy Sourcebook*, American Institute of Physics, New York, 1991, p. 274.

[15] Ref. 4, p. 403 (Original source: D. Deudney and C. Flavin, *Renewable Energy: The Power to Choose*, Worldwatch Institute, W.W. Norton, New York, 1983, p. 168).

This problem could be alleviated somewhat by allowing the incoming water to flow over the turbines, thus giving four power-producing cycles per day, or by a complicated scheme in which the collecting basin is divided into two or more separate chambers, and the flow is arranged so that there is always a tidal head between two chambers or one chamber and the ocean. However, the simple scheme with one basin and one flow per tide is the most economical. Another possibility for tidal plants is the use of electricity from other sources during times of low demand to pump water into the basin, to be utilized by the tidal turbines at peak-demand times.

Figure 16-13:   The tidal power plant at Annapolis Royal, Nova Scotia.

Environmental problems that can arise from tidal plants are sedimentation, flooding of upstream river areas if a plant is built at a river mouth, and destruction of tidal mudflats with consequent effects on animals and plants.

16.8 WAVE ENERGY

There is a great deal of energy in ocean waves —— for example, it has been estimated[16] that approximately 120 GW of wave power is dissipated on the west coast of the British Isles, about five times the British electrical power requirement. At the present time, there are only a few experimental projects investigating the use of wave energy to produce electricity, which is expected to cost about 9-15¢ per kW· h, well above the price of coal-fired or wind-produced electricity (cf. Fig. 16-9).

Most generators designed to use wave energy exploit the vertical motion. In a typical design, a rising wave compresses air in a chamber; the air passes through a one-way valve and turns a turbine. As the wave recedes, the air returns to the chamber through another one-way valve, again turning the turbine. About 1000 floating navigation buoys in oceans around the world are each powered by a 60-W wave-powered air turbine working on this principle[17]. A land-based pilot plant, recently constructed on the Scottish island of Islay, uses a similar (but larger) turbine to produce 40 kW of electrical power for a small nearby village.

Wave energy is renewable, and like the tides generates no air or water pollution and no radioactive waste. Large-scale exploitation of wave energy would undoubtedly affect local aquatic and shoreline ecosystems, and perhaps alter ocean currents as well.

16.9 BIOMASS ENERGY

Biomass energy refers to the energy that is available from plant material such as wood, or animal matter such as manure. Wood is the most common source of biomass energy, but as we discuss in this section, gaseous and liquid fuels can also be produced from biomass.

Wood

Around the world, fuelwood constitutes the single most common use of biomass energy. The forest-products industry frequently uses wood waste as an energy source, and residential wood-burning, even in North

------

[16] Ref. 14, p. 285 (Original source: D. Ross, *Energy from the Waves*, Pergamon Press, New York, 1979).

[17] Ref. 14, pp. 283-284.

America, is not uncommon. Wood shares the positive features of all biomass: it is renewable, contains little sulphur (and hence produces little SO_2), and if trees are replaced at the same rate that they are harvested, there is no net contribution to atmospheric CO_2. However, wood-burning generates particulate matter (PM) and NO_X, and perhaps most importantly, it produces a carcinogenic hydrocarbon: benzo(a)pyrene (abbreviated BaP). About 25% of BaP emissions in the U.S.A. in the year 1975 were estimated to come from burning wood. (The single largest source was industrial coke ovens, contributing almost 40%.) If a large fraction of residential heating were converted to wood-burning, pollution due to BaP and PM would be a serious problem. In areas of the U.S.A. where many homes have wood-heating, the outdoor concentration of BaP has been measured at 10-100 ng/m^3, compared with only 0.7 ng/m^3 as the average concentration in U.S. cities[18]. Unless BaP emissions can be controlled, large-scale burning of wood will not be acceptable as a major source of energy.

The precise meaning of "renewable" in the context of any forest use should be considered carefully. It does not mean, for example, that the primal rainforest of Canada's west coast will be regenerated in its original state. Rather, it means that new managed forests will replace it, with the type of trees selected for reasons that are in part economic. The overall ecology will be different.

Biofuels

Liquid and gaseous fuels can be produced from biomass. These fuels are produced by three methods, discussed separately below: fermentation, anaerobic digestion, and extraction of natural plant oils. Figure 16-9 shows the projected cost of electricity generated by burning liquid and gaseous fuels produced from biomass.

---

[18] N. Bunce, *Environmental Chemistry*, Wuerz Publ., Winnipeg, 1991, p. 113 (Original source: E. Calle and E. Zeighami, Ch. 3 in *Indoor Air Quality*, Ed. by P. Walsh, C. Dudney, and E. Copenhaver, CRC Press, Boca Raton, FL, 1984).

Fermentation

Certain bacteria and yeasts are able to break down the sugars in plant material to produce alcohol which can be used as a motor fuel, or blended with gasoline to produce "gasohol" (90% gasoline, 10% ethanol) to raise the octane rating of gasoline without lead additives. Critics of alcohol as a fuel point out that the production of alcohol is a very energy-intensive process; for every joule of recoverable energy from the alcohol about two joules of energy must be invested. Defenders of the process claim that the comparison is unfair because the present technology is based on the methods for producing drinking alcohol, a high-value product, where the cost of the energy is of small consideration. They argue that the technology could be much improved and that alcohol could compete in the liquid-fuels market. In addition, alcohol does not increase atmospheric CO_2, since the CO_2 is simply cycled through photosynthesis and burning.

It is not true, however, that alcohol has negligible environmental consequences. In Brazil, where gasoline is very expensive, the government has encouraged a vast plan to replace a large fraction of gasoline consumption with alcohol. To this end, large tracts of the virgin rain forest are being cleared to grow cane sugar and cassava for fermentation. The large-scale clearing of the rain forest is an event of great global concern. In addition, it has been discovered that although alcohol-burning reduces carbon monoxide and particulate emissions it produces more aldehydes, which react in the atmosphere to make peroxyacetyl nitrate, a compound which stunts plant growth and causes eye irritation. Aldehydes also help increase ozone levels and thus urban smog.

Some agricultural economists worry about large-scale development of bio-fuels because of their possible impact on the world food supply. At the present time there is no world-wide food shortage in spite of the existence of widespread famine. (It is a cruel irony that there is a world-wide food surplus, in the face of frequent, large-scale famine.) A very large excess of food can be, and is, grown and a small amount of this excess does find its way into the world's famine relief programs. Imagine what would be the effect on those countries with zero or small ability to pay for food if all of the surplus crops could be utilized to produce a high-value product like transportation fuel which again only the wealthy countries could afford to purchase. The argument has been made that the famine problem might worsen.

It is certainly true that gasohol using alcohol derived from farm products is the only new liquid fuel to make any impression in the marketplace; gasohol is available in a few service stations in North America but again the oil glut of the late 1980s has dampened enthusiasm.

Anaerobic Digestion

In the absence of oxygen, certain bacteria break down organic matter to produce methane. On a small scale this technology has been used for years to provide cooking and heating gas for single families, particularly in the Third World. Small digesters, usually using manure from a family's domestic animals, can supply a family with a clean source of cooking gas.

The production of biogas has been carried out on a somewhat larger scale, particularly in agriculture. Considerable development was carried out during the energy crises of the 1970s to design low-cost digesters suitable for use on a large farm, especially those with large manure production such as dairy farms. The digesters are of two types: The "conventional" digester is usually a vertical tank into which preheated (35 °C) manure is injected. The manure is stirred and maintained at this temperature, and biogas is produced and taken from the top of the tank. Plug-flow digesters are long concrete troughs having the manure continuously fed in at one end and removed at the other. The retention time varies from 14 days for cattle manure to 40 days for poultry. Typical figures for a 200-head dairy farm are:

Daily manure volume: 13 m^3
Volume of digester with water and gas storage : 360 m^3
Daily methane volume: 125 m^3

If 30% of the methane is required to maintain the temperature of the digester then there is a net production of 88 m^3 per day with an energy content of 3.3×10^9 J. This is the equivalent of 95 L of gasoline. Because of its lack of purity and difficulty of storage in a small volume, biogas is unsuitable as a transportation fuel but it is an excellent heating fuel.

On a large scale the technology has not been much developed although there is currently some interest in utilizing large landfill sites. When garbage is dumped in a landfill and covered with earth, the covering keeps oxygen out and anaerobic digestion starts to break down the organic material. The evolved methane seeps out of the ground into the atmosphere, and in the process adds to the greenhouse effect. If collecting wells are driven into the site the gas can be collected, cleaned and sold to gas distribution companies. Several installations of this type are operating.

Plant Oil

Oil doesn't come only from oil wells; the plant and animal kingdoms supply many useful and valuable oils. If oil, or oil by-products, are to be extracted from plants it is desirable, because of the argument given above, that the plants be grown on land that is not currently in production for food. For this reason some attention has been paid to crops which can be grown on arid and semi-desert lands, of which the earth has an abundance. Unfortunately, most of these plants have failed as yet to show great promise for producing fuel directly; however, their oils can be used to make products which now are made from petroleum. A few of the plant candidates are summarized in Table 16-1.

Table 16-1:   Oil-Producing Desert Plants

Plant	Products
Jojoba (*Simmodsia Chinensis*)	Wax (paper cups, candles etc.), cosmetics
Gopher plant (*Eophorbia Lathyris*)	Fuel Oil
Guayule (*Parthenium Argentatum*)	Rubber
Bladderpod (*Lesquerella*)	Plastics
Gumweed (*Grindelia*)	Plastics, Paper Coatings

Although there is some interest in these desert plants, none has been put into large-scale production and so far they have had no impact on world petroleum needs. The nearest that plant oils have come to making an impact on the transportation fuel market is the case of rape seed oil, which is a food product grown on prime agricultural land. Rape seed oil is extracted simply by crushing, yielding one tonne of oil for three tonnes of seed. Although the oil can be used directly in place of diesel fuel, the glycerine in the oil clogs engines. To prevent this, the oil is mixed with methanol which, in the presence of a catalyst, precipitates out the glycerine

378

leaving a clear clean fuel called "rape methyl ester" or RME.

RME is being used by many bus and taxi fleets in Europe, particularly in congested urban areas, because it produces almost no sulphur dioxide. Even when the petroleum-based fertilizers to grow the rape are taken into account, RME produces a net of only about 20% of the CO_2 released by fossil fuels. There is, however, an increase in the amount of methane and nitrogen oxides released to the atmosphere, so the fuel is not completely benign. The cost of RME is 5 to 15% higher than conventional diesel fuel, but this will probably decrease with volume production.

EXERCISES

16-1 The total electrical generating capacity in Canada[19] in 1990 was 104 GW. If all this power were to be provided by wind-electric turbines each having 500-kW power output, how many such turbines would be required? [Ans. 2.08×10^5 = 208 thousand]
(This seems like a huge number of turbines, but make an estimate of the number of automobiles in Canada; how does the number of turbines compare with the number of automobiles?)

16-2 What is the maximum possible theoretical efficiency of a thermal-electric plant operating at the Geysers thermal field, if the temperature of the steam is 200°C and the temperature of the condensed water at output is 20°C? What percentage of the total energy contained in the steam is discharged as waste heat? [Ans. 38%, 62%]

PROBLEMS

16-1 In the Toronto region, solar flux varies from 46 W/m^2 in November to 253 W/m^2 in June; these figures are 24 hour averages. If you cover your roof with Si solar cells of 10% efficiency, how much power will you

---

[19] *Electric Power in Canada 1990*, Energy, Mines and Resources Canada, Government of Canada, Ottawa, 1991, p. 46.

generate? Compare this to the consumption of your stove, TV, etc. What is the oil equivalent of one week of energy from the cells?
[Ans. (for 30 m^2) June: 760 W, 0.08 barrel]

16-2 Solar radiation could be used in a power tower for thermal dissociation of water to produce hydrogen. Taking a 75% collection efficiency into the boiler, calculate the H$_2$ produced daily per m^2 of collector in the Arizona desert. The dissociation energy of water is 2.4 × 10^5 J/mol. What amount of natural gas would 1 km^2 of collector replace? [Ans. 80 mol, 5 × 10^5 m^3 of gas per day]

16-3 Using Fig. 16-1 estimate the total solar radiation striking a horizontal plate at 44°N Lat. on March 20. Consider 1 m^2 of surface. How much water could be heated from 10°C to 40°C assuming that 50% of the sun's energy striking the plate can be converted into heat? [Ans. 50 kg]

16-4 The gas consumption profile for an older 6000-ft^2 house in Hamilton, Ontario, is shown in Fig. 16-14. Natural gas is used for space-heating, hot water and cooking.
(a) Estimate the yearly consumption of gas for the combined cooking and hot water, and for space-heating.
[Ans. 2000 m^3, 11500 m^3]
(b) What energy in kilowatt· hours is required to heat this house each year? [Ans. 120 000 kW· h]

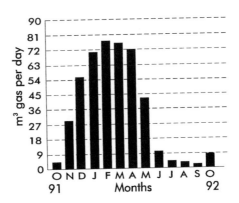

Figure 16-14: Problem 16-4.

(c) The house of Example 16-1 is two storeys plus a heated basement; the Hamilton house is three storeys plus a heated basement. What is the space-heating requirement per square metre of floor area? Which house is more efficient?

(d) Estimate the size of the roof of this dwelling and determine what fraction of its December space-heating requirement could be met by solar energy. [Ans. 40 m^2, about 12%]

16-5 (a) Derive Eqn. 16-1 for the power available from a horizontal-axis wind turbine. (Hints: Consider the kinetic energy of the air that strikes the windmill blades in time t. Think about how the volume of this air can be expressed mathematically.)

(b) Calculate the power in kilowatts for a reasonably sized windmill (R = 4.1 m) in a moderate wind (v = 5.6 m/s) for η = 23%. The density of air is 1.3 kg/m^3. [Ans. 1.4 kW]

16-6 Some years ago the Messerschmidt Co. of Germany revealed plans to construct a very large wind electric generating plant. The following data are from their press release.

> The mill will have a single blade 73 m in radius and will rotate at 17 r.p.m. It will operate at an electrical output of 5 MW, and will generate an average of 17 × 10^6 kW· h each year. This is enough to supply 350 houses and will save the burning of 31 000 barrels of oil.

(a) If the mill extracts 50% of the wind energy, what must the wind speed be to generate 5.0 MW if the generator operates at 90% efficiency? The density of air is 1.3 kg/m^3. [Ans. 2.2 m/s]

(b) What capacity factor do the designers hope to realize? Why is this? [Ans. 39%]

(c) Is the figure of 31 000 barrels of oil compatible with 17 × 10^6 kW· h per year, assuming a generating efficiency of 35% in an oil-burning plant? (Refer to the inside bookcover for the energy content of oil.) [Ans. yes]

(d) How many such plants would be required just to supply the homes of Guelph? [Ans. approx. 10^2]

16-7 The tidal power plant in Nova Scotia has one 20-MWe turbine. If it is operating at its rated power, what volume of water is flowing through it per second? Assume that the water is falling through a vertical distance of 4.0 m (the average tidal range in the Annapolis[20] area is 6.4 m), that the water density is 1.02×10^3 kg/m^3, and that the efficiency of conversion of gravitational potential energy to electrical energy is 95%.
[Ans. 5.3×10^2 m^3/s]

---

[20] Ref. 4, p. 380.

NEW ENERGY TECHNOLOGIES

In an effort to meet the world's ever-increasing demand for energy at reasonable cost, it will be necessary to find new sources if the present ones prove too damaging to the environment. At the same time, it is likely that the technologies of our present sources can be modified and improved. New and alternative sources will be covered in the next two chapters; some promising steps are being taken to improve the present sources and technologies in cleanliness, efficiency, or both.

17.1 Energy Currency

Many of the sources of energy are available at times and in places that are inconvenient or impractical. For example:

1. Water flows over Niagara Falls continuously, but during the night-time hours there is a reduced demand for electricity; the water however continues to flow.

2. The Sun shines for a longer time during summer days, but in temperate climates more energy is needed in the winter when the days are short.

3. There are more sunny hours over deserts — where few people live — than over more populated agricultural areas.

4. Populous areas have developed for historical and commercial reasons, with little regard for the proximity of energy sources. Energy, or its raw materials, must usually be brought to people, rather than the reverse.

As a general rule it is more economical and ecologically sound to deliver energy than to deliver the raw material. For example, the transportation of coal by rail is a greater strain on the economy and the environment than the delivery of electricity by high voltage transmission lines. Usually, however, when energy was cheap and apparently plentiful the method chosen for the energy distribution infrastructure was not the best one for later times. For example, most coal is delivered long distances (even inter-continentally) and little electricity is generated on-site.

One of the problems is that there is not enough variety in energy "currencies" compared to energy sources. In fact, there is only one currency in widespread use, and that is electricity. As the world's supply of petroleum decreases, there could develop a crisis in transportation due to lack of a convenient currency. The use of electricity in transportation as we now practise it is difficult. Of course we could all abandon our cars and trucks and revert to electric rail transport, but if there were another currency which could replace gasoline the present transportation infrastructure could continue. One currency which will probably be developed is hydrogen.

17.2 The Hydrogen "Currency"

Hydrogen is <u>not</u> a source of energy; there are no hydrogen mines or hydrogen wells. Hydrogen must be manufactured with the expenditure of energy, but by manufacturing hydrogen, a fuel is created that has almost no further environmental impact. This is because the energy is recovered from the hydrogen via the chemical reaction

$$H_2 + O \rightarrow H_2O \tag{17-1}$$

The waste product is simply water, and in some cases even oxygen is not used up since it can be created in the manufacture of hydrogen in the first place. As we saw in Section 8.10 hydrogen is an ideal fuel for fuel cells which can produce electricity directly via chemical reactions and are thus free from thermodynamic efficiency limitations. As a result, much effort has gone into research on methods of producing hydrogen from fossil fuels. There are several methods under investigation, such as steam reforming, thermal cracking, and partial oxidation.

Steam Reformation

Steam and fuel gas (e.g. methane) are combined in a reactor at temperatures between 750 and 900 °C where the reaction

$$CH_4 + H_2O \rightarrow CO + 3H_2 \qquad (17\text{-}2)$$

takes place. Reaction of the product gases with water converts the CO to CO_2 and the production of more H_2. Removal of the CO_2 results in hydrogen gas of 95 to 99% purity.

Thermal Cracking

In the presence of an appropriate catalyst, and at high temperature, it is possible to separate the carbon and hydrogen in methane according to the equation

$$CH_4 \rightarrow C + 2H_2 \qquad (17\text{-}3)$$

At a temperature of 1200 °C, methane from natural gas is blown through a bed (fluidized bed, Section 17.6) of catalyst pellets (7% Ni on alumina). The carbon is deposited as coke on the pellets, which must be continuously removed and regenerated in another fluidized bed with high-temperature air. The energy for the process is supplied by the burning of the coke. The process can produce 95% pure hydrogen.

Partial Oxidation

In the presence of steam and oxygen and with an appropriate catalyst, virtually any hydrocarbon (not only natural gas) can be converted into hydrogen and carbon monoxide. (The CO can also be converted with water to CO_2 and H_2.) The ability of this process to convert coal and heavy oils to the cleaner fuel has made it an attractive choice for development.

Of course, none of this does anything to reduce CO_2 in the atmosphere; so long as energy is extracted from hydrocarbons — by whatever method — CO_2 will be released. However, in principle, hydrogen can be manufactured from any energy source if the temperature is high enough. For some time there was enthusiasm for the production of hydrogen using the heat of nuclear reactors. Even if any future reactors are built, that possibility seems less likely as the trend is away from very high-temperature reactors (like breeders).

A further possibility is the production of hydrogen from the very high temperature of the receivers in large solar power plants, as described later.

It is possible to produce hydrogen by the electrolysis of water according to the equation

$$2H_2O + energy \rightarrow 2H_2 + O_2 \qquad (17\text{-}4)$$

The energy is provided by passing an electric current through the water, which has been made electrically conducting by the addition of a small amount of acid or a salt. Each electrolysis cell requires a DC voltage of about 2 V, so a battery of 100 cells in series can be operated from a 200 V DC source. The highest purity hydrogen is made in this way.

This process does not comprise a source of energy; it is simply an exchange of one energy currency (electricity) for another. The trade has a considerable cost since electrolysis cells operate at an efficiency of 50-60%; the energy is lost as resistive heating of the water. Hydrogen could nonetheless be of great importance in the future as a way of utilizing excess electrical energy production to make a fuel that is portable and suitable for transportation.

Storage of Hydrogen

One obvious way to store hydrogen is as a gas at high pressure using the well-established technologies that exist for handling gases at pressures up to 150 times atmospheric pressure. This technique, however, makes it a poor source for transportation fuel. The energy density of gasoline is 35×10^6 J/L whereas the 6600 L of hydrogen gas (pressurized to 44 L at 150 atm.) in a standard cylinder has only 69×10^6 J of energy, the equivalent of only 2L of gasoline! Clearly it is impractical to operate an automobile using hydrogen in cylinders. (Even this comparison neglects the energy cost of carrying around the cylinder which has a mass of about 40 kg.)

Hydrogen as a portable fuel will probably not be practical unless it is carried as a low-temperature liquid in a light well-insulated tank. Here however there is a large energy price to pay, as it takes about one third of the energy recoverable from the hydrogen when it is burned to liquefy it in the first place. In addition liquid hydrogen has a very low latent heat of vaporization, so the insulated containers must be very well insulated indeed or the boil-off rate becomes very large. Liquid hydrogen is used as a fuel in the space program rockets because the tanks are topped up until just before launch and it is all burned up in the first five minutes; even the storage tanks don't have to be very well insulated under those operating circumstances. A similar situation could exist in the context of take-off of large aircraft. A scenario can be envisaged in which liquid hydrogen powers the take-off, following which the aircraft reverts to a different fuel for cruising. Liquid hydrogen, while dangerous, is not as dangerous as gasoline and even has some safety advantages. In the case of a leak or spill it rises quickly and disperses unlike gasoline and propane which, being heavy, do not disperse well.

Hydrogen can also be stored within certain metals where it forms a hydride compound that, in some cases, can be made to release its hydrogen by the application of a modest amount of heat. Aluminum and titanium hydrides are examples which have been investigated extensively. One of the virtues of metal hydride storage is that the metal packs the hydrogen in an even smaller volume than for liquid hydrogen. This is because the liquid is a molecular liquid, and the molecules of H_2 occupy a large volume. As a hydride, the hydrogen is stored as atoms which have a much smaller volume. The downside of metal hydride storage is that the hydrides are heavy. Table 17-1 summarizes some of these properties for the fuel equivalent of 80 L of gasoline, a typical automobile load. The energy content of 80 L or 64 kg of gasoline is 2.8×10^9 J. Examining the table, it can be seen why gasoline is such an ideal portable fuel; it has remarkable properties of concentrated energy in a reasonable weight and volume, along with user-friendly handling properties.

The Hydrogen-Oxygen Fuel Cell

The operation of the H_2-O_2 fuel cell was described in Section 8.10. The perfection of such a fuel cell with its inherently high efficiency raises the possibility of a new kind of energy distribution infrastructure. The concept that has attracted most attention provides a method of storing the energy generated in off-peak times and using it efficiently later. Off-peak electrical supply may be used in several ways, as for instance to electrolyze water, or to produce hydrogen from coal, as above. The hydrogen can be easily stored and delivered by pipeline to localized fuel cells where it is used to generate electricity.

Table 17-1: Properties of Stored Hydrogen Equivalent to 80 L of Gasoline

Name	%H	Mass kg	Volume L
Gasoline	(na)	64	80
Liq. H_2	100	21	300
Lithium hydride	13	166	200
Calcium hydride	5	440	230
Titanium hydride	4	520	133
Iron titanium hydride	2	1123	205

From a transmission point of view this is very attractive; an underground gas pipeline leaves the surface land free of unsightly transmission lines and is cheaper in capital cost. The land area required is 1% of that required for conventional hydro-storage systems (see Section 17.5).

Since the 1970s, when there was intense interest in the "hydrogen economy", there has been a waning of interest with the precipitous decline in the world price of oil. The price of oil will surely rise again, sooner or later, and so the development of hydrogen may resume.

17.3 Synthetic Fuels

Coal can be used as a feedstock to produce methane from which, in principle, almost any organic fuel can be made. The process starts with the gasification reactions, described in the previous section, which produce carbon monoxide and hydrogen. In the presence of a suitable catalyst the following reaction can be made to occur:

$$CO + 3H_2 \rightarrow CH_4 + H_2 \tag{17-5}$$

The methane can be used directly as a fuel or as the feedstock in further catalyzed reactions even to the manufacture of gasoline.

Synthetic gasoline was made in this way in Germany during the Second World War and is currently made in South Africa by the German process. It has never been able to compete, however, with petroleum-derived gasoline and indeed no synthetic fuel yet competes economically with natural fuels. Estimates of the delivered price of synthetic fuels are always more than that of normal gasoline, sometimes by a factor of two. In addition, the conversion process is inefficient requiring much heat, and results in 40% more CO_2 being released to the atmosphere than for direct use of the original fuel. As well, the various chemical processes use large amounts of water, which constitutes a further environmental cost.

As a result of all these factors, research and development of synthetic fuels has not been attractive. Companies who embarked enthusiastically on these programs after the oil crises of the 1970s have since retrenched considerably, or withdrawn completely from activity.

17.4 The Electric Automobile

The petroleum crises of the 1970s prompted a great deal of activity in the United States and elsewhere to develop a practical electric automobile. The environmental concerns of the 1980s increased the interest in electric vehicles. Electrically powered vehicles have been available for many years but only in a form that requires low acceleration, relatively low speed and sufficient off-duty time to recharge the batteries. The electric milk-van is common in England and electrically powered warehouse trucks are common everywhere. The U.S. Post Office has ordered several thousand electric vehicles for extensive trial use. An average-use automobile, however, needs reasonable acceleration, short-term high power for emergency manoeuvring, and a cruising range of over 100 km between recharges.

The requirements for a viable electric car are threefold:
1. High torque, high speed electric motors of high efficiency.
2. Motor control circuitry of high reliability.
3. Batteries with a high power and energy density.

With the technology available in the 1970s, none of these three requirements could be met. Since that time, continued research has resulted in great improvements in the first two. DC motor efficiencies are now very high with reduced friction losses and efficient cooling. Motor control circuits that had to be made from discrete components can now be made with high reliability using integrated circuits. Only the battery remains a problem and, unfortunately, little has happened to improve it in the intervening quarter century. The Ragone plot (Fig. 8-15) illustrates clearly how almost all batteries fail to match the internal combustion engine in energy and power density. A few exotic systems involving expensive or hazardous materials (or both) are a possibility, but only the lead-acid battery is sufficiently cheap and reliable to consider.

With major improvements in two of the three areas, however, some manufacturers believe that a marketable electric automobile is possible. The first to achieve widespread distribution will probably be the General Motors "Impact" in the mid 1990s. Using light composite materials for the body, advanced aerodynamic design, high efficiency motors, low-rolling-friction tires (but lead-acid batteries) the Impact will have an acceleration of 0 to 95 km/hr in 8 seconds and a range of 200 km on a single charge. The car will probably sell in the range of $20,000 to $30,000.

The gasoline-powered automobile has been developed almost to the stage where no further environmental improvements will be significant. With modern pollution controls the modern engine is a marvel of clean burning; only very marginal improvements will be made. If further large gains are to be made in this area, it will be to change the fuel entirely — e.g. to hydrogen — or to transfer the burning of the fuel to another locality where further gains in environmental impact are possible. The electric car is an example of the latter. The burning of coal to generate electricity is one of the areas in which the greatest gains can be made in applying new technology. Of course adopting other forms of electric generation such as nuclear and solar energy would have an even greater impact on atmospheric cleanliness.

The gasoline automobile has also reached the stage where major improvements in efficiency will be very difficult to achieve. Again, the electric automobile offers an advantage. In a recent comparison study of the total of all transportation and manufacturing costs for all vehicles demonstrated that the electric automobile uses only 60 to 75% of the energy of its gasoline counterpart.

17.5 Energy Storage

Energy needs to be stored for a great variety of reasons, some of which have already been discussed. The most obvious one is to make it portable; this is usually the reason for the existence of batteries of all types. (The readily available stored energy in a lead-acid battery makes it easy to start an automobile and thereby begin to utilize the energy stored in the gasoline.) We must always realize, though, that the battery is not an energy source, it is simply a temporary energy repository. There is, however, another major reason for storing energy: the energy source might be available only in a continuous supply, while our energy needs vary with the time-of-day or the season. Rather than have the energy wasted ways must be found to store it for recovery when the demand rises.

Portability

Many methods have been devised to store energy and make it portable; indeed most of the transportation technology, with the exception of electric trains, relies on this technology in some form or other. So far, transportation has relied on rendering portable the source fuels such as coal and oil, perhaps after some industrial processing to forms like gasoline or jet fuel. Little progress has been made in the widespread use of stored energy currencies such as hydrogen, electricity, or fuel-cell fuels. As seen in previous sections, these technologies may be on the verge of development in the last years of the 20th century or the first decade of the 21st century.

One other method which has been proposed to make stored energy portable is to store it directly as mechanical energy in the form of a rotating flywheel. The limitation in this technology is in the tensile strength of the materials used in the flywheel. The energy of rotation of a rotating object is given by

$$KE_{rot} = (1/2)I\omega^2 = 2\pi^2If^2 \qquad (17\text{-}6)$$

where ω is the angular frequency (in radians·s^{-1}), f is the frequency (in Hz) and I is the *moment of inertia* of the rotating object. The moment of inertia of an object of circular cross-section of radius R and mass M is given by

$$I = pMR^2 \qquad (17\text{-}7)$$

where p is a number (usually of the order of 1) which depends on the geometry of the rotating object. Engineers define a quantity called the *"radius of gyration"*, k, as

$$k^2 = pR^2 \qquad (17\text{-}8)$$

and so for all rotating objects

$$KE_{rot} = 2\pi^2 Mk^2 f^2 \qquad (17\text{-}9)$$

The value of k^2 for a few geometries is given in Table 17-2.

Table 17-2:   Radius of Gyration of Some Rotating Objects

Object	k^2
Thin-walled, hollow cylinder rotating about its axis	R^2
Solid cylinder rotating about its axis	$(1/2)R^2$
Hollow solid cylinder of inner and outer radii, R_1 and R_2	$(1/2)(R_1{}^2 + R_2{}^2)$
Solid sphere rotating about its axis	$(2/5)R^2$

■

Example 17-1 A steel flywheel in the form of a solid cylinder 1.0 m in diameter and thickness h=10 cm spins at 120 revolutions per minute (2 Hz). How much energy is stored in the flywheel? The density of steel is $\rho = 8.0 \times 10^3$ kg/m³.

The mass of the flywheel is $\pi R^2 h\rho = \pi(0.50)^2(0.10)8.0 \times 10^3 = 630$ kg
The radius of gyration of the cylinder is given by (Table 17-3)
$k^2 = (1/2)R^2 = (1/2)(0.50)^2 = 0.125$ m²
Thus $KE_{rot} = 2\pi^2(630)(0.125)(2)^2 = 6.2 \times 10^3$ J ■

Example 17-2 *If the flywheel of Example 17-1 were installed in a vehicle whose gasoline engine of 100 horsepower (7.5 × 10⁴W) operates at 20% efficiency, how long would the vehicle run?*

Power required = 0.20(7.5 × 10⁴) = 1.5 × 10⁴W
Time = (6.2 × 10³ J)/(1.5 × 10⁴ J/s) = 0.4 s

The numbers in Examples 17-1 and 17-2 show that there is little promise for a practical method of storing energy in a flywheel made of today's materials. Of course a 630kg flywheel is far too heavy to be practical in an automobile in any case, and less than one second running time is useless. From Eqn. 17-9, if the mass is reduced the energy content can be restored by increasing the frequency. In fact, since the energy depends on the square of the frequency, research in this technology has been directed to designing light flywheels which turn very fast. Of course, the faster they turn the greater is the centrifugal stress at the rim of the flywheel trying to tear it apart. In past times this has been difficult since, in general, strong materials are also heavy. There is a criterion for how fast a wheel made of homogeneous material can be spun: The rim speed cannot be greater than the velocity of sound in the material from which the wheel is made (see Problem 17-8). In recent years, however, the aerospace industry has produced new synthetic materials such as Kevlar fibre which combine the strength of steel (or greater) with the lightness of plastics. As a result there is renewed interest in flywheel storage. It is now possible to build fibre flywheels that store energy of the order of 10^7 J.

A flywheel-driven car has several advantages over a battery-driven car: the energy can be extracted at any rate and so the car can have good acceleration. (This is a great technological limitation with batteries, as we have seen.) Furthermore, if the flywheel drives a dynamo which in turn drives the electric motors on the car's wheels, on braking or going downhill the wheel motors can be switched to dynamo action so that energy can be restored to the flywheel. Whereas batteries take a long time to recharge, a 30 kW·h (10^8J) flywheel can probably be spun up in about one hour using a 500 V, 75 A electric supply. Such a flywheel motor might have a mass in the 50 - 150 kg range and would spin in a vacuum chamber to eliminate friction.

From an environmental viewpoint it should be noted that the flywheel-electric motor combination is a more efficient user of primary fossil fuel than the internal combustion engine. If the flywheel is charged with electricity from an oil-fired power station, the generating efficiency is close to 40%; the flywheel converts the electricity to motive power with an efficiency close to 95% for an overall efficiency of 38%. Internal combustion engines operate at about 20% efficiency, so there is an improvement by almost a factor of two in the use of the primary fuel. In addition there is the ever-present environmental advantage of removing the fuel combustion back to a large plant rather than burning it at small distributed sites where pollution control is difficult and much more expensive per joule of produced energy.

Large-scale Storage

The water flowing over Niagara Falls does so continuously and most of it is used to generate electricity. The demand for electricity in the Ontario-New York area is variable, as has been discussed in Chapter 10. If the demand falls below that which can be supplied by the river flow, the energy will be wasted; this is particularly true at night. If large holding ponds are constructed at the top of the fall then, during periods of low demand, the water can be accumulated in the ponds. This water can be called upon during peak-load periods to augment the normal flow. In this way the energy can be stored very efficiently. At the large tidal power station at La Rance in France described in Section 16.7, where the tidal flow must be utilized or wasted, the off-peak generation is used to run electric pumps, which pump the water back to the head side of the dam to be used when needed. In effect the surplus electric energy is stored as gravitational potential energy.

In this case it would not be a good idea to store the energy as some form of fuel, e.g. hydrogen. (Doing so would cause some lost energy, i.e., thermodynamic efficiency loss.)

If the electricity demand is met by a wide-ranging mix of energy sources then this technology is not so important since, as discussed in Chapter 10, sources which are difficult to restart or which produce energy at awkward times can be included in the base-load, the peak being supplied by more costly but more flexible technologies.

394

Small-scale Heat Storage

A very large fraction of the world's energy supply is used to provide heat in the winter and cooling in the summer to our houses and workplaces. The ideal situation would be to have buildings so well insulated that their heat content would be a constant; heat could neither enter nor leave the building. In fact, if such a situation were possible it would be necessary to remove heat constantly since each occupant would be adding metabolic heat at a rate of at least 100 W. In very large buildings this load of metabolic heat is a significant part of the energy budget of the building. Its effect is usually to make the cooling energy requirement in the summer greater than the heating in the winter.

The next best situation would be to save the heat from air conditioners in the summer by storing it in some way to be used in the winter. Many schemes of this type have been designed and carried out in buildings ranging from the family home to large office complexes. The problem always is: what medium is cheap enough and has a sufficiently high specific heat capacity to serve as a storage medium? With cost being so important, the simplest answer has usually been to use rock or water.

Another method of heat storage which has attracted attention and some development is to utilize the latent heat of fusion of some substance which has a freezing point near the temperature of our normal living environment. A small-scale example of this is the phase-change hand warmers that can be purchased in any sporting goods shop. They consist of a plastic bag filled with a concentrated solution of sodium acetate $NaC_2H_3O_2 \cdot 3H_2O$. This substance melts at 58°C but can be supercooled to room temperature (20°C) without solidifying. When heat is required the contents are given a slight mechanical shock and the crystallization begins, releasing the latent heat and warming the bag to 58°C. The bag can be regenerated by boiling in water.

For heat storage in buildings the substance that has attracted the most attention is Glauber's salt (sodium sulphate decahydrate $Na_2SO_4 \cdot 10H_2O$) which melts at 30.5°C and has a latent heat of fusion of 1.64×10^5 J/kg. In Table 17-3 the relevant properties of water, rock and Glauber's salt are given. The largest latent heat of fusion shown is that of water, but its low melting temperature (0°C) makes it impractical for this use.

If the relative storage capacities and costs of these three are evaluated based on 1 m^3 of water, the values shown in Table 17-4 are obtained.

Table 17-3: Properties of Water, Rock and Glaubers Salt

Substance	Density kg/m^3	Specific Heat J·kg^{-1}·$^{\circ}$C^{-1}	Latent Heat of Fusion, J/kg
Water	1000	4190	3.35×10^5
Rock (Granite)	2600	~300	N.A.
Glauber's Salt	1464	N.A.	2.15×10^5

Table 17-4: Storage Capacity and Cost Relative to Water

Substance	Volume m^3	ΔT $^{\circ}$C	Energy J	Mass kg	Cost Relative to Water
Water	1	20	8.3×10^7	1000	1
Rock (Granite)	5.3	20	8.3×10^7	13800	4
Glauber's Salt	0.27	0	8.3×10^7	390	3

Examination of Table 17-4 would indicate that the poorest choice is rock and yet that is the technology that has been best developed. Glauber's salt is very attractive and a flurry of research in the 1970s solved many of the application problems such as dehydration of a fraction of the salt with every cycle. However, as with many of the attractive technologies investigated in the 1970s, interest has waned with the oil glut of the 1980s and 90s.

17.6 Coal Burning (The Fluidized Bed)

The one energy technology which produces the greatest number of environmental problems is the burning of fossil fuels, particularly coal. The atmospheric carbon dioxide problem, of course, cannot be overcome by any improvement of technology short of the abandonment of fossil fuel burning. However, burning fossil fuels has many other negative environmental impacts which have been discussed in Chapters 5 and 6.

When coal is burned on a large scale to produce electricity it is first crushed to a powder and injected into the burner much as a liquid fuel would be handled. Indeed, many such plants can burn powdered coal or oil with the same equipment. The powdered coal requires a large-volume reaction vessel in order that the high temperature gases can have sufficient surface to deposit their heat in tubes which produce steam. Such furnaces are physically quite large and cannot be cheaply built at a factory and moved to the site; they must be built on-site. The temperature of combustion is very high ($\sim 1700\,^\circ C$) with a resulting high production of nitrogen oxides. The sulphur content of the coal must be removed from the flue gas in the form of SO_2. Improved furnace design should address several problems:

1. Remove the sulphur content before or during the burning process to eliminate expensive scrubbers.

2. Lower the combustion temperature to decrease the production of nitrogen oxides.

3. Improve the efficiency of heat transfer to the water tubes to make the furnace physically smaller.

The fluidized bed furnace shown in Fig. 17-1 addresses all of these concerns.

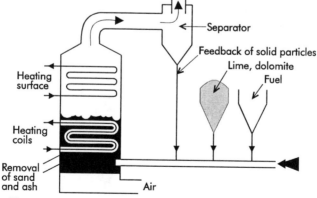

Figure 17-1:　　The Fluidized Bed Furnace.
F.D.Friedrich, Canmet Report 79-39

The greatest mass in the furnace at any time is a charge of limestone or dolomite pellets which are kept in violent agitation by upward jets of hot air. The turbulent motion of the pellets is much like that of a boiling liquid, whence the name "fluidized bed". The powdered coal is fed into this hot "boiling" mass where it reacts with the oxygen in the air to release its energy. The bulk of the energy is transferred to the pellets where they, in turn, transfer it to the steam tubes — not in the furnace walls, but mostly in the bed itself. The hot gases also give up their heat to tubes on surfaces higher in the furnace. The heat transfer in the bed is much more efficient and so the furnace can be significantly smaller. In addition the combustion temperature is lower ($\sim 900\,^\circ C$) and as a result there is much less production of nitrogen oxides.

The sulphur in the coal reacts with the limestone ($CaCO_3$) to form calcium sulphate ($CaSO_4$) as a crust on the stone. Between 150 and 300 kg of limestone are required to remove 90% of the sulphur from a tonne of coal. Of course this means that limestone must be mined as well as coal and transported to the plant site, with the attendant increase in environmental impact and strain on the transportation systems. The crusting of the pellets renders them inactive and so they must be constantly removed and replaced. Some research has been carried out on processes to remove the crust but generally the pellets are discarded with the coal ash.

A further elaboration of the method is the "Pressurized Fluidized Bed". Here the whole reactor is pressurized to about 10 atm. (1000 kPa). This permits more compact units for a given thermal output along with more efficient combustion of the fuel and removal of sulphur. The price paid is in a greater complexity of equipment to maintain the required pressure. The idea is, however, attractive as the high pressure, hot exhaust gases can, after cleaning, be fed directly to a turbine without generating steam.

In many cases the fluidized bed furnace is the most economical way to cope with the sulphur content of coal and the technology is now sufficiently well developed that many new installations or refittings are of this type. There is, however, an additional environmental price to pay. The formation of $CaSO_4$ is via the reaction

$$CaCO_3 + 3O + S \rightarrow CaSO_4 + CO_2 \qquad (17\text{-}10)$$

In other words, more atmospheric carbon dioxide is produced. One of the attractions of the fluidized bed process is its ability to handle high-sulphur coal which other furnaces cannot. But removing more sulphur means producing more carbon dioxide. In the 1970s when acid rain was the great and popular concern, fluidized bed combustion was hailed by many, including environmental organizations, as the desired technology. In the 1980s as attention shifted to atmospheric CO_2 and global warming, the technology is now seen by many as undesirable. This is an object lesson in the shifting fashions in environmental concerns. In the present example, one is faced with the stark facts of burning fossil fuels: either the environmentally dirty business must be accepted, or alternatives must be found (and they may be equally harsh if not more so).

Problems

17-1 Consider a standard cylinder of hydrogen gas compressed to a volume of 44 L at a pressure of 150 atmospheres.
a) What is the volume of this gas if it is released to the atmosphere? Assume hydrogen is an ideal gas. [Ans. 6600 L]
b) The density of liquid hydrogen is 71 g/L and gaseous hydrogen at 1 atm. and $20°C$ has a density of 0.084 g/L. What volume of liquid hydrogen can be made from one cylinder? [Ans. 7.8 L]
c) The energy content of liquid hydrogen is 120×10^6 J/kg. What is the energy content of 1 L of liquid hydrogen? [Ans. 8.5×10^6 J/L]
d) The energy content of gasoline is 35×10^6 J/L. What is the gasoline equivalent of one cylinder of gaseous hydrogen? [Ans. 1.9 L]

17-2 An automobile fuel tank holds 80 L of gasoline.
a) How large a tank would be required to hold enough liquid hydrogen to drive the car the same distance as the gasoline? [Ans. 330 L]
b) What would be the mass of each load of fuel (excluding their containers)? The specific gravity of gasoline is 0.8.
[Ans. gasoline 64 kg, H_2 23 kg]

17-3 The work done in compressing a gas isothermally (constant temperature) from volume V_1 to V_2 is given by

$$W = \int_{V_1}^{V_2} P \ dV$$

If the gas is ideal, then

$$PV = nRT$$

where n is the number of moles and R is the gas constant, equal to $(8.31 \ J \cdot mole^{-1} \cdot K^{-1})$.
a) Show that

$$W = nRT\ell n(V_2/V_1)$$

b) How much energy is required to fill a standard gas cylinder with hydrogen? (See Problem 17-1 for relevant data). [Ans. 3.7×10^6 J]
c) What percentage is this of its energy content? [Ans. 6%]

17-4 Instead of pressurizing the 6600 L of hydrogen from a gas cylinder it is desired to cool it and liquefy it. The specific heat of hydrogen gas is 14×10^3 J/kg and the latent heat of vaporization of the liquid is 32×10^3 J/L; the liquefaction temperature of hydrogen is 20 K.
a) What energy must be extracted from the gas to liquefy it?
    [Ans. 2.4×10^6 J]
b) What percentage is this of its energy content? [Ans. 4%]

17-5 The flywheel of Example 17-1 is hollowed out to remove one half of its mass.
   a) What is the inner radius? [Ans. 0.354 m]
   b) Spokes of negligible mass convert this hollow cylinder into a flywheel also spinning at 2 Hz. What is its energy content? Has much been sacrificed even though one half of the mass is removed?
       [Ans. 4.7×10^3 J]

17-6 a) How fast would the wheel of Example 17-1 have to be spun to operate the vehicle of Example 17-2 for 1 hr? [Ans. 190 Hz]
    b) What is the rim speed of the wheel in a)? [Ans. 590 m/s]

17-7 Using the data of Table 17-4 derive the data of Table 17-5 (except for the cost).

17-8 A flywheel of radius R and thickness h is made of a homogeneous material with a density ρ. Consider a thin rim of the flywheel which has a thickness dR.
  a) If the wheel turns with a rim speed v show that the centripetal force holding the rim in place is given by
$$F = 2\pi\rho v^2 h dR$$
  b) Show that the tensile stress (force per unit area) on the rim is given by
$$F/A = \rho v^2 dR/R$$
  c) Use the equation for elastic deformation
$$\text{stress/strain} = \text{Young's modulus}$$
  or $\qquad F/A = Y(\Delta\ell/\ell)$

where Y is Young's modulus. Assume that the wheel is disintegrating when the distortion $\Delta\ell$ is equal to the thickness of the rim and show that
$$v = \sqrt{Y/\rho}$$
The last formula is, in fact, the expression for the velocity of a sound wave in a material of Young's modulus Y and density ρ.

17-9
a) When you refuel your car you can transfer about 80 ℓ of gasoline from the pump into the gas tank in about 2 minutes. What power does this represent? [Ans. 25 MW]
b) If you owned an electric car with the same range as the automobile in a) it would need periodic refuelling (i.e. battery recharging) from an electricity source. A heavy duty household circuit such as the type that runs an electric stove can supply 30 A at a voltage of 220 V. How long would it take to refuel your electric car, remembering that a gasoline automobile operates at an efficiency of 20% whereas an electric car is nearly 100% efficient. Assume that the charging is done at an efficiency of 75%. [Ans. 29 h!]

18 *ENERGY CONSERVATION*

The previous two chapters have focused on new technologies, such as electric cars and coal gasification, which will help make better use of present energy sources, and alternative sources of energy — the wind and sun, for example — that can make an important contribution to energy supply where appropriate.

This chapter explores another "source" of energy, but not one which can be mined from the ground or captured in sunlight. The source is energy conservation — reducing energy consumption by a kW· h today means having an extra kW· h available tomorrow, in effect providing a source of 1 kW· h of energy. Energy conservation is achieved by increasing the efficiency of end-use devices, so that less energy is required to achieve the same level of heating, lighting, transportation, etc. (Energy conservation could also mean "doing without" — in other words, cutting back on heating, lighting, etc. — but this curtailment of services is not what is meant by energy conservation in this chapter.)

This chapter is divided into three broad sections, corresponding to sectors of society that use large amounts of energy: residential and commercial buildings (which consume 38% of world energy), industry (38%), and transportation (24%).

18.1 CONSERVATION IN THE RESIDENTIAL/COMMERCIAL SECTOR

Residential and commercial buildings include houses, apartment buildings, offices, stores, hotels, schools, and government edifices such as post offices, courthouses, etc. Many businesses and households have been conserving energy over the past couple of decades. In Canada from 1973 to 1987, the average household energy use declined[1] by about 32%, and energy consumption per m^2 of commercial floor space declined by almost

---

[1] *The State of Canada's Environment*, Government of Canada, Ottawa, 1991, p. 12-29.

18%. Some of this decrease was aided by government grants, for example to upgrade home insulation.

A considerable portion of the energy consumption in residential buildings goes into **lighting**. In a typical commercial structure, lighting consumes about 40% of the electricity used, plus another 10% for air conditioning to remove the unwanted heat generated by the lights[2]. In the U.S.A., lighting (in all sectors, not just residential and commercial) consumes about 25% of all electricity, 20% directly plus 5% for air conditioning[2].

Estimates of the potential reduction[2] in energy used for lighting in commercial buildings range from 55% to 85%. One obvious way to reduce energy for lighting is to change from incandescent bulbs, which convert only 5% of the electrical energy to light, to fluorescent bulbs, which have a conversion efficiency of 20%, as mentioned in Section 4.5. However, even if fluorescent bulbs are already being used, significant reductions are possible. A conventional four-tube fluorescent fixture consumes about 175 W of electrical power to produce 7000 lumens of light[3]. Replacing this fixture with a two-tube fixture using the most modern tubes and ballast[4], and installing a light-reflector above the tubes to direct light downward, can result in a light output of 7000 lumens with a power consumption of only about 70 W, providing a saving of 60% and a payback time of one to two years. Modern tubes have phosphors which can provide a wide range of "warm" and "cool" colours with higher efficiency than previous phosphors, and solid-state ballasts have been introduced with a resulting increase in efficiency[5] of about 20%.

In addition to upgrading lighting fixtures, there are a number of other methods to reduce energy consumption for lighting. Using existing daylight from windows or skylights can decrease lighting requirements, but attention has to be paid to the heating effect of direct sunlight, which is a

---

[2] A. Fickett, C. Gellings, and A. Lovins, Sci. Am. **263**, No. 3 (Sept. 1990), p. 66.

[3] A lumen is the SI unit of luminous flux.

[4] The "ballast" in a fluorescent fixture provides a high voltage to initiate the ionization of the mercury vapour in the tube, and also limits the current for safe operation.

[5] D. Hafemeister and L. Wall, "Energy Conservation," in *The Energy Sourcebook*, Amer. Inst. of Physics, New York, 1991, p. 455.

potential advantage in cold climates, but a problem in warm climates. Even in cold regions of the world, a large building in winter often requires air conditioning instead of heating because of the large amount of heat generated by people and equipment. Interior or exterior devices to diffuse the daylight can be used to reduce heating and glare produced by direct sunlight. It has been estimated[6] that the energy requirements of a typical commercial building can be reduced by about 10% by appropriate use of "daylighting." Another energy-saving step is to reduce the overall light level in a room, but provide compact fluorescent desklamps to illuminate work areas. As well, glass partitions in the upper part of walls between offices and rooms allow light to pass from room to room, and light-coloured furniture, carpets, and wall-coverings reflect light and hence reduce lighting requirements. Occupancy sensors can be used to turn off lights automatically when rooms are not in use. A number of these ideas can be used also in houses, apartments, etc.: conversion from incandescent bulbs to fluorescent bulbs (either as tubes or in compact form), use of daylight and light-coloured walls, turning off lights when not needed, etc.

As a specific example of the savings available from reducing lighting needs, the head office of Xerox Canada in North York, Ontario, is saving $187 000 annually as a result of installing energy-efficient lighting, occupancy sensors, and instituting a company policy to turn off equipment such as photocopiers, etc., when not in use[7]. The cost of the renovations was $550 000, giving a payback time of about three years. However, Ontario Hydro contributed half the cost of the renovations, and hence the payback time for the Xerox corporation is only about 1½ years.

Heating, Ventilating, and Air Conditioning

Another area of large potential reduction in energy consumption is the heating, ventilating, and air conditioning (HVAC) system in a building. Heating and air conditioning needs can clearly be reduced if insulation with a high R-value is used during construction or renovations, and if caulking around windows and weatherstripping around doors are installed

---

[6] Ref. 5, p. 451.

[7] Toronto *Globe and Mail*, "Energy Management Business Report," October 20, 1992, p. 1.

and maintained. Computer-control of air conditions in individual rooms, based on automatic monitoring of temperature, humidity, etc., can lead to greater efficiency. Proper maintenance, such as cleaning of coils in air conditioners and replacement of air filters in ventilating systems, also decrease energy consumption.

The use of high-efficiency motors in HVAC systems can play a significant role. In Canada, 50% of the electricity used in the commercial sector is consumed by electric motors[8], and in industry, the figure is 75%. In the U.S.A., more than half of all electricity generated is used by electric motors[9], and in industry they consume 65% to 70% of the electricity. High-efficiency motors have designs that reduce the magnetic, resistive, and mechanical losses to less than half the levels of a decade ago. For maximum effect, a high-efficiency motor is combined with an electronic variable-speed drive, which controls the speed of the motor. In a typical older motor installation, which might control a ventilation fan, for example, the motor operates at full speed all the time, and the amount of ventilation is controlled by a valve or damper which restricts the air flow. (This is somewhat like driving with one foot on the gas and the other on the brake.) Improved motor and drive systems use in total only about half the electricity of older systems[10].

Electrical needs for air conditioning can be reduced by use of thermal storage of cold water or ice, produced overnight when exterior temperatures are lower and hence the efficiency of air conditioning is greater. The cold water or ice can then be used during the daytime to supplement regular air conditioning. As well, overnight use of an air conditioner is usually less expensive because of lower commercial electricity rates for off-peak hours. Bell Canada in Ottawa has installed a state-of-the-art cool storage system, and will save $80 000 in energy costs annually[7]. Combined with other energy-efficiency improvements, including lighting upgrades, Bell will be saving $250 000 each year. The capital cost was $1.5 million, giving a payback time of about six years, shortened to three years by a financial contribution from Ontario Hydro.

---

[8] Ref. 7, p. 3.

[9] Ref. 2, p. 67.

[10] Ref. 2, p. 68.

Installation of energy-efficient modern windows can reduce heating needs, and as previously mentioned, wise placement of windows can also decrease electric-lighting requirements. It has been estimated that about 5% of the energy used in the U.S.A. is lost through windows[11]. A single-glazed window has an R-value (Section 5.4) of about 1, and double-glazed windows have R ≈ 2. Glass is a very strong absorber of the infrared (IR) radiation emitted by the interior of a building. A thin film of an IR-reflecting material (called a low-emissivity[12], or low-E material) can be used to coat the inside surface of a window to reflect the radiant heat back into the building. Whereas glass absorbs most of the IR, a low-E window absorbs only 10%, reflecting 90%. A low-E double-glazed window has an R-value of about 3. Adding a second low-E film and filling the space between the double glazing with a poorly conducting gas such as argon increases R to about 4. Thermal resistance values as high as R-6 to R-10 can be achieved by filling the space between a double-glazed low-E window with a transparent, poorly conducting material such as aerogel, which is a sparse skeleton of tiny glass particles (5%) and air (95%).

In warm climates, and in large buildings in cold climates, heat gained through windows is more of a problem than heat loss. For large office buildings, it is common practice to place reflective coatings on the exterior of windows to transmit visible light, but reflect solar IR radiation. Still under development are photosensitive windows which darken when the ambient light levels get too high.

Appliances and Office Equipment

Efficiency improvements in lighting and motors have been matched by various appliances. For example, present-day refrigerators and freezers consume about half the energy of the models from 1972. Between 1950 and 1972, energy consumption by new refrigerators in the U.S.A.[13]

---

[11] Ref. 6.

[12] The IR-reflective materials are poor absorbers of IR and also poor emitters of IR —— hence the phrase "low emissivity."

[13] Ref. 5, pp. 456-457.

increased from approximately 600 kW· h per year to almost 1800 kW· h per year (Fig. 18-1), because of increased size (from about 7 ft³ to 17 ft³), a decrease of 40% in actual efficiency, and the addition of features such as automatic defrosting. The "energy crisis" of 1974 led to the establishment of tighter standards on refrigerator efficiency by the California and U.S. Governments, resulting in a continuing decline in electricity usage by refrigerators[14]. The efficiency of refrigerators has been increased by improving the insulation (both in thickness and quality), using more efficient motors and compressors, and increasing the area of the heat exchangers.

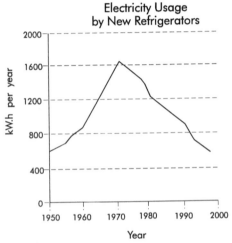

Figure 18-1: Energy usage by new refrigerators in the U.S.A. increased from 1950 to 1972, and has declined gradually since then. Data after 1990 are projections based on new standards.

Other devices can show large efficiency improvements. Efficient televisions can consume 75% less electricity than their inefficient counterparts[15], photocopiers can save 90%, and computers 95%. Clearly a consumer who wishes to reduce energy consumption can do so by shopping wisely.

---

[14] Ref. 13; and Ref. 2, p. 74.

[15] Ref. 2, p. 68.

Additional Savings, Especially for Houses

Owners of houses can take advantage of many of the energy-saving strategies mentioned above, such as upgrading insulation and using low-E windows, and there are many other ways to effect energy savings, some involving no cost at all. In cold climates, deciduous trees can be planted on the south and west sides of a house to provide summer shade, but when the leaves are absent in the winter, the sun provides light and warmth. Coniferous trees can be planted on the north side of a house to block winter winds, and ideally, this side of a house should have fewer windows to reduce heat loss. Turning down the furnace thermostat a few degrees at night and when no one is in the house results in a saving of heat energy of about 10-15%. The temperature of the water heater can also be turned down from $140\,°F$ to $120\,°F$ to reduce heat loss from the tank and hot-water pipes. With modern detergents, using cold water for the laundry is possible.

A detailed study[16] on a typical home in Northern California showed that its total energy consumption could be reduced from about 70 MW·h to 30 MW·h per year, with only a modest investment. Energy-reducing steps included upgrading ceiling insulation from R-11 to R-19 and wall insulation from R-0 to R-11, installing a fluorescent light fixture in the kitchen, using a low-flow showerhead, and decreasing the hot-water temperature. The payback time is about three years.

18.2 DISTRICT HEATING

In Sweden, Denmark, and many Eastern European cities, the waste heat from electrical power stations is used to heat surrounding communities. This district heating has the obvious advantage of using thermal energy that would otherwise be discarded. In addition, controlling pollution from one central plant is easier than control of individual furnaces. In at least one Swedish city, there is even enough waste heat to melt snow from the roads,

---

[16] A. Rosenfeld and A. Meier, "Energy Demand," in *Physics Vade Mecum*, H. Anderson, Ed., American Institute of Physics, New York, 1981, p. 170.

thereby saving on snow removal costs. District heating is easier to implement in European cities than in North America because of the denser housing patterns in Europe. In addition, some North Americans would undoubtedly object to their lack of choice in home heating if district heating were mandatory, as it is in some European cities.

18.3 CONSERVATION IN INDUSTRY

Industry can avail itself of all the energy-conserving devices discussed for residential and commercial buildings. In particular, appreciable savings are possible from the use of high-efficiency motors, improved lighting, and upgrading insulation and HVAC systems. At least as important, however, are improvements in industrial processes. For example, the introduction of a new steel-making procedure in the 1960s reduced the energy requirements by 50% relative to the open-hearth methods that were previously used. The newer technique involves blowing oxygen through the molten metal to burn much of the carbon that is present; the heat thus generated provides energy to remove other impurities in the form of slag. In the open-hearth process, all the heat was provided from outside the steel. Another general area of process improvement is the use of computers to monitor and control complex manufacturing procedures, reducing waste and inefficiency.

From 1973 to 1987, Canadian industry reduced its energy consumption[17] per unit of industrial output by approximately 6%. In the U.S.A. from 1971 to 1986, energy consumption per unit output declined by 35%.[18] However, it has been estimated that about 40% of the U.S. energy reduction is not due to conservation measures and process improvements, but rather to a shift toward the manufacturing of products that happen to require less energy per unit of output[19].

Industries are usually careful to recycle materials, because the energy needed to manufacture a product from virgin materials is typically about

---

[17] Ref. 1.

[18] M. Ross and D. Steinmeyer, Sci. Am. **263**, No. 3 (Sept. 1990), p. 89.

[19] Ref. 18, p. 94B.

double that required when recycled materials are used. Recycling consumer products also saves energy; Figure 18-2 shows the energy required[20] to produce a consumer container (glass bottle, etc.) from various materials, either virgin or recycled. Aluminum clearly offers large savings in recycling. (The value given for a recycled plastic container is hypothetical, because plastic containers are not now made of recycled material.) Of course, even recycling consumes more energy than using refillable containers.

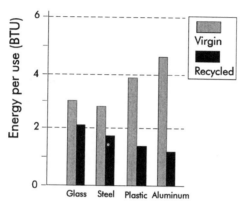

Figure 18-2: Producing a consumer container (glass bottle, steel can, etc.) from recycled material requires less energy than producing it from virgin material.

18.4 COGENERATION

Industries often are in need of heat for various industrial processes, as well as electricity for machinery, lighting, etc. Since most electricity is generated by thermal plants, which produce a great deal of waste heat, why not supply industry with both electricity and heat? This cogeneration of electricity and heat can occur in two ways: an electric power plant can sell waste heat to industry (or the heat could be used for district heating, as described in Section 18.2), or an industry which now produces heat for itself can convert to generation of both electricity and heat, possibly selling

---

[20] Ref. 18, p. 96.

excess electricity to a local utility.

However, there are barriers to cogeneration. If an electric plant is to provide heat to industry, the plant and industry must be physically close together. As well, the times when an industry requires the most heat might not correspond to times when electricity generation is high. Industries are often hampered from considering cogeneration themselves because they consider the payback time for the electrical generating system to be too long. In addition, if surplus electricity is to be sold to a utility, the industry will likely not be paid a good price for it (because the surplus would probably not be supplied on a regular basis) and the industry is apt to be subject to stricter environmental laws regarding the generation of the electricity.

Although cogeneration is clearly a good idea in principle, the barriers have prevented effective development of this concept.

18.5 CONSERVATION IN TRANSPORTATION

"That the automobile has reached the limit of its development is suggested by the fact that during the past year no improvements of a radical nature have been introduced."
— *Scientific American*, 1909

Perhaps there were no significant changes in automobiles in 1909, but in the past couple of decades there have been a great many improvements in the energy-efficiency of automobiles. In Canada[21], the number of automobiles increased from 8.9 million in the year 1975 to 13.2 million in 1990, but during this period the total consumption of gasoline decreased from 3.45×10^7 m^3 to 3.39×10^7 m^3. In 1975, the average car in Ontario required 18.6 L of gasoline to travel 100 km in combined city and highway driving; by 1989, 9.6 L was required. (Nevertheless, it is interesting to note that the 1989 figure is greater than the comparable figure from 1985, 9.3 L.)

Road vehicles account for about half of the world's oil consumption

---

[21] Toronto *Globe and Mail*, January 25, 1992, p. B1-B4.

each year, and the number of registered vehicles is steadily growing[22] (Fig. 18-3). Many people in industrialized countries have become heavily dependent on the use of automobiles, and urban design is often based on the convenience of automobile-drivers. In Canada in 1987, 77% of all households owned one or more cars, and in 1983, 73% of journeys to work in Canada were made by car[23]. Many people travel a great distance to work; in Toronto in 1980 about 25% of commuters[24] had a one-way commuting distance greater than 19 km.

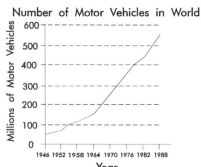

Number of Motor Vehicles in World

Figure 18-3:   The global fleet of motor vehicles grew by almost a factor of 11 from 1946 to 1988.

In the U.S.A. in 1988, about 27% of the energy consumed was devoted to transportation[25]. A breakdown of the transportation energy[26] used in 1986 in the U.S. is provided in Figure 18-4. Notice that automobiles consumed almost 50% of the energy in the transportation sector, and cars and trucks together accounted for 85% of the consumption. Aircraft took up most of the remainder.

---

[22] D. Bleviss and P. Walzer, Sci. Am. **263**, No. 3 (Sept. 1990), p. 109.

[23] Ref. 1, p. 13-21.

[24] Ref. 1, p. 13-25.

[25] A. Chachich, "Energy and Transportation," in *The Energy Sourcebook*, American Institute of Physics, New York, 1991, p. 358.

[26] Ref. 25, p. 359.

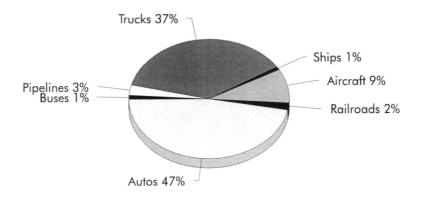

Figure 18-4:   Transportation energy consumption in the U.S.A. in 1986.

As a result of the energy crises in the 1970s and a growing awareness of the pollution problems resulting from the use of motor vehicles, the U.S. and California Governments passed laws requiring a gradual increase in fuel efficiency of automobiles and light trucks.   Much of the resulting improvement has been due to a reduction in weight. A typical midsize car in the early 1970s had a large, iron engine, rear-wheel drive which required a heavy driveshaft and differential gearbox, and a metal chassis.   A comparable car now has a smaller, aluminum engine, front-wheel drive, and no chassis. Many of the parts which used to be metallic — such as the dashboard and bumpers — are now made of plastic, thus reducing the weight. A typical car 20 years ago had a mass of about 1800 kg; in the 1980s, the mass had decreased to about 1400 kg, and it is anticipated that another 400 kg can be shed[27] by replacing steel and cast-iron with high-strength low-alloy steel, aluminum, composite plastic, and magnesium.

Automotive engines tend to be smaller and more fuel-efficient than their gas-guzzling predecessors in the 1970s, and on average, the North American car is now smaller. (However, in comparison with typical car sizes in Europe, North American cars are still larger.)

Other changes that have led to a decrease in fuel consumption include the

---

[27] Ref. 25, p. 370.

following:

- Body shapes that are more aerodynamic, thus reducing air friction.
- The use of radial tires, which can have over 30% less rolling resistance than bias-ply tires[28], producing fuel savings of 15% on the highway, and 10% overall.
- Electronic engine control, which regulates air intake, exhaust recirculation, fuel injection, spark timing, and idle speed; fuel savings of up to 10% can result.
- Lower highway speed limits.
- Computerized traffic control in cities, thus reducing time spent waiting at lights and accelerating and decelerating.

Future changes in automobiles will probably include further reduction in weight, even better aerodynamics, improvements in lubricants, and use of a continuously variable transmission that permits the engine to work at its most efficient speed almost all the time.

Thus far in this section we have been discussing automobiles (and light trucks), which consume a large fraction of the energy devoted to transportation. However, fuel consumed by transportation could be reduced significantly if people would use mass transit instead of automobiles. Table 18-1 shows the energy used per person·km for various forms of urban transportation[29]. Many urban travellers drive alone in their cars, but this is the most energy-inefficient way to travel. Unfortunately, car-drivers are unlikely to use more efficient types of transit unless there is considerable intervention by governments through subsidies, introduction of special traffic lanes for buses, or development of cities in which access to the downtown areas by private automobiles is limited.

---

[28] Ref. 25, p. 386.

[29] Ref. 1, p. 13-20 (Original source: Ontario Ministry of Transportation and Communications); and J. Fowler, *Energy and the Environment (2nd Ed.)*, McGraw-Hill, New York, 1984, p. 466.

Table 18-1: Energy Consumption in Urban Transportation

Transportation Mode	Typical Energy Used (MJ· person^{-1}· km^{-1})
Automobile (1 person)	8
Automobile (5 people)	2
Commuter Rail	0.8
Bus	0.6
Walking	0.2
Bicycle	0.1

Notice in Table 18-1 that the most efficient mode of transportation (even better than walking) is the bicycle. In the Netherlands, a large fraction of commuters use bicycles on an extensive network of bicycle paths that even includes separate bicycle traffic lights.

For intercity transportation, the energy used per person·km decreases. Buses offer the most energy-efficient transportation, followed by railroads, automobiles, and airplanes. A typical airplane consumes about six times as much energy per person·km than does a bus.

Long-Haul Trucks

No fuel-efficiency standards have been mandated for large trucks, and hence progress in this area has been due only to cost savings to operators. Various energy-saving (and hence cost-saving) measures undertaken so far include installation of air deflectors above cabs to reduce aerodynamic drag (by 15-25%), design of new cabs with rounded corners, angled windshields, etc., and use of radial tires and diesel engines (which are more energy-efficient than their gasoline counterparts). As time passes, we can expect to see engine improvements such as electronic engine control, and further aerodynamic changes such as skirts around tractor-trailers.

Although trucks carry most of the freight in North America, ships, pipelines (for oil and gas), and railroads offer more energy-efficient ways of moving goods[30] (Table 18-2). Only airplanes have a greater energy

---

[30] Adapted from J. Fowler, *Energy and the Environment (2nd Ed.)*, McGraw-Hill, New York, 1984, p. 466 (Original source: R. Hemphill, "Energy Conservation in the Transportation Sector," in *Energy Conservation and Public Policy*, J. Sawhill, ed., Prentice-Hall, Englewood Cliffs, N.J., 1979, p. 92).

cost for freight transport than do trucks.

Table 18-2:   Energy Used in Freight Transport

Transportation Mode	Typical Energy Used ($MJ \cdot tonne^{-1} \cdot km^{-1}$)
Waterway	0.07 - 0.4
Pipeline	0.07 - 0.9
Railroad	0.1 - 0.7
Truck	0.8 - 1.4
Airplane	6 - 18

Airplanes

Airplanes have also undergone improvements in design to decrease their fuel consumption.   Engines have become more efficient, lighter-weight materials such as graphite fibres in epoxy resin are being utilized in new aircraft in place of aluminum, lighter carpets and seats are being used, and less paint is being applied (thus reducing weight). Operational factors are also important: changes in airport design, airspace routing, and flight scheduling can all make improvements in fuel efficiency. In the future, it is expected that new wing designs which create less turbulence and hence save fuel will come into common use, and that there will be a return to propeller-driven aircraft (having better fuel efficiency) for short-haul routes.

EXERCISES

18-1 Make a numerical estimate of the total amount of energy that you use in a typical day. Include transportation, heating, lighting, appliances, etc. For each category, how could your consumption be reduced (without significantly affecting your lifestyle), and how much of an effect would such reductions have on your total consumption?

18-2 In Table 18-1, the energy used in urban transportation is given in units of MJ· person^{-1}· km^{-1}. In the U.S.A., another common unit for this quantity would be Btu· person^{-1}· mi^{-1}. Convert the energy consumption given in Table 18-1 for an automobile containing 5 people to Btu· person^{-1}· mi^{-1}. (1 mi = 1.609 km)   [Ans. 3×10^3 Btu· person^{-1}· mi^{-1}]

PROBLEMS

18-1 Electrical consumption in Canada[31] in 1990 was 466 TW· h in total, of which roughly 20% was used for lighting. If conservation measures were to reduce energy consumption for lighting by 50%, what would be the power saving in megawatts, and what could be the reduction in the number of 1000-MWe plants in operation?   (Assume that the power consumed by lighting is constant throughout each day of the year.)
[Ans. 5.3×10^3 MW; 5 power plants]

18-2 Suppose that an electric pump is being used to drive a fluid through a pipe. If the pump is run at a slower speed, so that the speed of the fluid is reduced by a factor of two, by what factor is the power requirement of the pump reduced? Hint: consider a quantity of fluid (having mass m) to be at rest prior to passing through the pump, and travelling with speed v when leaving the pump a short time later. [Ans. 8]

18-3 Figure 18-5 shows a semilog plot of the number of registered motor vehicles in the world vs. time. The graph has essentially two straight-line portions, one from the year 1946 to 1976, and another (with a smaller slope) from 1976 to 1988. (Use a straight-edge to confirm this.)

(a) Determine the average annual percentage growth rate between 1946 and 1976, and between 1976 and 1988.
[Ans. approx. 6.7% and 3.7%]

(b) If the growth rate remains constant at the 1976-88 value, how many motor vehicles will there be in the year 2030?
[Ans. 2.5 billion]

---

[31] *Electric Power in Canada 1990*, Energy, Mines and Resources Canada, Government of Canada, Ottawa, 1991, p. 29.

(c) Assume that the average mass of a motor vehicle is 1400 kg. If the growth rate were to remain constant at the 1976-88 value, when would the total mass of motor vehicles equal the mass of the earth (5.98×10^{24} kg)? [Ans. in the year 2810 (approx.)]

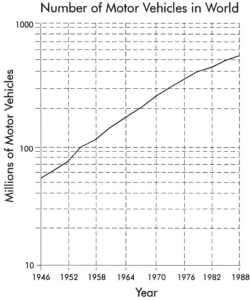

Figure 18-5:   Problem 18-3.

EPILOGUE

BACK TO THE FUTURE

LESSONS FROM THE PAST

It is extremely difficult to predict what lies ahead in the energy field. The direction that will be followed depends on political and economic situations, prices of various forms of energy, new developments in technology, and the level of concern about the environment.

Much of the interest in energy conservation and alternative sources of energy in the late 1970s and early 80s was the result of an increase in the price of crude oil[1] between 1973 and 1982 from $3 (U.S.) per barrel to $35 (in money-of-the-day). Most of this increase occurred in two steps: from $3 to $11 in a six-month period between 1973 and 1974, and from $13 to $35 between 1979 and 1982. These price increases were precipitated by political events: in late 1973 there was the so-called Yom Kippur war between Egypt and Israel, and the Arab nations who controlled the Organization of Petroleum Exporting Countries (OPEC) embargoed oil exports to the U.S.A. and other Western countries in an attempt to put indirect pressure on Israel to settle territorial disputes. The price of crude oil roughly quadrupled, gasoline was in short supply (producing long lines of cars at gasoline pumps in the U.S.), and an atmosphere of crisis surrounded the energy industry. Some people argue that it was not a coincidence that this embargo and the huge price increases occurred shortly after U.S. domestic oil production had begun to decline after peaking in 1970. In 1979, a revolution in Iran led to temporary suspension of Iranian oil production and roughly a tripling of oil prices, with gasoline rationing in parts of the U.S.A.

In early 1986, oil prices plummeted to about $15 a barrel, and have fluctuated between $15 and $22 since then. Why this huge drop in oil prices? We quote a newspaper[2] of the day: "The Organization of Petroleum Exporting Countries said late last year that it wanted a 'fair

---

[1] *BP Statistical Review of World Energy*, June 1985, p. 13, and July 1989, p. 14.

[2] *Toronto Globe and Mail*, January 23, 1986.

market share' of the world oil market, even if it meant price-cutting and the risk of a price war. OPEC, which has seen its market share cut dramatically as North Sea output has increased, has urged non-OPEC producers to cut production." As a result of the low price of oil, interest in alternative energy sources such as solar energy and wind energy waned quickly. Only now, in the mid 1990s, has there been a rekindling of interest in these alternative technologies, since they are now on the verge of being economically competitive. It is the belief of some people that OPEC has held prices around the $20 level in order to suppress development of alternative energy sources.

The point is that what happens in the energy arena depends not only on science, technology, and environmental issues, but on many other political, economic, and social factors over which an individual country (let alone an individual person) might have very little control. This does not mean that we should all just sit back and "let it happen," but we should all be aware of the complexity of the situation.

With this high level of uncertainty planted firmly in our minds, we now explore two general directions which individual countries (or regions of countries) might follow —— the "hard path" and the "soft path," phrases created in the mid-1970s by a young physics graduate student, Amory Lovins, who became a crusader and guru for the soft path.

HARD PATH VS. SOFT PATH

The hard path can be summarized as "more of the same:" that is, fossil fuels will dominate energy supplies, and electricity will be provided by large utilities employing huge distribution systems and centralized power plants using coal, fission, and perhaps fusion. Much of the energy domain will continue to be controlled by big-business interests. This approach has the advantage of providing certain economies of scale. (As power-plant size increases, the price per unit energy decreases.) Hard-path conservation is produced only by reaction to high prices, i.e., by supply and demand forces in the economy, not by government incentives.

The soft path involves reliance on renewable energy, such as solar and wind energy. Many people who advocate the soft path contend that all energy technologies should be understandable by the general public, and hence fission and fusion are not possible energy sources. There is an emphasis on decentralized production of energy, with many small diverse energy plants, and hence no large distribution system. Improvements in efficiency and conservation measures are an integral feature of the soft path, with one focus being on improvement of end-use devices rather than production of more energy.

Of course there are many approaches that can involve aspects of both the hard path and soft path. For example, in Ontario at present, about 50% of the electricity comes from nuclear plants (hard path), and at the same time Ontario Hydro is awarding grants to business and industry to assist them in energy conservation measures (soft path).

Around the world, specific countries, states, and communities will be faced with energy decisions in the future, and will come to different conclusions. At the moment, France is committed to heavy use of nuclear energy, whereas Sweden is planning to phase out its nuclear program. As well, the needs and resources of developing countries differ from those of highly industrialized countries.

Whatever energy path is followed by individual regions of the world, one thing remains certain: as long as human population continues to grow, there will eventually be too many of us to extract energy from the limited resources available without doing irreversible environmental damage to the Earth.

APPENDIX I   SIGNIFICANT DIGITS

The number of significant digits in the answer to a calculation will depend on the number of significant digits in the given data, as discussed in the rules below. *Approximate* calculations (order-of-magnitude estimates) always result in answers with only one or two significant digits.

When Are Digits Significant?

Non-zero digits are always significant. Thus, 22 J has two significant digits, and 22.3 J has three significant digits.

With zeroes, the situation is more complicated:

(a) Zeroes placed before other digits are not significant; 0.046 J has two significant digits.
(b) Zeroes placed between other digits are always significant; 4009 kg has four significant digits.
(c) Zeroes placed after other digits behind a decimal are significant; 7.90 has three significant digits.
(d) Zeroes at the end of a number are significant only if they are behind a decimal as in (c). Otherwise, it is impossible to tell if they are significant. For example, in the number 8200, it is not clear if the zeroes are significant or not. The number of significant digits in 8200 is at least two, but could be three or four. To avoid uncertainty, use scientific notation to place zeroes behind a decimal:

8.200×10^3 has four significant digits
8.20×10^3 has three significant digits
8.2×10^3 has two significant digits

Significant Digits in Multiplication, Division, Trig. Functions, etc.:

In a calculation involving multiplication, division, trigonometric functions, etc., the number of significant digits in an answer should equal the least number of significant digits in any one of the numbers being multiplied, divided, etc.

Thus, in evaluating sin(kx), where $k = 0.097$ m^{-1} (two significant digits) and $x = 4.73$ m (three significant digits), the answer should have two significant digits.

Note that whole numbers have essentially an unlimited number of significant digits. As an example, if 1 hairdryer uses 1.2 kW of power, then 2 identical hairdryers use 2.4 kW:

1.2 kW {2 sig. digits} $\times$ 2 {unlimited sig. digits} = 2.4 kW {2 sig. digits}

Significant Digits in Addition and Subtraction:

When quantities are being added or subtracted, the number of *decimal places* (not significant digits) in the answer should be the same as the least number of decimal places in any of the quantities being added or subtracted.

Example: 5.67   J   (two decimal places)
             1.1   J   (one dec. pl.)
          + 0.9378 J   (four dec. pl.)
            7.7   J   (one dec. pl.)

Keep One Extra Digit in Intermediate Answers

When doing multi-step calculations, *keep one more significant digit in intermediate results* than needed in your final answer.

For instance, if a final answer requires two significant digits, then carry three significant digits in calculations. If you round off all your intermediate answers to only two digits, you are discarding the information contained in the third digit, and as a result the *second* digit in your answer might be incorrect. (This phenomenon is known as "round-off error.")

The Two Biggest Sins Regarding Significant Digits

1. Writing more digits in an answer (intermediate or final) than justified by the number of digits in the data;
2. rounding-off to, say, two digits in an intermediate answer, and then writing three digits in the final answer.

Exercises

1. e^{kt} = ?, where k = 0.0189 yr^{-1}, and t = 25 yr.
   [Ans. 1.6]

2. ab/c = ?, where a = 483 J, b = 73.67 J, and c = 15.67
   [Ans. 2.27 x 10^3 J^2]

3. x + y + z = ?, where x = 48.1, y = 77, and z = 65.789
   [Ans. 191]

4. m - n - p = ?, where m = 25.6, n = 21.1, and p = 2.43
   [Ans. 2.1]

APPENDIX II   PROPERTIES OF FOSSIL FUELS

ENERGY CONTENT OF FOSSIL FUELS

Fuel	Energy Content
Ethyl alcohol	3×10^6 J/kg
Coal	$(17\text{-}29) \times 10^6$ J/kg
Gasoline	48×10^6 J/kg, or
	35×10^6 J/L
Natural gas	38×10^6 J/m^3
Wood	$(17\text{-}19) \times 10^6$ J/kg
Kerosene (jet fuel)	40×10^6 J/kg
Oil	40×10^6 J/kg
1 barrel of oil	5500×10^6 J

PROPERTIES OF COAL

	Anthracite	Bituminous	Sub-bituminous	Lignite
Moisture %	4.4	2-6	~20	~40
Volatiles %	4.8	20-40	~30	~30
Carbon %	82	65-45	~40	~30
Ash %	9	12-4	10-8	~5
Energy J/kg	30×10^6	30×10^6	22×10^6	16×10^6

APPENDIX III   SEMILOGARITHMIC GRAPH PAPER

APPENDIX IV   RECOMMENDED FURTHER READING

The publication 'Annual Reviews of Energy' offers a selection of in-depth articles by specialists. While the level of technical difficulty in some cases exceeds that of this book, many articles can be appreciated in full or in part, and they in turn supply many references. Journals such as Scientific American, Endeavour, New Scientist, Science and American Scientist offer a wide variety of articles on energy topics. These range from refereed original works by experts to articles that are essentially news stories on current developments. A selection is given below. In addition, some journals on occasion present a special issue with extended coverage of a particular subject. In September 1990, Scientific American published a special issue (volume 263 #3) entitled 'Energy for Planet Earth'; this is excellent supplementary reading material.

CHAPTER 1
N. Smith, The origins of the water turbine. *Scientific American* 242, 138-148 (1980).
C. Starr, M.F. Searl and S. Alpert, Energy sources: a realistic outlook. *Science* 256, 981-987 (1992).

CHAPTER 6
V.A. Mohnen, The challenge of acid rain. *Scientific American* 259, 30-38 (1988).

CHAPTER 7
R.A. Houghton and G.M. Woodwell, Global climate change. *Scientific American* 260, 36-44 (1989).
T.E. Graedel and P.J. Crutzen, The changing atmosphere. *Scientific American* 261, 58-69 (1989).
R.M. White, The great climate debate. *Scientific American* 263, 36-45 (1990).
P.D. Jones and T.M.L. Wigley, Global warming trends. *Scientific American* 263, 84-91 (1990).

CHAPTER 8
J. Manassen, The new role of rechargeable batteries. *Endeavour* 16, 164-166 (1992).

CHAPTER 9
J.W. Coltmann, The transformer. *Scientific American* 258, 86-96 (1988).

CHAPTER 10
T.H. Geballe and J.K.Hulm, Superconductors in electric power technology. *Scientific American* 243, 138-172 (1980).
A.M. Wolsky, R.F. Giese and E.J. Daniels, The new superconductors: prospects for applications. *Scientific American* 260, 60-69 (1989).
R. Stone, Polarized debate: EMFs and cancer. *Science* 258, 1724-1725 (1992).
Health effects of low-frequency electric and magnetic fields, in *Environmental Science and Technology* 27, 42-58 (1993): this feature includes the executive summary of the Oak Ridge Associated Universities Report, followed by commentary by D.A.Savitz and T.S. Tenforde.

CHAPTERS 12 and 15
H.W. Lewis, The safety of fission reactors. *Scientific American* 242, 53-65 (1980).
A.M. Weinberg and I. Spiewak, Inherently safe reactors and a second nuclear era. *Science* 224, 1398-1402 (1984).
J.J. Taylor, Improved and safer nuclear power. *Science* 244, 318-325 (1989).
M.W. Golay and N.E. Todreas, Advanced light-water reactors. *Scientific American* 262, 82-89 (1990).
J.F. Ahearne, The future of nuclear power. *American Scientist* 81, 24-35 (1993).

CHAPTER 13
M.G. Haines, Tokamak physics. *Contemporary Physics* 25, 331-353 (1984).
R.S. Craxton, R.L. McCrory and J.M. Soures, Progress in laser fusion. *Scientific American* 255, 68-79 (1986).
R.W. Conn, V.A. Chuyanov, N. Inoue and D.R. Sweetman, The international thermonuclear experimental reactor. *Scientific American* 266, 103-110 (1992).

CHAPTER 14
R.S. Yalow, Radiation and Society. *Interdisciplinary Science Reviews* 16, 351-356 (1991).
A.C. Upton, Health effects of low-level ionizing radiation. *Physics Today* 44, 34-39 (1991).

CHAPTER 16
R. Winston, Non-imaging optics. *Scientific American* 264, 76-81 (1991).
I. Dostrovsky, Chemical fuels from the sun. *Scientific American* 265, 102-107 (1991).
A.J. McEvoy, Outlook for solar photovoltaic electricity. *Endeavour* 17, 17-20 (1993).
P.M. Moretti and L.V. Divone, Modern windmills. *Scientific American* 254, 110-118 (1986).
J.G. McGowan, Tilting towards windmills. *Technology Review*, July 1993, 40-46.
T.R. Penney and D. Bharathan, Power from the sea. *Scientific American* 256, 86-92 (1987).

CHAPTER 17
C.L. Gray, Jr. and J.A. Alson, The case for methanol. *Scientific American* 261, 108-114 (1989).
E. Corcoran, Cleaning up coal. *Scientific American* 264, 106-116 (1991).
The future for coal. A multi-author survey published in *New Scientist* 137, 20-41 (1993).
S.R. Ovshinsky, M.A.Fetcenko and J. Ross. A nickel metal hydride battery for electric vehicles. *Science* 260, 176-181 (1993).
R.S. Claassen and L.A. Girifalco, Materials for energy utilization. *Scientific American* 255, 103-118 (1986).

CHAPTER 18
R.C. Marlay, Trends in industrial use of energy. *Science* 226, 1277-1283 (1984).
A.H. Rosenfeld and D. Hafemeister. Energy-efficient buildings. *Scientific American* 258, 78-85 (1988).

EPILOGUE
P.H. Abelson, Energy futures. *American Scientist* 75, 584-593 (1987).
J.H. Gibbons, P.D. Blair and H.L. Gwin, Strategies for energy use. *Scientific American* 261, 136-143 (1989).
H.M. Hubbard, The real cost of energy. *Scientific American* 264, 36-42 (1991).

Index

Index

Index

Index

Index

Index

About the Authors

Ernest L McFarland was born in Toronto, Ontario, Canada. He received a B Sc degree (1968) from the U of Western Ontario, and a M Sc (1970) from McMaster U in Hamilton. In 1974 he became a member of the Physics Department of the University of Guelph, where is main interest is in undergraduate teaching. He has been very active in the American Association of Physics Teachers and the Ontario Association of Physics Teachers (of which he is Founding President). Dr McFarland has also served as a physics-education consultant to TV-Ontario, the Ontario Ministry of Education, and the Ontario Natural Gas Association.

James L Hunt was born in Guelph, Ontario, Canada. He earned a B Sc degree (1955) from Queen's U, Kingston, and a M A (1956) and Ph D (1959) from the U of Toronto. He joined the Physics Department of the Memorial U of Newfoundland in 1959, and the U of Guelph in 1964. Dr Hunt's research is in the area of the induced spectra of simple molecules and in particular the spectrum of solid hydrogen irradiated with high-energy protons.

John L Campbell was born in Kilmarnock, Scotland. He holds the degrees B Sc (1963), Ph D (1967), and D Sc (1982) from the U of Glasgow. In 1968 he joined the University of Guelph, where he is professor of Physics and Dean of the College ofg Physical and Engineering Science. Dr Campbell's research area is atomic physics, and he is one of the developers of proton-induced X-ray emission (PIXE) as a quantitative microprobe technique for elemental analysis of environmental and other materials. He is co-author of a text on PIXE.

Acronyms

AC — alternating current
AGR — advanced gas-cooled reactor
BE — binding energy
BaP — benzo(a)pyrene
BWR — boiling water reactor
C.O.P. — coefficient of performance
CANDU — Canadian deuterium uranium reactor
DC — direct current
E-M — electro-magnetic
ECCS — emergency core cooling system
HC — hydrocarbons
HDD — heating degree days
HTGR — high temperature gas-cooled reactor
HVAC — heating, ventilating, and air conditioning
ICRP — International Commission on Radiation Protection
IR — infrared
KE — kinetic energy
LOCA — loss of cooling accident
PE — potential energy
PM — particulate matter
PWR — pressurized water reactor
QF — quality factor
RBE — relative biological effectiveness
RME — rape methyl ester
rms — root mean square
RSI — thermal resistance in SI units
TMI — Three Mile Island (Pennsylvania)

Symbols

α — alpha particle or helium nucleus
α — planetary albedo
β — beta particle (electron or positron)
γ — gamma ray
ϵ — (epsilon) average emissivity
η — (eta) efficiency
λ — (lambda) wavelength
λ — radioactive decay constant
λ_b — biological decay constant
λ_e — effective decay constant
λ_p — physical decay constant
μ — (mu) linear attenuation coefficient
μ_m — mass attenuation coefficient
ρ — (rho) electrical resistivity
ρ — density
σ — (sigma) Stefan's constant
σ — standard deviation of Gaussian function
σ — nuclear absorption cross-section
ϕ — (phi) magnetic flux
A — atomic mass number
A — radioisotope activity
B — magnetic field strength
Btu — British thermal unit
c — speed of light
c — specific heat
cal — calorie
D — dose
e — (absolute value of) electronic charge
E — effective dose
E — energy
eV — electron-volt
f — frequency
g — gravitational acceleration
h — Planck's constant
h — conduction-convection parameter
H_T — equivalent dose
I — intensity
I — electric current
I — moment of inertia
J — heat flux

k — Boltzmann's constant
k — reactor constant
k — radius of gyration
k — exponential-growth constant
k — thermal conductivity
kW•h — kilowatt•hour
L — litre
$L_{1/2}$ — half thickness
L_F — latent heat of fusion
L_V — latent heat of vaporization
m — mass
MW_e — megawatts of electrical power
N_M — maximum value of Gaussian function
NO_x — nitrogen oxides
P — power
Q — amount of heat
Q,q — electric charge
Q_T — resource remaining
Q_∞ — total quantity of resource
quad — quadrillion British thermal units
R — thermal resistance
R — molar gas constant
R — electric resistance
R — percent growth Rate
S — entropy
t — time
T — absolute temperature
T — resource lifetime
$T_{1/2}$ — half life
$_bT_{1/2}$ — biological half life
$_eT_{1/2}$ — effective half life
$_pT_{1/2}$ — physical half life
T_2 — doubling time
T_M — time at which Gaussian function is maximum
u — atomic mass unit
U — electric potential energy
v — speed
V — electric potential
V — volume
W — work
W — work
w_R — radiation weighting factor
w_T — tissue weighting factor
Z — atomic number